D0138277

To Hannah
With love

Eukaryotic Transcription Factors

Fourth Edition

David S. Latchman

*Master, Birkbeck, University of London and
Professor of Genetics, Birkbeck College and
University College London*

ELSEVIER
ACADEMIC
PRESS

Amsterdam Boston Heidelberg London New York Oxford Paris
San Diego San Francisco Singapore Sydney Tokyo

Elsevier Academic Press
84 Theobald's Road, London WC1X 8RR, UK
http://books.elsevier.com

Elsevier Academic Press
525 B Street, Suite 1900 San Diego, California 92101-4495, USA
http://books.elsevier.com

A catalogue record for this book is available from the British Library

Library of Congress Catalog Number: 2003105489

ISBN 0-12-437178-7

Typeset by Keyword Typesetting Services Ltd, Keyword House,
30 Stafford Road, Wallington, Surrey, SM6 9AA
Printed and bound in Italy

04 05 06 07 PT 9 8 7 6 5 4 3 2 1

CONTENTS

List of Tables   xiii
About the author   xv
Preface   xvii
Preface to the third edition   xix
Preface to the second edition   xxi
Preface to the first edition   xxiii
Acknowledgements   xxv

1   DNA SEQUENCES, TRANSCRIPTION FACTORS AND
    CHROMATIN STRUCTURE   1
    1.1   The importance of transcription   1
    1.2   Chromatin structure and its remodelling   2
          1.2.1   Chromatin structure and gene regulation   2
          1.2.2   Chromatin remodelling factors   3
          1.2.3   Histone acetylation   5
    1.3   DNA sequence elements   7
          1.3.1   The gene promoter   7
          1.3.2   Sequences involved in the basic process of
                  transcription   8
          1.3.3   Sequences involved in regulated transcription   9
          1.3.4   Sequences which act at a distance   13
          1.3.5   Negatively acting DNA sequences   16
          1.3.6   Interaction between factors bound at various sites   17
    1.4   Conclusions   19

2   METHODS FOR STUDYING TRANSCRIPTION FACTORS   23
    2.1   Introduction   23
    2.2   Methods for studying DNA–protein interactions   23
          2.2.1   DNA mobility shift assay   23
          2.2.2   DNAseI footprinting assay   27
          2.2.3   Methylation interference assay   30
          2.2.4   *In vivo* footprinting assay   32
    2.3   Methods for purifying and/or cloning transcription factors   34

	2.3.1	Protein purification	34
	2.3.2	Gene cloning	38
2.4	Use of cloned genes		41
	2.4.1	Domain mapping of transcription factors	41
	2.4.2	Determining the DNA binding specificity of an uncharacterized factor	45
	2.4.3.	Identification of target genes for transcription factors	47
2.5	Conclusions		50

3 RNA POLYMERASES AND THE BASAL TRANSCRIPTIONAL COMPLEX 55

3.1	RNA polymerases		55
3.2	The stable transcriptional complex		57
3.3	RNA polymerase I		58
3.4	RNA polymerase III		60
3.5	RNA polymerase II		63
	3.5.1	Stepwise assembly of the RNA polymerase II basal transcriptional complex	63
	3.5.2	The RNA polymerase holoenzyme	66
3.6	TBP, the universal transcription factor?		67
3.7	Conclusions		73

4 FAMILIES OF DNA BINDING TRANSCRIPTION FACTORS 77

4.1	Introduction		77
4.2	The homeodomain		78
	4.2.1	Transcription factors in *Drosophila* development	78
	4.2.2	The homeobox	79
	4.2.3	DNA binding by the helix-turn-helix motif in the homeobox	81
	4.2.4	Regulation of DNA binding specificity by interactions between different homeobox proteins	88
	4.2.5	Homeodomain transcription factors in other organisms	90
	4.2.6	POU proteins	93
	4.2.7	Pax proteins	100
4.3	The two cysteine two histidine zinc finger		102
	4.3.1	Transcription factors with the two cysteine two histidine finger	102
	4.3.2	DNA binding by the two cysteine two histidine finger	104
4.4	The multi-cysteine zinc finger		107
	4.4.1	Steroid receptors	107
	4.4.2	DNA binding by the multi-cysteine zinc finger	109

4.5   The basic DNA binding domain   117
    4.5.1   The leucine zipper and the basic DNA binding
        domain   117
    4.5.2   The helix-loop-helix motif and the basic DNA
        binding domain   121
    4.5.3   Dimerization of basic DNA binding domain-containing
        factors   122
4.6   Other DNA-binding motifs   124
4.7   Conclusions   125

5   ACTIVATION OF GENE EXPRESSION BY TRANSCRIPTION FACTORS   135
5.1   Activation domains   135
5.2   Nature of activation domains   136
    5.2.1   Acidic domains   137
    5.2.2   Glutamine-rich domains   139
    5.2.3   Proline-rich domains   140
    5.2.4   Functional relationship of the different
        activation domains   140
5.3   Interaction of activation domains with the basal
    transcriptional complex   142
    5.3.1   Activators and the basal transcriptional complex   142
    5.3.2   Stimulation of factor binding   143
    5.3.3   Stimulation of factor activity   146
5.4   Interaction of activation domains with other regulatory
    proteins   149
    5.4.1   The mediator complex   149
    5.4.2   TAFs   151
    5.4.3   CBP and other co-activators   157
    5.4.4   A multitude of targets for transcriptional activators   161
5.5   Other targets for transcriptional activators   162
    5.5.1   Modulation of chromatin structure   162
    5.5.2   Stimulation of transcriptional elongation   170
5.6   Conclusions   174

6   REPRESSION OF GENE EXPRESSION BY TRANSCRIPTION FACTORS   183
6.1   Repression of transcription   183
6.2   Indirect repression   183
    6.2.1   Inhibition of activator binding by masking of its
        DNA binding site   183
    6.2.2   Inhibition of activator binding by formation of a
        non-DNA binding complex   187

	6.2.3	Quenching of an activator	189
	6.2.4	Degradation of the activator	190
6.3	Direct repression		191
	6.3.1	Mechanisms of transcriptional repression	191
	6.3.2	Direct repression by DNA binding transcription factors	192
	6.3.3	Direct repression by factors binding to the basal transcriptional complex	198
6.4	Other targets for transcriptional repressors		200
	6.4.1	Modulation of chromatin structure	200
	6.4.2	Inhibition of transcription elongation	204
6.5	Conclusions		205

7   REGULATION OF TRANSCRIPTION FACTOR SYNTHESIS   211

7.1	Transcription factor regulation		211
7.2	Regulated synthesis of transcription factors		211
	7.2.1	The MyoD transcription factor	212
	7.2.2	Homeobox transcription factors	219
7.3	Mechanisms regulating the synthesis of transcription factors		229
	7.3.1	Regulation of transcription	230
	7.3.2	Regulation of RNA splicing	232
	7.3.3	Regulation of translation	238
7.4	Conclusions		240

8   REGULATION OF TRANSCRIPTION FACTOR ACTIVITY   245

8.1	Evidence for the regulated activity of transcription factors		245
8.2	Regulation by protein–ligand binding		247
	8.2.1	Examples of regulation by ligand binding	247
	8.2.2	The nuclear receptors	248
8.3	Regulation by protein–protein interactions		255
	8.3.1	Inhibition of transcription factor activity by protein–protein interaction	255
	8.3.2	Activation of transcription factors by protein–protein interaction	262
	8.3.3	Alteration of transcription factor function by protein–protein interaction	263
8.4	Regulation by protein modification		265
	8.4.1	Transcription factor modification	265
	8.4.2	Phosphorylation	266
	8.4.3	Acetylation	272
	8.4.4	Methylation	274

    8.4.5   Ubiquitination   275
  8.5   Regulation by protein degradation and processing   281
  8.6   Role of regulated activity   283
  8.7   Conclusions   285

9   TRANSCRIPTION FACTORS AND HUMAN DISEASE   293
  9.1   Diseases caused by transcription factor mutations   293
  9.2   Cancer   297
  9.3   Cellular oncogenes and cancer   299
    9.3.1   Fos, Jun and AP1   299
    9.3.2   *v-erbA* and the thyroid hormone receptor   304
    9.3.3   The *myc* oncogene   307
    9.3.4   Other oncogenic transcription factors   312
  9.4   Anti-oncogenes and cancer   314
    9.4.1   Nature of anti-oncogenes   314
    9.4.2   p53   315
    9.4.3   The retinoblastoma protein   322
    9.4.4   Other anti-oncogenic transcription factors   328
  9.5   Conclusions   331

10   CONCLUSIONS AND FUTURE PROSPECTS   341

Index   345

COLOUR PLATES ARE LOCATED BETWEEN PP. 166–7

LIST OF TABLES

2.1 Purification of transcription factor NFI from HeLa cells

3.1 Eukaryotic RNA polymerases

4.1 Transcriptional regulatory proteins containing Cys_2-His_2 zinc fingers

4.2 Relationship of various hormone response elements

4.3 Transcriptional regulatory proteins with multiple cysteine fingers

4.4 DNA binding motifs

ABOUT THE AUTHOR

David S. Latchman graduated with a BA (First Class Honours) in Natural Sciences from Cambridge University and subsequently obtained a PhD in Genetics from Cambridge University.

Following a period of post-doctoral research at Imperial College London, he was appointed to a lectureship at University College London in 1984. Subsequent appointments at UCL included Director of the Medical Molecular Biology Unit (1988), Professor of Molecular Pathology and Head of the Department of Molecular Pathology (1991), Director of the Windeyer Institute of Medical Sciences (1996) and in 1999, Dean of the Institute of Child Health and Professor of Human Genetics.

At the beginning of 2003, he took up a new appointment as Master of Birkbeck, University of London, leading this multi-faculty College comprising over 1000 staff and 16 000 students studying at levels from short courses to PhDs. He continues to retain an active research laboratory at the Institute of Child Health, UCL and in recognition of this he is now Professor of Genetics at Birkbeck and UCL.

Since he set up his own laboratory in 1984, Professor Latchman's interests have focused particularly on the regulation of gene expression in mammalian cells by specific transcription factors. Initial studies involved the eukaryotic virus herpes simplex virus (HSV) and were aimed at elucidating why this virus was able to replicate in epithelial cells, while establishing silent latent infections in neuronal cells.

The identification of a cellular POU family transcription factor, Oct-2, which was expressed in neuronal cells and blocked the virus-lytic cycle, led to an interest in the role of specific POU family transcription factors in regulating cellular as well as viral gene expression in neuronal cells. As well as studies of the role of Oct-2, these investigations resulted in the characterization of a small sub-family of POU family transcription factors which comprised three members: Brn-3a, Brn-3b and Brn-3c. In particular, the key role of Brn-3a in promoting neuronal differentiation and enhancing the survival of neuronal cells was defined by Professor Latchman's laboratory and has important therapeutic potential for the treatment of human neurodegenerative diseases involving losses of neuronal cells.

More recently, Professor Latchman's group has also studied the role of specific transcription factors in the heart and, in particular, in its response to damage caused by ischaemia, as in a heart attack. They have demonstrated that a specific transcription factor, STAT-1, becomes activated during ischaemia and induces a number of genes which enhance the programmed cell death of cardiac cells, resulting in the loss of these irreplaceable cells. Thus, inhibition of STAT-1 represents a potential therapeutic mechanism for decreasing the damage caused by interruption of blood flow to the heart.

On the basis of his scientific research, Professor Latchman was awarded a DSc by London University in 1994 and was appointed a Fellow of the Royal College of Pathologists (FRCPath) in 1999. He serves on a number of Committees, including the Department of Health, Genetics and Insurance Committee, the National Biological Standards Board and its Scientific Policy Advisory Committee, the Health Protection Agency and the Research Strategy Committee of the Higher Education Funding Council for England.

It is now over ten years since the first edition of *Eukaryotic Transcription Factors* was published. It is obvious that in that time an enormous amount of information about transcription factors has accumulated and this has been reflected in subsequent editions of the book. However, over the past years, we have moved from a situation where only a few transcription factors had been characterized in any detail, to a situation where a very large number of transcription factors have been extensively characterized. This has led to the decision in this new edition to abandon the dual structure of previous editions in which the role of a few transcription factors in inducible, cell-type specific and developmental gene regulation was extensively discussed, followed by chapters dealing with the mechanistic aspects of transcription factors.

In the new edition therefore, the book adopts a single approach of dealing in turn with the specific properties of transcription factors, using a range of examples including those which were extensively discussed in previous editions but also others as appropriate. This has allowed a much more detailed analysis of various mechanistic aspects which have become of increasing importance in recent years.

As before, the work begins with a chapter on DNA sequences and chromatin structure in which the section on the modulation of chromatin structure by chromatin remodelling complexes and histone modifying enzymes has been considerably expanded to reflect recent work. This is followed, as before, by a chapter describing the methods used to analyse the properties of transcription factors which now has an additional section dealing with the methods of identifying target genes for previously uncharacterized transcription factors. As before, this is followed by a chapter dealing with RNA polymerase enzymes and the basal transcriptional complex.

Following these three initial chapters, however, the format of the book has dramatically changed. Thus, Chapter 4 now deals extensively with specific transcription factor families. Moreover, since these families are defined primarily on the basis of their DNA binding domain, this chapter also deals with the features which allow these various factors to bind to DNA. Subsequently, separate chapters deal with activation and repression of transcription respectively, replacing the single chapter which previously dealt with both these processes. This has allowed a considerable expansion of the discussion of these topics, allowing subjects such as the mediator complex, co-activators and the activation or repression of transcription by alterations in chromatin structure, to be discussed in much greater depth.

Similarly, the single chapter in the previous edition dealing with the regulation of transcription factor synthesis and activity, has now been split into two chapters dealing respectively with the regulation of transcription factor synthesis and the regulation of transcription factor activity. Again, this has allowed a number of topics, such as the regulation of transcription factor activity by a variety of different post-translational modifications, to be discussed in greater depth. As part of these changes, the chapter on transcription factors and human disease has been moved to the end of the work and is followed by a final conclusion chapter.

It is hoped that these changes will avoid the increasing duplication which would have been necessary if the initial approach had been maintained and will allow the work to build on the success of its predecessors, by providing an up-to-date account of this critically important topic.

Finally, I would like to thank Miss Maruschka Malacos for typing the text and coping with the necessity to move around large and small sections, to reflect the change in the structure of the book. I am also most grateful to Dr Tessa Picknett and the staff at Elsevier Academic Press for commissioning this new edition and producing it with their customary efficiency.

David S. Latchman

PREFACE TO THE THIRD EDITION

As in previous years, the period between the publication of the second and third editions of this book has been marked by a considerable further accumulation of information about individual transcription factors and the manner in which they act. This new edition has therefore been extensively updated to reflect this and several sections have been completely rewritten.

As well as such increased general understanding of transcription factors, a major new theme unifying much of this information has emerged. This involves the role of co-activator molecules such as CBP in the action of a number of different activating transcription factors as well as the finding that such co-activators frequently possess histone acetyltransferase activity indicating that they may act by modulating chromatin structure. In addition to discussion of co-activators in the appropriate sections on individual transcription factors, the new edition of this work now includes specific new sections dealing with this important topic. Thus the role of chromatin structure and histone acetylation in the regulation of gene expression is now introduced in Chapter one (section 1.4), the role of CBP in cyclic AMP mediated gene activation where it was originally discovered is discussed in Chapter four together with other aspects of this signalling pathway (section 4.3) and the interaction of transcriptional activators with co-activators is discussed in a separate section of Chapter nine (section 9.2.4).

In addition to these new sections on this aspect, other new sections have been added describing topics which are now of sufficient importance to merit a separate section. These are the methods used to determine the DNA binding specificity of an uncharacterized transcription factor (Chapter two, section 2.3.4), the Pax family transcription factors (Chapter six, section 6.3.2), anti-oncogenic transcription factors other than p53 or Rb (Chapter seven, section 7.3.4) and the regulation of transcription factor activity by protein degradation and processing (Chapter ten, section 10.3.5). Similarly, Chapter seven now includes an extensive discussion of the role of transcription factors in diseases other than cancer and its title has therefore been changed to 'Transcription factors and human disease' (from 'Transcription factors and cancer').

As well as these changes in the text, we have been able to include, for the first time, a special section of colour illustrations illustrating various aspects of transcription factor structure which are being progressively elucidated. It is hoped that all these changes will allow this new edition, like its predecessors, to provide an up-to-date overview of the important area of transcription factors and their vital role in regulating transcription in different cell types, during development and in disease.

Finally, I would like to thank Mrs Sarah Franklin for her efficiency in producing the text and dealing with the need to make numerous changes from the previous edition as well as Mrs Jane Templeman for continuing to use her outstanding skills in the preparation of the numerous new illustrations in this edition. Thanks are also due to Tessa Picknett and the staff at Academic Press for producing this new edition with their customary efficiency.

David S. Latchman

PREFACE TO THE SECOND EDITION

In the four years since the first edition of this work was published, the explosion of information about transcription factors has continued. The genes encoding many more transcription factors have been cloned and this information used to analyse their structure and function culminating in many cases with the use of inactivating mutations to prepare so called 'knock out' mice, thereby testing directly the role of these factors in development. Nonetheless, the examples used in the first part of this book to illustrate the role of transcription factors in processes as diverse as inducible gene expression and development still remain among the best understood. The discussion of these factors has therefore been considerably updated to reflect the progress made in the last few years. In addition new sections have been added on topics such as TBP; the *myc* oncogene and anti-oncogenes where the degree of additional information now warrants a separate section.

Even greater changes have been necessary in the second part of the book which deals with the mechanisms by which transcription factors act. Thus, for example, the sections in Chapter nine on the mechanisms of transcriptional activation and on transcriptional repression have been completely rewritten. In addition, the increasing emphasis on transcriptional repression discussed in Chapter nine has led to a change in the title of Chapter ten to 'What regulates the regulators?' (from 'What activates the activators?'). Moreover, this chapter now includes a much more extensive section on the interaction between different factors which is another major theme to have emerged in the last few years. It is hoped that these changes will allow the new edition to build on the success of the first edition in providing an overview of these vital factors and the role they play in gene regulation.

Finally I would like to thank Jane Templeman who has prepared a large number of new illustrations to complement the excellent ones she provided for the first edition and Sarah Chinn for coping with the necessity of adding, deleting or amending large sections of the first edition. I am also grateful to Tessa Picknett and the staff at Academic Press for commissioning this new edition and their efficiency in producing it.

David S. Latchman

PREFACE TO THE FIRST EDITION

In my previous book, *Gene Regulation: A Eukaryotic Perspective* (Unwin-Hyman, 1990); I described the mechanisms by which the expression of eukaryotic genes is regulated during processes as diverse as steroid treatment and embryonic development. Although some of this regulation occurs at the post-transcriptional level, it is clear that the process of gene transcription itself is the major point at which gene expression is regulated. In turn this has focused attention on the protein factors, known as transcription factors, which control both the basal processes of transcription and its regulation in response to specific stimuli or developmental processes. The characterization of many of these factors and, in particular, the cloning of the genes encoding them has resulted in the availability of a bewildering array of information on these factors, their mechanism of action and their relationship to each other. Despite its evident interest and importance, however, this information could be discussed only relatively briefly in *Gene Regulation* whose primary purpose was to provide an overview of the process of gene regulation and the various mechanisms by which this is achieved.

It is the purpose of this book therefore to discuss in detail the available information on transcription factors emphasizing common themes and mechanisms to which new information can be related as it becomes available. As such it is hoped the work will appeal to final year undergraduates and postgraduate students entering the field as well as to those moving into the area from other scientific or clinical fields who wish to know how transcription factors may regulate the gene in which they are interested.

In order to provide a basis for the discussion of transcription factors, the first two chapters focus respectively on the DNA sequences with which the factors interact and on the experimental methods that are used to study these factors and obtain the information about them provided in subsequent chapters. The remainder of the work is divided into two distinct portions. Thus Chapters three to seven focus on the role of transcription factors in particular processes. These include constitutive and inducible gene expression, cell type-specific and developmentally regulated gene expression and the role of transcription factors in cancer. Subsequently Chapters eight to ten adopt a more mechanistic approach and consider the features of transcription factors which allow them to fulfil their function. These include the ability to bind to DNA and modulate transcription either positively or negatively as well as the ability to respond to specific stimuli and thereby activate gene expression in a regulated manner.

Although this dual approach to transcription factors from both a process-oriented and mechanistic point of view may lead to some duplication, it is the most efficient means of providing the necessary overview both of the nature of transcription factors and the manner in which they achieve their role of modulating gene expression in many diverse situations.

Finally I would like to thank Mrs Rose Lang for typing the text and coping with the continual additions necessary in this fast moving field and Mrs Jane Templeman for her outstanding skill in preparing the illustrations.

David S. Latchman

ACKNOWLEDGEMENTS

I would like to thank all the colleagues, listed below, who have given permission for material from their papers to be reproduced in this book and have provided prints suitable for reproduction.

Figures 4.1 and 7.9, photographs kindly provided by Professor W.J. Gehring from Gehring, Science 236, 1245 (1987) by permission of the American Association for the Advancement of Science.

Figure 4.14, photograph kindly provided by Dr P. Holland from Holland and Hogan, Nature 321, 251 (1986) by permission of Macmillan Magazines Ltd.

Figure 4.25 redrawn from Redemann *et al.*, Nature 332, 90 (1988) by kind permission of Dr H. Jackle and Macmillan Magazines Ltd.

Figures 4.31 and 4.35 redrawn from Schwabe *et al.*, Nature 348, 458 (1990) by kind permission of Dr D. Rhodes and Macmillan Magazines Ltd.

Figure 4.40 redrawn from Abel and Maniatis, Nature 341, 24 (1989) by kind permission of Professor T. Maniatis and Macmillan Magazines Ltd.

Figures 7.4 and 7.7, photographs kindly provided by Dr R.L. Davis from Davis *et al.*, Cell 51, 987 (1987) by permission of Cell Press.

Figures 7.12 and 7.13, photographs kindly provided by Dr R. Krumlauf from Graham *et al.*, Cell 57, 367 (1989) by permission of Cell Press.

Figure 8.5, photograph kindly provided by Professor M. Beato from Willmann and Beato, Nature 324, 688 (1986) by permission of Macmillan Magazines Ltd.

Figure 8.12, photograph kindly provided by Dr C. Wu from Zimarino and Wu, Nature 327, 727 (1987), by permission of Macmillan Magazines Ltd.

I am also especially grateful to the colleagues who have provided colour prints of transcription factor structures, allowing us to include this feature.

Plate 1, kindly provided by Dr J. H. Geiger.

Plate 2 kindly provided by Dr T. Li and Professor C. Wolberger.

Plate 3 kindly provided by Professor P. E. Wright from Lee *et al.*, Science 245, 635 (1989) by permission of the American Association for the Advancement of Science.

Plate 4 kindly provided by Dr R. J. Fletterick.

Plate 5 kindly provided by Professor R. Kaptein from Hard *et al.*, Science 249, 157 (1990) by permission of the American Association for the Advancement of Science.

Plate 6 kindly provided by Dr D. Rhodes from Schwabe *et al.*, Cell 75, 567 (1993) by kind permission of Cell press.

Plate 7 kindly provided by Professor D. Moras.

CHAPTER 1

DNA SEQUENCES, TRANSCRIPTION FACTORS AND CHROMATIN STRUCTURE

1.1 THE IMPORTANCE OF TRANSCRIPTION

The fundamental dogma of molecular biology is that DNA produces RNA which, in turn, produces protein. Hence if the genetic information that each individual inherits as DNA (the genotype) is to be converted into the proteins which produce the corresponding characteristics of the individual (the phenotype), it must first be converted into an RNA product. The process of transcription, whereby an RNA product is produced from the DNA, is therefore an essential element in gene expression. The failure of this process to occur will obviously render redundant all the other steps that follow the production of the initial RNA transcript in eukaryotes, such as RNA splicing, transport to the cytoplasm or translation into protein (for review of these stages see Nevins, 1983; Latchman, 2002).

The central role of transcription in the process of gene expression also renders it an attractive control point for regulating the expression of genes in particular cell types or in response to a particular signal. Indeed, it is now clear that, in the vast majority of cases, where a particular protein is produced only in a particular tissue or in response to a particular signal, this is achieved by control processes which ensure that its corresponding gene is transcribed only in that tissue or in response to such a signal (for reviews see Darnell, 1982; Latchman, 2002). For example, the genes encoding the immunoglobulin heavy and light chains of the antibody molecule are transcribed at high level only in the antibody-producing B cells while the increase in somatostatin production in response to treatment of cells with cyclic AMP is mediated by increased transcription of the corresponding gene. Therefore, while post-transcriptional regulation affecting for example, RNA splicing or stability plays some role in the regulation of gene expression (for reviews see Bashirullah *et al.*, 2001; Graveley, 2001) the major control point lies at the level of transcription.

1.2 CHROMATIN STRUCTURE AND ITS REMODELLING

1.2.1 CHROMATIN STRUCTURE AND GENE REGULATION

The central role of transcription, both in the basic process of gene expression and its regulation in particular tissues, has led to considerable study of this process. Initially such studies focused on the nature of the DNA sequences within individual genes which were essential for either basal or regulated gene expression. These sequences will be discussed in section 1.3. It is now clear, however, that the accessibility of these DNA sequences and hence their ability to regulate gene expression is controlled by the manner in which they are packaged in the cell. The packaging of DNA will therefore be discussed in this section.

It has been known for some time that the DNA in eukaryotic cells is packaged by association with specific proteins, such as the histones, into a structure known as chromatin (for reviews see Wolffe, 1995; Latchman, 2002; Felsenfeld and Groudine, 2003). The fundamental unit of this structure is the nucleosome in which the DNA is wrapped twice around a unit of eight histone molecules (two each of histones H2A, H2B, H3 and H4) (for review see Kornberg and Lorch, 1999). This structure is compacted further into the so-called solenoid structure in genes which are not transcriptionally active or about to become active. In contrast, active or potentially active genes exist in the simple nucleosomal structure. Moreover, in the regulatory regions of these genes nucleosomes are either removed altogether or undergo a structural alteration which facilitates the binding of specific transcription factors to their binding sites in these regions (Fig. 1.1).

Interestingly, the tightly packed solenoid structure can be compacted even further, by extensive looping, to form the chromosomes which are visible during cell division. These loops are linked at their bases to a protein scaffold known as the nuclear matrix, with such linkage occurring via specific DNA sequences, known as matrix attachment regions (MARs) (Fig. 1.2; for review see Horn and Peterson, 2002).

Figure 1.1

Levels of chromatin structure in active or inactive DNA.

Higher order structure of inactive DNA

Nucleosome-free regulatory region

Active gene in nucleosome structure

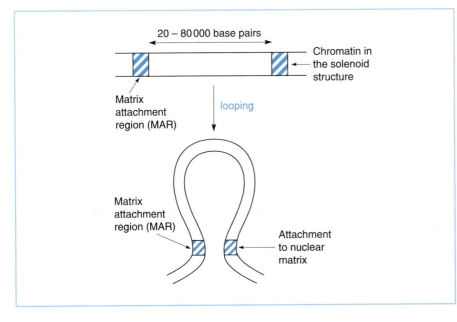

Figure 1.2

The tightly packed solenoid structure can be further compacted by the formation of loops. These loops (which contain approximately 20–80 000 bases of DNA) are attached to the nuclear matrix via specific DNA elements known as matrix attachment regions.

Clearly, the access of a transcription factor to its appropriate binding site will be affected by the manner in which that site is packaged within the chromatin structure. Evidently, therefore, genes that are about to be transcribed must undergo changes in chromatin structure which facilitate such transcription by allowing access of activating transcription factors to their binding sites. Although a detailed discussion of these changes is beyond the scope of this book (for reviews see Aalfs and Kingston, 2000; Wu and Grunstein, 2000; Bradbury, 2002; Latchman, 2002; Richards and Elgin, 2002; Felsenfeld and Groudine, 2003), at least two mechanisms which can alter chromatin structure are of particular importance in terms of transcription factor regulation and these will be discussed in turn.

1.2.2 CHROMATIN REMODELLING FACTORS

A number of studies have identified protein complexes which are capable of binding to DNA, hydrolysing ATP and using the energy generated to disrupt the nucleosomal structure. The best characterized of these is the SWI/SNF complex which contains a number of different polypeptides. It was originally defined in yeast but has now been identified in a range of organisms including humans (for review see Aalfs and Kingston, 2000; Sudarsanam and Winston, 2000). The critical role of this complex in regulating gene expression is indicated by the phenotype of the brahma mutation in *Drosophila* which inactivates the SWI2 component of the complex. Thus, in this mutant the genes encoding several homeobox-containing genes, which control the correct patterning of the body (see Chapter 4, section 4.2), remain in an inactive chromatin structure and are hence not transcribed. This results in a mutant fly with a grossly abnormal body structure (for review see Simon, 1995).

It is likely that SWI/SNF and other chromatin remodelling complexes can act by three different methods to alter the accessibility of the DNA. Thus, they may act by altering the association of the histone molecules within the nucleosome so that the nucleosome structure is changed in such a way as to allow other factors to bind to DNA (nucleosome remodelling: Fig. 1.3a). Secondly, they may act by causing the nucleosome to move along the DNA, so exposing a particular DNA sequence (nucleosome sliding: Fig. 1.3b). Finally, they may act by displacing a nucleosome so that it leaves the target DNA and binds to another DNA molecule (nucleosome displacement: Fig.1.3c). All these methods have in common the use of ATP hydrolysis to alter the nucleosome in some way so as to allow a particular region of DNA to become more accessible and hence bind specific regulatory factors.

Evidently, these mechanisms beg the question of how the SWI/SNF complex is itself recruited to the genes which need to be activated. This can occur via its association with the RNA polymerase complex or by its association with other transcription factors which can bind to their specific DNA binding sites even in tightly packed, non-remodelled chromatin. These processes are discussed in subsequent chapters.

Interestingly, it has recently been shown that chromatin remodelling complexes can also be recruited to the DNA by the SATBI protein which is involved in the looping of the chromatin into a highly compact structure

Figure 1.3

The SWI/SNF complex can allow a regulatory protein access to its binding site (X) by (a) producing an altered structure of the nucleosome in a process known as nucleosome remodelling; (b) inducing nucleosome sliding to a different position on the DNA; or (c) displacing the nucleosome onto another DNA molecule.

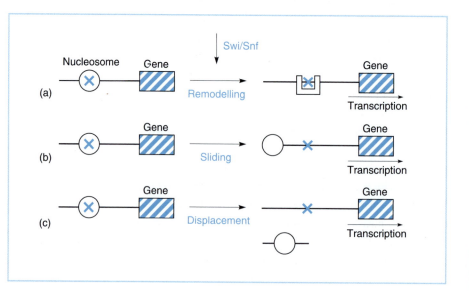

(Yasui *et al.*, 2002) (see section 1.2.1). This provides a link between the looping process and chromatin remodelling/gene regulation and suggests that such remodelling processes can target the large regions of DNA (20–80 000 bases of DNA) contained in individual loops.

1.2.3 HISTONE ACETYLATION

The histone molecules which play a key role in chromatin structure are subject to a number of post-translational modifications such as phosphorylation, ubiquitination or acetylation (for reviews see Strahl and Allis, 2000; Wu and Grunstein, 2000; Jenuwein and Allis, 2001; Felsenfeld and Groudine, 2003). In particular, the addition of an acetyl group to a free amino group in lysine residues in the histone molecule reduces its net positive charge. Such acetylated forms of the histones have been found preferentially in active or potentially active genes where the chromatin is less tightly packed. Moreover, treatments which enhance histone acetylation, such as addition of sodium butyrate to cultured cells, result in a less tightly packed chromatin structure and the activation of previously silent cellular genes. This suggests that hyperacetylation of histones could play a causal role in producing the more open chromatin structure characteristic of active or potentially active genes.

Hence, activation of gene expression could be achieved by factors with histone acetyltransferase activity which were able to acetylate histones and hence open up the chromatin structure, whereas inhibition of gene expression would be achieved by histone deacetylases which would have the opposite effect (Fig. 1.4). Most interestingly, recent studies have identified both components of the basal transcriptional complex and specific activating transcription factors with histone acetyltransferase activity as well as specific inhibitory transcription factors with histone deacetylase activity (for review see Brown *et al.*, 2000). These findings, which link studies on modulation of chromatin structure with those on activating and inhibitory transcription factors, are discussed further in later chapters.

It is clear, therefore, that histone acetylation plays a key role in regulating chromatin structure. However, in the last few years it has become increasingly clear that other histone modifications, such as methylation, phosphorylation or the addition of the small protein, ubiquitin (ubiquitination) are also involved in this process and that these modifications interact with one another and with acetylation. Thus, for example, demethylation of the lysine amino acid at position 9 in histone H3 facilitates phosphorylation of serine 10 and acetylation of lysine 14 of H3 leading to opening of the chromatin and gene activation (Paro, 2000) (Fig. 1.5). Such interaction can also occur between modifications on one histone molecule and those on another. Thus, ubiqui-

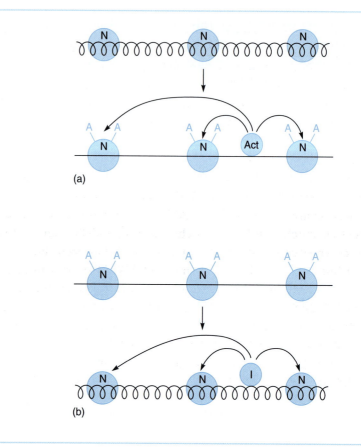

(a)

(b)

Figure 1.4

(a) An activating molecule (Act) can direct the acetylation of histones in the nucleosome (N) thereby resulting in a change in chromatin structure from a tightly packed (wavy line) to a more open (solid line) configuration. (b) An inhibitory molecule can direct the deacetylation of histones thereby having the opposite effect on chromatin structure.

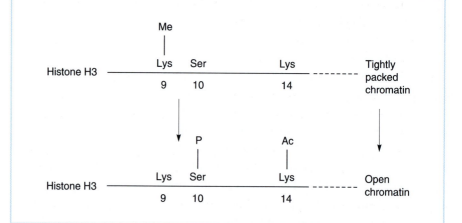

Figure 1.5

Demethylation of the lysine amino acid at position 9 in histone H3 facilitates phosphorylation of serine 10 and acetylation of lysine 14 leading to a more open chromatin structure.

tination of histone H2B facilitates subsequent methylation of histone H3 on the lysines at positions 4 and 79 leading to a tightly packed chromatin structure and gene silencing (Briggs *et al.*, 2002; Sun and Allis, 2002).

This complex pattern of modification has led to the idea of a 'histone code' in which the chromatin structure of a particular gene is specified by the pattern of different modifications of the histones that package it (for reviews see Strahl and Allis, 2000; Berger, 2001; Goll and Bestor, 2002; Turner, 2002).

Hence, both ATP-dependent chromatin remodelling complexes and alterations in histone acetylation/modification play a vital role in regulating the chromatin structure of specific genes. Although these two processes have been discussed separately, it is likely that chromatin remodelling and histone modification enzymes cooperate. Thus, for example, it has been shown that acetylation of histones can allow recruitment of SWI/SNF to a promoter (Agalioti *et al.*, 2002) as well as preventing it from dissociating once it has bound (Hassan *et al.*, 2001). Hence, it appears that these two processes act together to ensure that the DNA sequences involved in transcription control become accessible at the correct time in development or in response to appropriate signals (for review see Wu and Grunstein, 2000; Narlikor *et al.*, 2002). The nature of these DNA sequences is discussed in the next section.

1.3 DNA SEQUENCE ELEMENTS

1.3.1 THE GENE PROMOTER

The primary aim of chromatin remodelling processes is to expose specific DNA sequences so that these can be targeted by transcription factors involved in the process of gene transcription. In prokaryotes, such sequences are found immediately upstream of the start site of transcription and form part of the promoter directing expression of the genes. Sequences found at this position include both elements found in all genes which are involved in the basic process of transcription itself and those found in a more limited number of genes which mediate their response to a particular signal (for review see Muller-Hill, 1996).

Early studies of cloned eukaryotic genes, therefore, concentrated on the region immediately upstream of the transcribed region where, by analogy, sequences involved in transcription and its regulation should be located. Putative regulatory sequences were identified by comparison between different genes and the conclusions reached in this way confirmed either by destroying these sequences by deletion or mutation or by transferring them to another gene in an attempt to alter its pattern of regulation.

This work carried out on a number of different genes encoding specific proteins identified many short sequence elements involved in transcriptional control (for reviews see Davidson *et al.*, 1983; Jones *et al.*, 1988). The elements of this type present in two typical examples, the human gene encoding the 70 kd heat inducible (heat shock) protein (Williams *et al.*, 1989) and the human metallothionein IIA gene (Lee *et al.*, 1987), are illustrated in Figure 1.6.

Comparisons of these and many other genes revealed that, as in bacteria, their upstream regions contain two types of elements. First, sequences found in very many genes exhibiting distinct patterns of regulation which are likely to be involved in the basic process of transcription itself and secondly those found only in genes transcribed in a particular tissue or in response to a specific signal which are likely to produce this specific pattern of expression. These will be discussed in turn.

Figure 1.6

Transcriptional control elements upstream of the transcriptional start site in the human genes encoding hsp70 (panel a) and methallothionein IIA (panel b). The TATA, Sp1 and CCAAT boxes bind factors that are involved in constitutive transcription while the glucocorticoid response element (GRE), metal response element (MRE), heat shock element (HSE) and the AP1 and AP2 sites bind factors involved in the induction of gene expression in response to specific stimuli.

1.3.2 SEQUENCES INVOLVED IN THE BASIC PROCESS OF TRANSCRIPTION

Although they are regulated very differently, the hsp70 and metallothionein genes both contain a TATA box. This is an AT-rich sequence (consensus TATAA/TAA/T) which is found about thirty base pairs upstream of the transcriptional start site in very many but not all genes. Mutagenesis or relocation of this sequence has shown that it plays an essential role in accurately positioning the start site of transcription (Breathnach and Chambon, 1981). The region of the gene bracketed by the TATA box and the site of transcriptional initiation (the Cap site) has been operationally defined as the gene promoter or core promoter (Goodwin *et al.*, 1990). It is likely that this region binds several proteins essential for transcription, as well as RNA polymerase II itself which is the enzyme responsible for transcribing protein coding genes.

Although the TATA box is found in most eukaryotic genes, it is absent in some genes, notably housekeeping genes expressed in all tissues and in some tissue-specific genes (for reviews of the different classes of core promoters see Smale, 2001; Butler and Kadonaga, 2002). In these promoters, a sequence known as the initiator element, which is located over the start site of transcription itself, appears to play a critical role in determining the initiation point and acts as a minimal promoter capable of producing basal levels of transcription (see Chapter 3, section 3.6 for a discussion of transcription from promoters containing or lacking a TATA box).

In promoters which contain a TATA box and in those which lack it, the very low activity of the promoter itself is dramatically increased by other elements located upstream of the promoter. These elements are found in a very wide variety of genes with different patterns of expression indicating that they play a role in stimulating the constitutive activity of promoters. Thus inspection of the hsp70 and metallothionein IIA genes reveals that both contain one or more copies of a GC-rich sequence, known as the Sp1 box, which is found upstream of the promoter in many genes both with and without TATA boxes (for review see Lania *et al.*, 1997).

In addition, the hsp70 promoter but not the metallothionein promoter contains another sequence, the CCAAT box, which is also found in very many genes with disparate patterns of regulation. Both the CCAAT box and the Sp1 box are typically found upstream of the TATA box as in the metallothionein and hsp70 genes. Some genes, as in the case of hsp70 may have both of these elements, whereas others such as the metallothionein gene have single or multiple copies of one or the other. In every case, however, these elements are essential for transcription of the genes and their elimination by deletion or mutation abolishes transcription. Hence these sequences play an essential role in efficient transcription of the gene and have been termed upstream promoter elements (UPE: Goodwin *et al.*, 1990).

1.3.3 SEQUENCES INVOLVED IN REGULATED TRANSCRIPTION

Inspection of the hsp70 promoter (see Fig.1.6) reveals several other sequence elements which are only shared with a much more limited number of other genes and which are interdigitated with the upstream promoter elements discussed above. Indeed, one of these, which is located approximately ninety bases upstream of the transcriptional start site, is shared only with other heat shock genes whose transcription is increased in response to elevated temperature. This suggests that this heat shock element may be essential for the regulated transcription of the hsp70 gene in response to heat.

To prove this directly, however, it is necessary to transfer this sequence to a non-heat-inducible gene and show that this transfer renders the recipient gene heat inducible. Pelham (1982) successfully achieved this by linking the heat shock element to the non-heat-inducible thymidine kinase gene of the eukaryotic virus herpes simplex. This hybrid gene could be activated following its introduction into mammalian cells by raising the temperature (Fig. 1.7). Hence the heat shock element can confer heat inducibility on another gene, directly proving that its presence in the hsp gene promoters is responsible for their heat inducibility.

Moreover, although these experiments used a heat shock element taken from the hsp70 gene of the fruit fly *Drosophila melanogaster*, the hybrid gene was introduced into mammalian cells. Not only does the successful functioning of the fly element in mammalian cells indicate that this process is evolutionarily conserved, but it permits a further conclusion about the way in which the effect operates. Thus, in the cold-blooded *Drosophila*, 37°C represents a thermally stressful temperature and the heat shock response would normally be active at this temperature. The hybrid gene was inactive at 37°C in the mammalian cells, however, and was only induced at 42°C, the heat shock

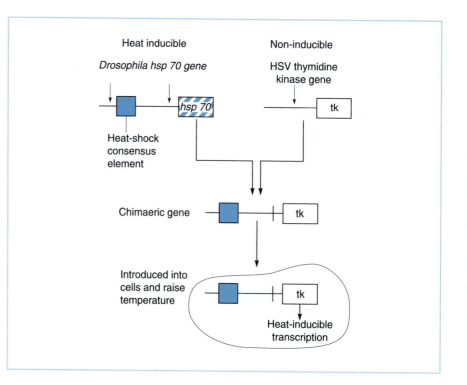

Figure 1.7

Demonstration that the heat shock element mediates heat inducibility. Transfer of this sequence to a gene (thymidine kinase) which is not normally inducible renders this gene heat inducible.

temperature characteristic of the cell into which it was introduced. Hence this sequence does not act as a thermostat, set to go off at a particular temperature since this would occur at the *Drosophila* heat shock temperature (Fig. 1.8a). Rather, this sequence must act by being recognized by a cellular protein which is activated only at an elevated temperature characteristic of the mammalian cell heat shock response (Fig.1.8b).

This experiment therefore not only directly proves the importance of the heat shock element in producing the heat inducibility of the hsp70 gene, but also shows that this sequence acts by binding a cellular protein which is activated in response to elevated temperature. The binding of this transcription factor then activates transcription of the hsp70 gene. The manner in which this factor activates transcription of the hsp70 gene and the other heat shock genes is discussed further in Chapter 8 (section 8.3.1).

The presence of specific DNA sequences which can bind particular proteins, will therefore confer on a specific gene the ability to respond to particular stimuli. Thus, the lack of a heat shock element in the metallothionein IIA gene (see Fig.1.6) means that this gene is not heat inducible. In contrast, however, this gene, unlike the hsp70 gene, contains a glucocorticoid response element (GRE). Hence it can bind the complex of the glucocorticoid receptor and the hormone itself, which forms following treatment of cells with glucocorticoid. Its transcription is therefore activated in response to glucocorticoid

Figure 1.8

Predicted effects of placing the *Drosophila* heat shock element in a mammalian cell if the element acts as a thermostat detecting elevated temperature directly (panel a) or if it acts by binding a protein which is activated by elevated temperature (panel b). Note that only possibility (b) can account for the observation that the *Drosophila* heat shock element only activates transcription in mammalian cells at the mammalian heat shock temperature of 42°C and not at the *Drosophila* heat shock temperature of 37°C.

whereas that of the hsp70 gene is not (see Chapter 4, section 4.4). Similarly, only the metallothionein gene contains metal response elements (MRE) allowing it to be activated in response to treatment with heavy metals such as zinc and cadmium (Thiele, 1992). In contrast both genes contain binding sites for the transcription factor AP2 which mediates gene activation in response to cyclic AMP and phorbol esters.

Similar DNA sequence elements in the promoters of tissue specific genes play a critical role in producing their tissue specific pattern of expression by binding transcription factors which are present in an active form only in a particular tissue where the gene will be activated. For example, the promoters of the immunoglobulin heavy and light chain genes contain a sequence known as the octamer motif (ATGCAAAT) which can confer B-cell specific expression on an unrelated promoter (Wirth et al., 1987). Similarly, the related sequence ATGAATAA/T is found in genes expressed specifically in the anterior pituitary gland, such as the prolactin gene and the growth hormone gene, and binds a transcription factor known as Pit-1 which is expressed only in the anterior pituitary (for review see Andersen and Rosenfeld, 1994). If this short sequence is inserted upstream of a promoter, the gene is expressed only in pituitary cells. In contrast the octamer motif which differs by only two bases will direct expression only in B cells when inserted upstream of the same promoter (Elsholtz et al., 1990; Fig. 1.9). Hence small differences in control element sequences can produce radically different patterns of gene expression.

Figure 1.9

Linkage of the octamer binding motif ATGCAAAT (1) and the related Pit-1 binding motif ATGAATAT (2) to the prolactin promoter and introduction into B cells and pituitary cells (panel a). Only the octamer containing construct 1 directs a high level of activity in B cells, whereas only construct 2 containing the Pit-1 binding site directs a high level of gene activity in pituitary cells (panel b). Data from Elsholtz et al. (1990).

1.3.4 SEQUENCES WHICH ACT AT A DISTANCE

(a) Enhancers

One of the characteristic features of eukaryotic gene expression is the existence of sequence elements located at great distances from the start site of transcription which can influence the level of gene expression. These elements can be located upstream, downstream or within a transcription unit and function in either orientation relative to the start site of transcription (Fig. 1.10). They act by increasing the activity of a promoter, although they

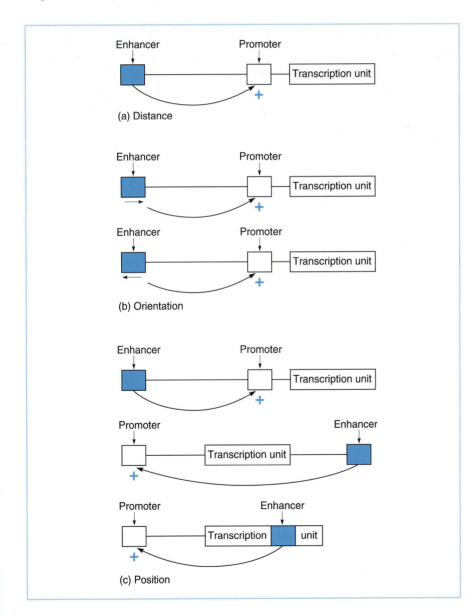

Figure 1.10

Characteristics of an enhancer element which can activate a promoter at a distance (a); in either orientation relative to the promoter (b); and when positioned upstream, downstream, or within a transcription unit (c).

lack promoter activity themselves and are hence referred to as enhancers (for reviews see Hatzopoulos *et al.*, 1988; Muller *et al.*, 1988). Some enhancers are active in all tissues and increase the activity of a promoter in all cell types, while others function as tissue specific enhancers which activate a particular promoter only in a specific cell type. Thus the enhancer located in the intervening region of the immunoglobulin genes is active only in B cells and the B-cell-specific expression of the immunoglobulin genes is produced by the interaction of this enhancer and the immunoglobulin promoter which, as we have previously seen, is also B-cell specific (Garcia *et al.*, 1986).

As with promoter elements, enhancers contain multiple binding sites for transcription factors which interact together to mediate enhancer function. This has led to the idea that a multi-protein complex, known as the enhanceosome, assembles on the enhancer and induces transcriptional activation of the target gene (for reviews see Merika and Thanos, 2001; Strahl, 2001). In many cases the elements within enhancers are identical to those contained immediately upstream of gene promoters. Thus, the immunoglobulin heavy chain enhancer contains a copy of the octamer sequence (Sen and Baltimore, 1986) which is also found in the immunoglobulin promoters (section 1.3.3). Similarly multiple copies of the heat shock element are located far upstream of the start site in the *Xenopus* hsp70 gene and function as a heat inducible enhancer when transferred to another gene (Bienz and Pelham, 1986).

Enhancers therefore consist of sequence elements which are also present in similarly regulated promoters and may be found within the enhancer associated with other control elements or in multiple copies.

(b) Locus control regions

The genes encoding the β-globin component of haemoglobin and other related molecules are found clustered together in the genome with five functional genes located adjacent to one another. All of these genes are expressed in erythroid (red blood cell) precursors and not in other cell types and this pattern of expression is dependent on an element located 10–20 kilo-bases upstream of the gene cluster which is known as a locus control region (LCR) (Fig. 1.11). In the absence of this element, none of the genes is expressed in the correct erythroid-specific manner (for reviews of LCRs see Bulger and Groudine, 1999; Li *et al.*, 1999).

It is likely that the LCR functions by regulating chromatin structure so that the entire region of the genome containing β-globin-like genes is opened up in red blood cell precursors. Each of the genes within the region can then be individually regulated in the red blood cell lineage by their own individual enhancer and promoter elements with, for example, the ϵ-globin gene being

Figure 1.11
The locus control region (LCR) in the β-globin gene cluster directs the correct pattern of chromatin opening in erythroid cells. Regulatory processes acting on each gene in the cluster then allow it to be expressed at the correct time in erythroid development with the ε-globin gene being expressed in the early embryo, the Gγ and Aγ-globin genes in the fetus and the β and δ-globin genes in the adult.

expressed in the embryo and the β- and δ-globin genes in the adult (see Fig. 1.11).

Since its original identification in the β-globin locus, LCRs have been found regulating the expression of a number of other gene clusters expressed in different cell types. Interestingly, in several cases, LCRs contain matrix attachment regions (see section 1.2.1). This suggests that a region controlled by an LCR, such as the β-globin cluster, may form a single large loop attached to the nuclear matrix whose chromatin structure is regulated as a single unit.

The typical eukaryotic gene will therefore consist of multiple distinct transcriptional control elements (Fig. 1.12). These are first, the promoter itself, secondly upstream promoter elements (UPE) located close to it, which are required for efficient transcription in any cell type, thirdly, other elements adjacent to the promoter which are interdigitated with the UPEs and which activate the gene in particular tissues or in response to particular stimuli and, lastly, elements such as enhancers or locus control regions which act at a distance to regulate gene expression.

Such sequences often act by binding positively acting factors which then stimulate transcription (Fig. 1.13a). As will be discussed in later chapters, this could involve the DNA binding protein either altering chromatin structure to make the DNA more accessible to other positively-acting regulatory factors or

Figure 1.12
Structure of a typical gene with a TATA box-containing promoter, upstream promoter elements such as the CCAAT and Sp1 boxes, regulatory elements inducing expression in response to treatment with substances such as glucocorticoid (GRE) and cyclic AMP (CRE) and other elements within more distant enhancers. Note that as discussed in the text and illustrated in Figure 1.6, the upstream promoter elements are often interdigitated with the regulatory elements while the same regulatory elements can be found upstream of the promoter and in enhancers.

Figure 1.13

Panel (a) A specific DNA sequence (X) can act to stimulate transcription by binding a positively acting factor (Y). Panel (b) In contrast, binding of the negatively acting factor (W) to the DNA sequence Z inhibits transcription.

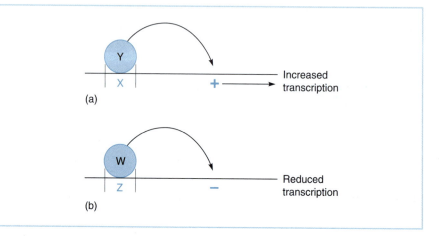

direct stimulation of transcription by the DNA binding protein interacting with RNA polymerase or its associated molecules. Interestingly, however, although most sequences act in such a positive way, some sequences do appear to act in a negative manner to inhibit transcription and these are discussed in the next section.

1.3.5 NEGATIVELY ACTING DNA SEQUENCES

(a) Silencers

Silencer elements, which act to inhibit gene transcription, have been defined in a number of genes including the cellular oncogene c-*myc* (Chapter 7, section 7.2.3) and those encoding proteins such as growth hormone or collagen type II. As with activating sequences, some silencer elements are constitutively active while others display cell-type specific activity. Thus, for example, the silencer in the gene encoding the T-lymphocyte marker CD4 represses its expression in most T cells where CD4 is not expressed but is inactive in a subset of T cells allowing these cells actively to express the CD4 protein (Sawada *et al.*, 1994). In many cases silencer elements have been shown to act by binding regulatory factors which then act to reduce the rate of transcription (Fig. 1.13b) either by promoting a more tightly packed chromatin structure or by interacting with RNA polymerase and its associated molecules in an inhibitory manner.

(b) Insulators

The ability of sequences such as enhancers or LCRs to act over large distances evidently begs the question of how their activity is limited to the genes that they need to regulate and does not affect other genes in adjacent regions. This is achieved by DNA elements known as insulators which act to block the spread of enhancer or silencer activity (Fig. 1.14) (for reviews see Bell *et al.*, 2001; Labrador and Corces, 2002; West *et al.*, 2002).

It is likely that insulators act by blocking the alterations in DNA structure induced by enhancers or silencers. In some cases this involves a direct effect on chromatin structure preventing the opening of chromatin structure induced by enhancers (or the production of a more tightly packed chromatin structure induced by silencers) from spreading to a particular region of chromatin. In other cases an insulator may prevent the looping of DNA which is required to bring together regulatory proteins bound at the enhancer/silencer with their target proteins bound to the promoter (see below section 1.3.6).

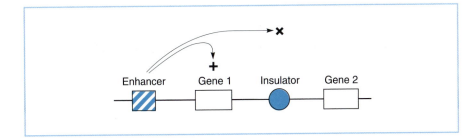

Figure 1.14
An insulator sequence can limit the action of an enhancer to genes located between the enhancer and the insulator.

1.3.6 INTERACTION BETWEEN FACTORS BOUND AT VARIOUS SITES

Obviously the balance between positively and negatively acting transcription factors which bind to the regulatory regions of a particular gene will determine the rate of gene transcription in any particular situation. In some cases binding of the RNA polymerase and associated factors to the promoter and of other positive factors to the UPEs will be sufficient for transcription to occur and the gene will be expressed constitutively. In other cases, however, such interactions will be insufficient and transcription of the gene will occur only in response to the binding, to another DNA sequence, of a factor which is activated in response to a particular stimulus or is present only in a particular tissue. These regulatory factors will then interact with the constitutive factors allowing transcription to occur. Hence their binding will result in the observed tissue specific or inducible pattern of gene expression.

Such interaction is well illustrated by the metallothionein IIA gene. As illustrated in Figure 1.6 this gene contains a binding site for the transcription factor AP1 which produces induction of gene expression in response to phorbol ester treatment. The action of AP1 on the expression of the metallothionein gene is abolished, however, both by mutations in its binding site and by mutations in the adjacent Sp1 motif which prevent this motif binding its corresponding transcription factor Sp1 (Lee *et al.*, 1987). Although these mutations in the Sp1 motif do not abolish AP1 binding they do prevent its action, indicating that the inducible AP1 factor interacts with the constitutive Sp1 factor to activate transcription.

Clearly such interactions between bound transcription factors need not be confined to factors bound to regions adjacent to the promoter but can also involve the similar factors bound to more distant enhancers. It is likely that this is achieved by a looping out of the intervening DNA allowing contact between factors bound at the promoter and those bound at the enhancer (Fig. 1.15) (for further discussion see Bulger and Groudine, 1999; Latchman, 2002).

This need for transcription factors to interact with one another to stimulate transcription means that transcription can also be stimulated by a class of factors which act indirectly by binding to the DNA and bending it so that other DNA bound factors can interact with one another (Fig. 1.16). Thus, the LEF-1 factor, which is specifically expressed in T lymphocytes, binds to the

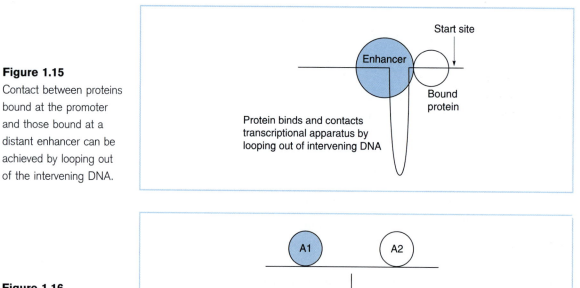

Figure 1.15

Contact between proteins bound at the promoter and those bound at a distant enhancer can be achieved by looping out of the intervening DNA.

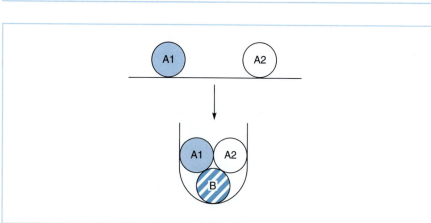

Figure 1.16

A factor which bends the DNA (B) can indirectly activate transcription by facilitating the interaction of two activating transcription factors (A1 and A2).

enhancer of the T-cell receptor α gene and bends the DNA so that other constitutively expressed transcription factors can interact with one another, thereby allowing them to activate transcription. This results in the T-cell specific expression of the gene even though the directly activating factors are not expressed in a T-cell specific manner (for review see Werner and Burley, 1997). Similarly, the DNA binding transcription factor HMGI (Y) plays a critical role in the multi-protein enhanceosome which assembles on the interferon β gene enhancer and is essential for the inducibility of this gene following viral infection (for review see Merika and Thanos, 2001) (for further discussion of the processes involved in the activation of this promoter see Chapter 5, section 5.6).

1.4 CONCLUSIONS

It is clear that both the process of transcription itself and its regulation in particular tissues or in response to particular signals are controlled by short DNA sequence elements located adjacent to the promoter or in enhancers. In turn such sequences act by binding proteins which are either active constitutively or are present in an active form only in a specific tissue or following a specific inducing signal. Such DNA bound transcription factors then interact with each other and the RNA polymerase itself in order to produce constitutive or regulated transcription. The nature of these factors, the manner in which they function and their role in different biological processes form the subject of this book.

REFERENCES

Aalfs, J. D. and Kingston, R. E. (2000) What does 'chromatin remodelling' mean? Trends in Biochemical Sciences 25, 548–555.

Agalioti, T., Chen, G. and Thanos, D. (2002) Deciphering the transcriptional histone acetylation code for a human gene. Cell 111, 381–392.

Andersen, B. and Rosenfeld, M.G. (1994) Pit-1 determines cell types during the development of the anterior pituitary gland. Journal of Biological Chemistry 269, 29335–29338.

Bashirullah, A., Cooperstock, R. L. and Lipshitz, H. D. (2001) Spatial and temporal control of RNA stability. Proceedings of the National Academy of Sciences USA 98, 7025–7028.

Bell, A. C., West, A. G. and Felsenfeld, G. (2001) Insulators and boundaries: versatile regulatory elements in the eukaryotic genome. Science 291, 447–450.

Berger, S. L. (2001) The histone modification circus. Science 292, 64–65.

Bienz, M. and Pelham, H.R.B. (1986) Heat shock regulatory elements function as an inducible enhancer when linked to a heterologous promoter. Cell 45, 753–760.

Bradbury, E. M. (2002) Chromatin structure and dynamics: state-of-the-art. Molecular Cell 10, 13–19.

Breathnach, R. and Chambon, P. (1981) Organization and expression of eukaryotic split genes coding for proteins. Annual Review of Biochemistry 50, 349–383.

Briggs, S. D., Xiao, T., Sun, Z-W. et al. (2002) *Trans*-histone regulatory pathway in chromatin. Nature 418, 498.

Brown, C. E., Lechner, T., Howe, L. and Workman, J. L. (2000) The many HATs of transcription coactivators. Trends in Biochemical Sciences 25, 15–19.

Bulger, M. and Groudine, M. (1999) Looping versus linking: toward a model for long-distance gene activation. Genes and Development 13, 2465–2477.

Butler, J. E. F. and Kadonaga, J. T. (2002) The RNA polymerase II core promoter: a key component in the regulation of gene expression. Genes and Development 16, 2583–2592.

Darnell, J.E. (1982) Variety in the level of gene control in eukaryotic cells. Nature 297, 365–371.

Davidson, E. H., Jacobs, H. T. and Britten, R. J. (1983) Very short repeats and co-ordinate induction of genes. Nature 301, 468–470.

Elsholtz, H.P., Albert, V.R., Treacy, M. N. and Rosenfeld, M.G. (1990) A two-base change in a POU factor binding site switches pituitary-specific to lymphoid-specific gene expression. Genes and Development 4, 43–51.

Felsenfeld, G. and Groudine, M. (2003) Controlling the double helix. Nature 421, 448–453.

Garcia, J.V., Bich-Thuy, L., Stafford, J. and Queen, C. (1986) Synergism between immunoglobulin enhancers and promoters. Nature 322, 383–385.

Goll, M. G. and Bestor, T. H. (2002) Histone modification and replacement in chromatin activation. Genes and Development 16, 1739–1742.

Goodwin, G.H., Partington, G.A. and Perkins, N.D. (1990) Sequence specific DNA binding proteins involved in gene transcription. In: Chromosomes: eukaryotic, prokaryotic and viral. Adolph, K.W. (ed) vol. 1 pp. 31–85, CRC Press.

Graveley, B. R. (2001) Alternative splicing: increasing diversity in the proteomic world. Trends in Genetics 17, 100–107.

Hassan, A.H., Neely, K.E. and Workman, J.L. (2001) Histone acetyltransferase complexes stabilise SWI/SNF binding to promoter nucleosomes. Cell 104, 817–827.

Hatzopoulos, A.K., Schlokat, U. and Gruss, P. (1988) Enhancers and other cis-acting sequences In: Transcription and splicing, Hames, B.D. and Glover, D.M. (eds), pp. 43–96, IRL Press.

Horn, P. J. and Peterson, C. L. (2002) Chromatin higher order folding: wrapping up transcription. Science 297, 1824–1827.

Jenuwein, T. and Allis, C. D. (2001) Translating the histone code. Science 293, 1074–1080.

Jones, N.C., Rigby, P.W.J. and Ziff, E.G. (1988) Trans-acting protein factors and the regulation of eukaryotic transcription. Genes and Development 2, 267–281.

Kornberg, R. D. and Lorch, Y. (1999) Twenty-five years of the nucleosome, fundamental particle of the eukaryote chromosome. Cell 98, 285–294.

Labrador, M. and Corces, V.G. (2002) Setting the boundaries of chromatin domains and nuclear organisation. Cell 111, 151–154.

Lania, L., Majello, B. and de Luca, P. (1997) Transcriptional regulation by the Sp family proteins. International Journal of Biochemistry and Cell Biology 29, 1313–1323.

Latchman, D.S. (2002) Gene Regulation: A eukaryotic perspective. Fourth Edition. Nelson Thornes Ltd.

Lee, W., Haslinger, A., Karin, M. and Tjian, R. (1987) Activation of transcription by two factors that bind promoter and enhancer sequences of the human metallothionein gene and SV40. Nature 325, 369–372.

Li, Q., Harju, S. and Peterson, K. R. (1999) Locus control regions coming of age at a decade plus. Trends in Genetics 15, 403–408.

Maniatis, T., Goodboun, S. and Fischer, J.A. (1987) Regulation of inducible and tissue specific gene expression. Science 236, 1237–1245.

Merika, M. and Thanos, D. (2001) Enhanceosomes. Current Opinion in Genetics and Development 11, 205–208.

Muller, M.M., Gerster, T. and Schaffner, W. (1988) Enhancer sequences and the regulation of gene transcription. European Journal of Biochemistry 176, 485–495.

Muller-Hill, B.W. (ed) (1996) The lac operon: a short history of a genetic paradigm. de Grayler Co. Berlin.

Narlikor, G. J., Fan, H-Y. and Kingston, R. E. (2002) Co-operation between complexes that regulate chromatin structure and transcription. Cell 108, 475–487.

Nevins, J.R. (1983) The pathway of eukaryotic mRNA transcription. Annual Review of Biochemistry 52, 441–446.

Paro, R. (2000) Formatting genetic text. Nature 406, 579–580.

Pelham, H.R.B. (1982) A regulatory upstream promoter element in the Drosophila hsp70 heat-shock gene. Cell 30, 517–528.

Richards, E. J. and Elgin, S. C. R. (2002) Epigenetic codes for heterochromatin formation and silencing: rounding up the usual suspects. Cell 108, 489–500.

Sawada, S., Scarborough, J.D., Kileen, N. and Littman, D.R. (1994) A lineage-specific transcriptional silencer regulates CD4 gene expression during T lymphocyte development. Cell 77, 917–929.

Sen, R. and Baltimore, D. (1986) Multiple nuclear factors interact with the immunoglobulin enhancer sequences. Cell 46, 705–716.

Simon, J. (1995) Locking in stable states of gene expression: transcriptional control during *Drosophila* development. Current Opinion in Cell Biology 7, 376–385.

Smale, S.T. (2001) Core promoters: active contributors to combinatorial gene regulation. Genes and Development 15, 2503–2508.

Strahl, B. D. and Allis, C. D. (2000) The language of covalent histone modifications. Nature 403, 41–45.

Strahl, K. (2001) A paradigm for precision. Science 293, 1054–1055.

Sudarsanam, P. and Winston, F. (2000) The Swi/Snf family nucleosome-remodelling complexes and transcriptional control. Trends in Genetics 16, 345–351.

Sun, Z-W. and Allis, C. D. (2002) Ubiquitination of histone H2B regulates H3 methylation and gene silencing in yeast. Nature 418, 104–108.

Thiele, D.J. (1992) Metal regulated transcription in eukaryotes. Nucleic Acids Research 20, 1183–1191.

Turner, B.M. (2002) Cellular memory and the histone code. Cell 111, 285–291.

Wang, J. and Manley, J.L. (1997) Regulation of pre-mRNA splicing in metazoa. Current Opinion in Genetics and Development 7, 205–211.

Werner, M.H. and Burley, S.K. (1997) Architectural transcription factors: proteins that remodel DNA. Cell 88, 733–736.

West, A. G., Gaszner, M. and Felsenfeld, G. (2002) Insulators: many functions, many mechanisms. Genes and Development 16, 271–288.

Williams, G.T., McClanahan, T.K. and Morimoto, R. I. (1989) E1a transactivation of the human hsp70 promoter is mediated through the basal transcriptional complex. Molecular and Cellular Biology 9, 2574–2587.

Wirth, T., Staudt, L. and Baltimore, D. (1987) An octamer oligonucleotide upstream of a TATA motif is sufficient for lymphoid specific promoter activity. Nature 329, 174–178.

Wolffe, A. (1995) Chromatin: structure and function, Second Edition. Academic Press.

Wu, J. and Grunstein, M. (2000) 25 years after the nucleosome model: chromatin modifications. Trends in Biochemical Sciences 25, 619–623.

Yasui, D., Miyano, M., Cal, S. *et al.* (2002) SATB1 targets chromatin remodelling to regulate genes over long distances. Nature 419, 641–645.

CHAPTER 2

METHODS FOR STUDYING TRANSCRIPTION FACTORS

2.1 INTRODUCTION

The explosion in the available information on transcription factors that has occurred in recent years has arisen primarily because of the availability of new or improved methods for studying these factors. Initially such studies may focus on identifying a factor that interacts with a particular DNA sequence and characterizing this interaction and the methods for doing this are discussed in section 2.2. Subsequently, the protein identified in this way is further characterized and purified and its corresponding gene isolated. The methods involved in the purification and/or cloning of transcription factors are considered in section 2.3, while section 2.4 analyses the methods used to characterize such cloned transcription factors including the methods for determining the DNA binding site or gene targets of a transcription factor which is initially identified by means other than its DNA binding characteristics. (For details of the methodologies involved see Latchman, 1999.)

2.2 METHODS FOR STUDYING DNA–PROTEIN INTERACTIONS

2.2.1 DNA MOBILITY SHIFT ASSAY

As discussed in Chapter 1 (section 1.3), the initial stimulus to identify a transcription factor frequently comes from the identification of a particular DNA sequence that confers a specific pattern of expression on a gene which carries it. The next step therefore following the identification of such a sequence will be to define the protein factors that bind to it. This can be readily achieved by the DNA mobility shift or gel retardation assay (Fried and Crothers, 1981; Garner and Revzin, 1981).

This method relies on the obvious principle that a fragment of DNA to which a protein has bound will move more slowly in gel electrophoresis than the same DNA fragment without bound protein. The DNA mobility shift assay is carried out therefore by first radioactively labelling the specific DNA sequence whose protein binding properties are being investigated. The labelled DNA is then incubated with a nuclear (Dignam *et al.*, 1983) or whole cell (Manley *et al.*, 1980) extract of cells prepared in such a way as to contain the DNA binding proteins. In this way DNA–protein complexes are allowed to form. The complexes are then electrophoresed on a non-denaturing polyacrylamide gel and the position of the radioactive DNA visualized by autoradiography. If no protein has bound to the DNA, all the radioactive label will be at the bottom of the gel, whereas if a protein–DNA complex has formed, radioactive DNA to which the protein has bound will migrate more slowly and hence will be visualized near the top of the gel (Fig. 2.1). (For methodological details see Smith *et al.*, 1999.)

This technique can be used therefore to identify proteins which can bind to a particular DNA sequence in extracts prepared from specific cell types. Thus,

Figure 2.1

DNA mobility shift or gel retardation assay. Binding of a cellular protein (B) to the radioactively-labelled DNA causes it to move more slowly upon gel electrophoresis and hence results in the appearance of a retarded band upon autoradiography to detect the radioactive label.

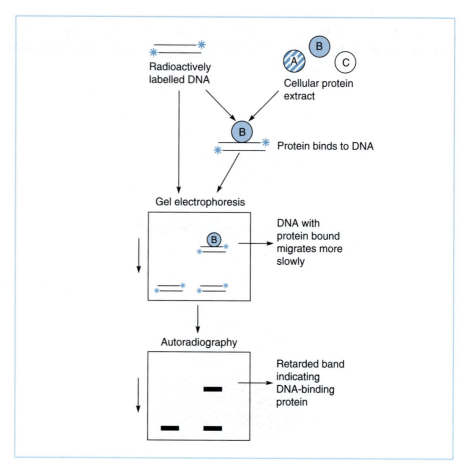

for example, in the case of the octamer sequence discussed in Chapter 1, (section 1.3.3), a single retarded band is detected when this sequence is mixed, for example, with a fibroblast extract. In contrast, when an extract from immunoglobulin-producing B cells is used, two distinct retarded bands are seen (Fig. 2.2). Since each band is produced by a distinct protein binding to the DNA, this indicates that, in addition to the ubiquitous octamer binding protein Oct-1 which is present in most cell types, B cells also contain an additional octamer binding protein, Oct-2, which is absent in many other cells.

As well as defining the proteins binding to a particular sequence, the DNA mobility shift assay can also be used to investigate the precise sequence specificity of this binding. This can be done by including in the binding reaction a large excess of a second DNA sequence which has not been labelled. If this DNA sequence can also bind the protein bound by the labelled DNA, it will do so. Moreover, binding to the unlabelled DNA will predominate since it is present in large excess. Hence the retarded band will not appear in the presence of the unlabelled competitor since only protein−DNA complexes containing labelled DNA are visualized on autoradiography (Fig. 2.3b). In contrast if the competitor cannot bind the same sequence as the labelled

Figure 2.2
DNA mobility shift assay using a radioactively-labelled probe containing the binding site for octamer binding proteins (ATGCAAAT) and extracts prepared from fibroblast cells (1) or B cells (2). Note that fibroblast cells contain only one protein Oct-1 (01) capable of producing a retarded band, whereas B cells contain both Oct-1 and an additional tissue-specific protein Oct-2 (02). The complexes formed by Oct-1 and Oct-2 on the labelled oligonucleotide in the absence of unlabelled oligonucleotide (track a) are readily removed by a one hundredfold excess of unlabelled octamer oligonucleotide (track b). They are not removed, however, by a similar excess of a mutant octamer oligonucleotide (ATAATAAT) which is known not to bind octamer binding proteins (track c), or of the binding site for the unrelated transcription factor Sp1 (track d: Dynan and Tjian, 1983). This indicates that the retarded bands are produced by sequence specific DNA binding proteins which bind specifically to the octamer motif and not to mutant or unrelated motifs.

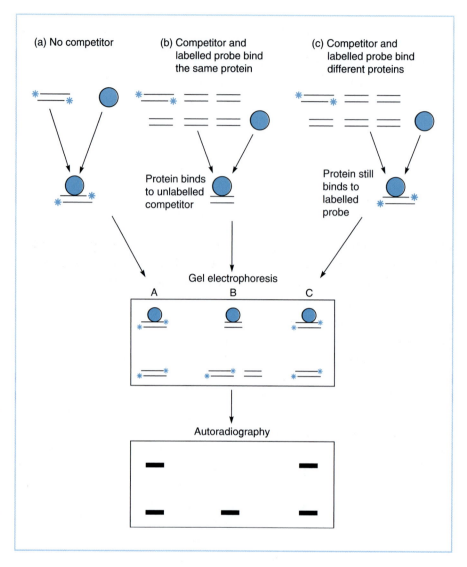

Figure 2.3

Use of unlabelled competitor DNAs in the DNA mobility shift assay. If an unlabelled DNA sequence is capable of binding the same protein as is bound by the labelled probe, it will do so (B) and the radioactive retarded band will not be observed, whereas if it cannot bind the same protein (C), the radioactive retarded band will form exactly as in the absence of competitor (A).

DNA, the complex with the labelled DNA will form and the labelled band will be visualized as before (Fig. 2.3c).

Thus, by using competitor DNAs which contain the binding sites for previously described transcription factors, it can be established whether the protein detected in a particular mobility shift experiment is identical or related to any of these factors. Similarly, if competitor DNAs are used which differ in only one or a few bases from the original binding site the effect of such base changes on the efficiency of the competitor DNA and hence on binding of the transcription factor can be assessed. Figure 2.2 illustrates an example of this type of competition approach showing that the octamer binding proteins Oct-

1 and Oct-2 are efficiently competed away from the labelled octamer probe by an excess of identical unlabelled competitor but not by a competitor containing three base changes in this sequence which prevents binding (ATGCAAAT to ATAATAAT). Similarly no competition is observed, as expected, when the binding site of an unrelated transcription factor Sp1 is used as the competitor DNA.

The DNA mobility shift assay therefore provides an excellent means of initially identifying a particular factor binding to a specific sequence and characterizing both its tissue distribution and its sequence specificity.

2.2.2 DNAseI FOOTPRINTING ASSAY

Although the mobility shift assay provides a means of obtaining information on DNA–protein interaction, it cannot be used directly to localize the area of the contact between protein and DNA. For this purpose, the DNAseI footprint assay is used (Galas and Schmitz, 1978; Dynan and Tjian, 1983).

In this assay, DNA and protein are mixed as before, the DNA being labelled however, only at the end of one strand of the double-stranded molecule. Following binding, the DNA is treated with a small amount of the enzyme deoxyribonuclease I (DNAseI) which will digest DNA. The digestion conditions are chosen, however, so that each molecule of DNA will be cut once or a very few times by the enzyme. Following digestion, the bound protein is removed and the DNA fragments separated by electrophoresis on a polyacrylamide gel capable of resolving DNA fragments differing in size by only one base. This produces a ladder of bands representing the products of DNAseI cutting either one or two or three or four etc., bases from the labelled end. Where a particular piece of the DNA has bound a protein, however, it will be protected from digestion and hence the bands corresponding to cleavage at these points will be absent. This will be visualized on electrophoresis as a blank area on the gel lacking labelled fragments and is referred to as the footprint of the protein (Fig. 2.4). Similar labelling of the other strand of the DNA molecule will allow the interaction of the protein with the other strand of the DNA to be assessed.

The footprinting technique therefore allows a visualization of the interaction of a particular factor with a specific piece of DNA. By using a sufficiently large piece of DNA, the binding of different proteins to different DNA sequences within the same fragment can be assessed. An analysis of this type is shown in Figure 2.5. This shows the footprints (A and B) produced by two cellular proteins binding to two distinct sequences within a region of the human immunodeficiency virus (HIV) control element which has an inhibiting effect on promoter activity (Orchard et al., 1990). Interestingly some

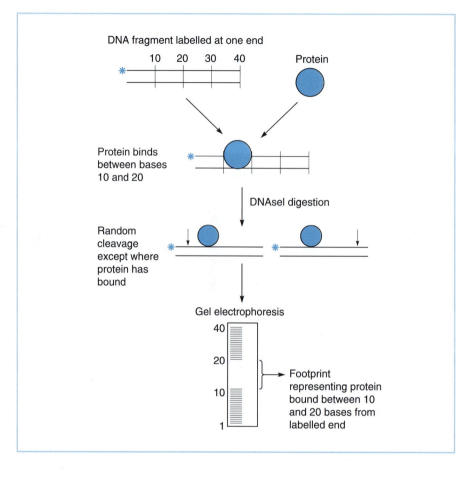

Figure 2.4

DNAseI footprinting assay. If a protein binds at a specific site within a DNA fragment labelled at one end, the region of DNA at which the protein binds will be protected from digestion with DNAseI. Hence this region will appear as a footprint in the ladder of bands produced by the DNA being cut at all other points by DNAseI.

insights into the topology of the DNA–protein interaction are also obtained in this experiment since bands adjacent to the protected region appear more intense in the presence of the protein. These regions of hypersensitivity to cutting are likely to represent a change in the structure of the DNA in this region when the protein has bound rendering the DNA more susceptible to enzyme cleavage.

As with the mobility shift assay, unlabelled competitor sequences can be used to remove a particular footprint and determine its sequence specificity. In the HIV case illustrated in Figure 2.5, short DNA competitors containing the sequence of one or other of the footprinted areas were used specifically to remove each footprint without affecting the other, indicating that two distinct proteins produce the two footprints.

As well as footprinting using DNAseI, other footprinting techniques have been developed which rely on the protection of DNA which has bound protein from cleavage by other reagents that normally cleave the DNA. These include hydroxyl radical footprinting and phenanthroline-copper footprinting

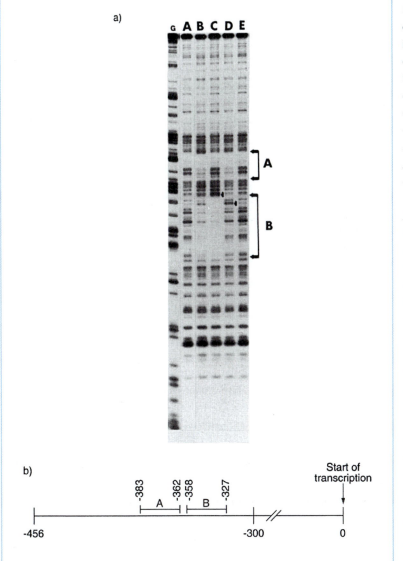

a)

b)

Figure 2.5

Panel (a): DNAseI footprinting assay carried out on a region of the human immunodeficiency virus (HIV) control element. The two footprints (A and B) are not observed when no cell extract is added to the reaction (track A) but are observed when cellular extract is added in the absence of competitor (track B). Addition of unlabelled oligonucleotide competitor containing the DNA sequence of site A removes the site A footprint without affecting site B (track C) while an unlabelled oligonucleotide containing the site B DNA sequence has the opposite effect (track D). Both footprints are removed by a mixture of unlabelled site A and B oligonucleotides (track E). Arrows indicate the position of sites at which cleavage with DNAseI is enhanced in the presence of protein bound to an adjacent site indicating the existence of conformational changes induced by protein binding. The track labelled G represents a marker track consisting of the same DNA fragment chemically cleaved at every guanine residue. Panel (b): Position of sites A and B within the HIV control element. The arrow indicates the start site of transcription.

which, like DNAseI footprinting, rely on the ability of the reagents to cleave the DNA in a non-sequence specific manner (for further details see Kreale, 1994; Papavassilou, 1995).

Of greater interest, however, is the technique of dimethyl sulphate (DMS) protection footprinting since it can provide information on the exact bases within the binding site which are contacted by the protein. Thus, this method relies on the ability of DMS to specifically methylate guanine residues in the DNA. These methylated G residues can then be cleaved by exposure to piperidene, whereas no cleavage occurs at unmethylated G residues (Maxam and Gilbert, 1980). A protein bound to the DNA will protect the guanine residues which it contacts from methylation and hence they will not be cleaved upon subsequent piperidene treatment. As in the other footprinting techniques, therefore, specific bands produced by such treatment of naked DNA are absent in the protein–DNA sample. Unlike the other methods, however, because cleavage occurs at specific guanine residues, this method identifies specific bases within the DNA that are contacted by the transcription factor protein.

These footprinting techniques therefore offer an advance on the mobility shift assay, allowing a more precise visualization of the DNA–protein interaction. (For methodological details see Spiro and McMurray, 1999.)

2.2.3 METHYLATION INTERFERENCE ASSAY

The pattern of DNA–protein interaction can also be studied in more detail using the methylation interference assay (Siebenlist and Gilbert, 1980). Like methylation protection, this method relies on the ability of DMS to methylate G residues which can then be cleaved with piperidene. However, methylation interference is based on assessing whether the prior methylation of specific G residues in the target DNA affects subsequent protein binding. Thus, the target DNA is first partially methylated using DMS so that on average only one G residue per DNA molecule is methylated (Maxam and Gilbert, 1980). Each individual DNA molecule will therefore contain some methylated G residues with the particular residues which are methylated being different in each molecule. These partially methylated DNAs are then used in a DNA mobility shift experiment with an appropriate cell extract containing the DNA binding protein. Following electrophoresis the band produced by the DNA which has bound protein and that produced by the DNA which has not, are excised from the gel and treated with piperidine to cleave the DNA at the methylated G residues and not at unmethylated Gs. Clearly, if methylation of a particular G prevents protein binding then cleavage at this particular methylated G will be observed only in the DNA which failed to bind the protein. Conversely, if a particular G residue plays no role in binding, then cleavage at this G residue will be observed equally in both the DNA which bound the protein and that which failed to do so (Fig. 2.6).

Figure 2.7 shows this type of analysis applied to the protein binding to site B within the negatively acting element in the human immunodeficiency virus promoter (for the footprint produced by the binding of this protein see Fig. 2.5). In this case the footprinted sequence was palindromic (Fig. 2.7) suggesting that the DNA–protein interaction may involve similar binding to

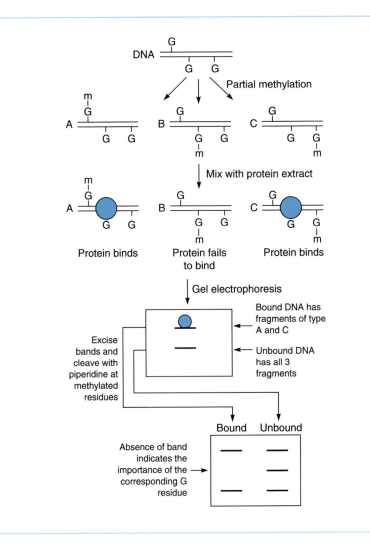

Figure 2.6
Methylation interference assay. Partially methylated DNA is used in a DNA mobility shift assay and both the DNA that has failed to bind protein and that which has bound protein and formed a retarded band are subsequently cleaved at methylated G residues with piperidine. If methylation at a specific G residue has no effect on protein binding (types A and C) the bound and unbound DNA will contain equal amounts of methylated G at this position. In contrast, if methylation at a particular G prevents binding of the protein (type B), only the unbound DNA will contain methylated G at this position.

the two halves of the palindrome. The methylation interference analysis of site B confirms this by showing that methylation of equivalent G residues in each half of the palindrome interferes with binding of the protein, indicating that these residues are critical for binding.

Although the DMS method only studies contacts of the protein with G residues, interference analysis can also be used to study the interaction of DNA binding proteins with A residues in the binding site. This can be done either by methylating all purines to allow study of interference at A and G residues simultaneously (see for example Ares *et al.*, 1987) or by using diethyl-pyrocarbonate specifically to modify A residues (probably by carboxyethyla-tion) rendering them susceptible to piperidine cleavage (see for example

Figure 2.7

Panel (a): Methylation interference assay applied to the DNA of site B in the HIV control element as defined in the footprinting experiment shown in Figure 2.5. Both the upper (tracks A and B) and lower (tracks C and D) strands of the double-stranded DNA sequence were analysed. Tracks B and C show the methylation pattern of the unbound DNA that failed to bind protein, whereas tracks A and D show the methylation pattern of DNA that has bound protein. The arrows show G residues whose methylation is considerably lower in the bound compared to the unbound DNA and which are therefore critical for binding the specific cellular protein that interacts with this DNA sequence. Panel (b): DNA sequence of site B. The extent of the footprint region is indicated by the square brackets and the critical G residues defined by the methylation interference assay in panel (a) are asterisked. Note the symmetrical pattern of critical G residues within the palindromic DNA sequence.

a)

b)

```
     -370                    **              *            -320
       TACACACCAGGGCCAGGGGTCAGATATCCACTGACCTTTGGATGG

       ATGTGTGGTCCCGGTCCCCAGTCTATAGGTGACTGGAAACCTACC
                                *              **
```

Sturm *et al.*, 1988). These techniques are of particular value when studying sequences such as the octamer motif in which there are relatively few G residues, hence limiting the information which can be obtained by studying interference at G residues alone (Sturm *et al.*, 1987; Baumruker *et al.*, 1988). Chemical interference techniques can therefore be used to supplement footprinting methodologies and identify the precise DNA–protein interactions within the footprinted region. (For methodological details see Spiro and McMurray, 1999.)

2.2.4 IN VIVO FOOTPRINTING ASSAY

Although the methods described so far can provide considerable information about DNA–protein contacts they all suffer from the deficiency that the DNA–protein interaction occurs *in vitro* when cell extract and the DNA are mixed.

Hence they indicate what factors can bind to the DNA rather than whether such factors actually do bind to the DNA in the intact cell where a particular factor may be sequestered in the cytoplasm or where its binding may be impeded by the association of DNA with other proteins such as histones.

These problems are overcome by the technique of *in vivo* footprinting, which is an extension of the *in vitro* DMS protection footprinting technique described in section 2.2.2. Thus intact cells are freely permeable to DMS which can therefore be used to methylate the DNA within its native chromatin structure in such cells. Exactly as in the *in vitro* technique, G residues, to which a protein has bound, will be protected from such methylation and will therefore not be cleaved when the DNA is subsequently isolated and treated with piperidene. Hence the bands produced by cleavage at these residues will be absent when the pattern produced by intact chromatin is compared to that produced by naked DNA (Fig. 2.8).

Figure 2.8

In vivo footprinting using the methylation protection assay in which specific G residues are protected by bound protein (X) from methylation by DMS treatment of intact cells. Hence following DNA isolation, cleavage of methylated G residues with piperidine and subsequent amplification by the polymerase chain reaction (PCR), the band corresponding to cleavage at this protected residue will be absent. In contrast, cleavage at this position will be observed in naked DNA where no protein protects this residue from methylation.

Obviously the amounts of any specific DNA sequence obtained from total chromatin in this procedure are vanishingly small compared to when a cloned DNA fragment is used in the *in vitro* procedure. It is thus necessary to amplify the DNA of interest from within total chromatin by the polymerase chain reaction in order to obtain sufficient material for analysis by this method. When this is done, however, *in vivo* footprinting provides an excellent means for analysing DNA–protein contacts within intact cells *in vivo* as well as determining the changes in such contacts which occur in response to specific treatments (see Herrera *et al.*, 1989; Mueller and Wold, 1989 for examples of this approach and Spiro and McMurray, 1999 for a full description of the methodologies involved).

Taken together, therefore, the three methods of DNA mobility shift, footprinting and methylation interference can provide considerable information on the nature of the interaction between a particular DNA sequence and a transcription factor. They serve as an essential prelude to a detailed study of the transcription factor itself.

2.3 METHODS FOR PURIFYING AND/OR CLONING TRANSCRIPTION FACTORS

2.3.1 PROTEIN PURIFICATION

As discussed above, once a particular DNA sequence has been shown to be involved in transcriptional regulation, a number of techniques are available for characterizing the binding of transcription factors to this sequence. Although such studies can be carried out on crude cellular extracts containing the protein, ultimately they need to be supplemented by studies on the protein itself. This can be achieved by purifying the transcription factor from extracts of cells containing it. Unfortunately, however, conventional protein purification techniques such as conventional chromatography and high pressure liquid chromatography (HPLC) result in the isolation of transcription factors at only 1–2% purity (Kadonaga and Tjian, 1986).

To overcome this problem and purify the transcription factor Sp1, Kadonaga and Tjian (1986) devised a method involving DNA affinity chromatography. In this method (Fig. 2.9), a DNA sequence containing a high affinity binding site for the transcription factor is synthesized and the individual molecules joined to form a multimeric molecule. This very high affinity binding site is then coupled to an activated sepharose support on a column and total cellular protein passed down the column. The Sp1 protein binds specifically to its corresponding DNA sequence while all other cellular proteins do

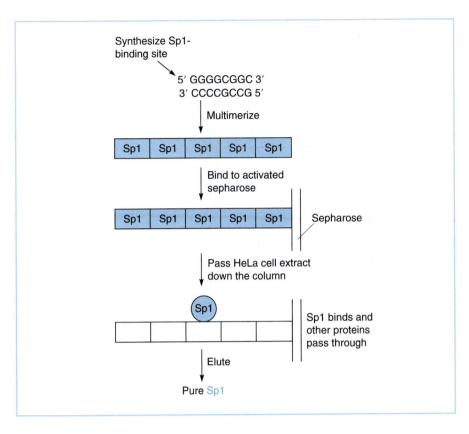

Synthesize Sp1-
binding site

5′ GGGGCGGC 3′
3′ CCCCGCCG 5′

Multimerize

| Sp1 | Sp1 | Sp1 | Sp1 | Sp1 |

Bind to activated
sepharose

| Sp1 | Sp1 | Sp1 | Sp1 | Sp1 | Sepharose

Pass HeLa cell extract
down the column

Sp1 binds and
other proteins
pass through

Elute

Pure Sp1

Figure 2.9
Purification of
transcription factor Sp1
on an affinity column in
which multiple copies of
the DNA sequence
binding Sp1 have been
coupled to a sepharose
support (Kadonaga and
Tjian, 1986).

not bind. The bound Sp1 can be eluted simply by raising the salt concentration. Two successive affinity chromatography steps of this type successfully resulted in the isolation of Sp1 at 90% purity, 30% of the Sp1 in the original extract being recovered, representing a 500–1000-fold purification (Kadonaga and Tjian, 1986).

Although this simple one step method was successful in this case, it relies critically on the addition of exactly the right amount of non-specific DNA carrier to the cell extract. Thus this added carrier acts to remove proteins which bind to DNA in a non-sequence specific manner and which would hence bind non-specifically to the Sp1 affinity column and contaminate the resulting Sp1 preparation. This contamination will occur if too little carrier is added. If too much carrier is added, however, it will bind out the Sp1 since, like all sequence specific proteins, Sp1 can bind with low affinity to any DNA sequence. Hence in this case no Sp1 will bind to the column itself (Fig. 2.10).

To overcome this problem Rosenfeld and Kelley (1986) devised a method in which proteins capable of binding to DNA with high affinity in a non-sequence specific manner are removed prior to the affinity column. To do this the bulk of cellular protein was removed on a Biorex 70 high capacity ion

Figure 2.10

Consequences of adding different amounts of non-specific carrier DNA to the protein passing through the Sp1 affinity column. If the correct amount of non-specific carrier is added it will bind proteins which interact with DNA in a non-sequence specific manner allowing Sp1 to bind to the column (A). However, addition of too little carrier will result in non-sequence specific proteins binding to the column thereby preventing the binding of Sp1 (B), whereas in the presence of too much carrier both the non-specific proteins and Sp1 will bind to the carrier (C).

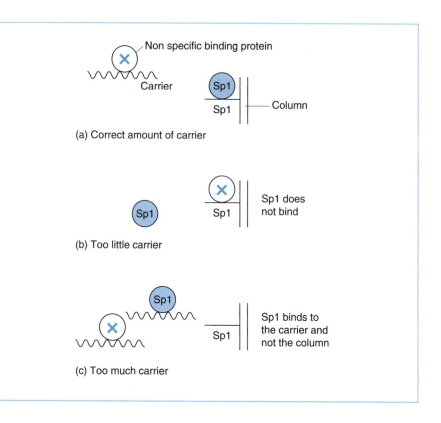

exchange column and proteins which can bind to any DNA with high affinity were then removed on a cellulose column to which total bacterial DNA had been bound. Subsequently the remaining proteins which had bound to non-sequence specific DNA only with low affinity were applied to a column containing a high affinity binding site for transcription factor NF-1 (Fig. 2.11). NF-1 bound to this site with high affinity and could be eluted in essentially pure form by raising the salt concentration (Table 2.1). It should be noted that in this and other purification procedures the fractions containing the transcription factor can readily be identified by carrying out a DNA mobility shift or footprinting assay with each fraction using the specific DNA binding site of the transcription factor.

The purified protein obtained in this way can obviously be used to characterize the protein, for example, by determining its molecular weight or by raising an antibody to it to characterize its expression pattern in different cell types. Similarly the activity of the protein can be assessed by adding it to cellular extracts and assessing its effect on their ability to transcribe an exogenously added DNA in an *in vitro* transcription assay. Unfortunately, however, because of the very low abundance of transcription factors in the cell, these purification procedures yield very small amounts of protein. For exam-

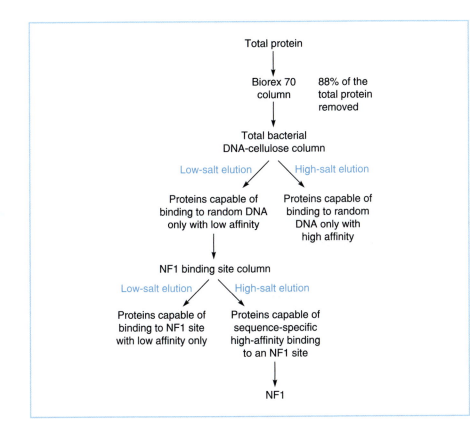

Figure 2.11
Purification of transcription factor NF-1 (Rosenfeld and Kelley, 1986). Following removal of most cellular proteins on a Biorex 70 ion exchange column, proteins that bind to all DNA sequences with high affinity were removed on a bacterial DNA-cellulose column. Subsequent application of the remaining proteins to a column containing the NF-1 binding site results in the purification of NF-1 since it is the only protein which binds with low affinity to random DNA but with high affinity to an NF-1 site.

Table 2.1
Purification of transcription factor NF1 from HeLa cells

	Total protein (mg)	Specific binding of ^{32}P DNA (fmol/ mg protein) $\times 10^{-3}$	Purification (fold)	Yield (%)
HeLa cell extract*	4590	3.1	1.0	100
Biorex 70 column	550	27.1	8.7	104
E. coli DNA cellulose	65.2	181	58.4	83
NF1 affinity matrix 1st passage 2nd passage	2.1 1.1	4510 7517	1455 2425	67 57

* Prepared from 6×10^{10} cells or 120 g cells

ple Treisman (1987) succeeded in purifying only 1.6 μg of the serum response factor starting with 2×10^{10} cells or 40 g of cells. Such difficulties clearly limit the experiments that can be done with purified material. Indeed, the primary use of purified factor in most cases has simply been to provide material to isolate the gene encoding the protein. This gene can then be expressed either *in vitro* or in bacteria to provide a far more abundant source of the corresponding protein than could be obtained from cells that naturally express it.

2.3.2 GENE CLONING

Several methods are available for cloning the gene encoding a particular transcription factor and these will be discussed in turn.

(a) Use of oligonucleotide probes predicted from the protein sequence of the factor

If a particular transcription factor has been purified, it is possible to obtain portions of its amino acid sequence. In turn, such sequences can be used to predict oligonucleotides containing DNA sequences capable of encoding these protein fragments. Due to the redundancy of the genetic code, whereby several different DNA codons can encode a particular amino acid, there will be multiple different oligonucleotides capable of encoding a particular amino acid sequence. All these possible oligonucleotides are synthesized chemically, made radioactive and used to screen a cDNA library prepared from mRNA isolated from a cell type expressing the factor. The oligonucleotide in the mixture which does correspond to the transcription factor amino acid sequence will hybridize to the corresponding sequence in a cDNA clone derived from mRNA encoding the factor. Hence such a clone can be readily identified in the cDNA library (Fig. 2.12).

In cases where purified protein is available as in those discussed in the previous section, this approach represents a relatively simple method for isolating cDNA clones. It has therefore been widely used to isolate cDNA clones corresponding to purified factors such as Sp1 (Kadonaga *et al.*, 1987: Fig. 2.12), NF1 (Santoro *et al.*, 1988) and the serum response factor (Norman *et al.*, 1988) (for methodological details see Nicolas *et al.*, 1999).

(b) Use of oligonucleotide probes derived from the DNA binding site of the factor

Although relatively simple, the use of oligonucleotides derived from protein sequences does require purified protein. As we have seen, purification of a transcription factor requires a vast quantity of cells and is technically difficult.

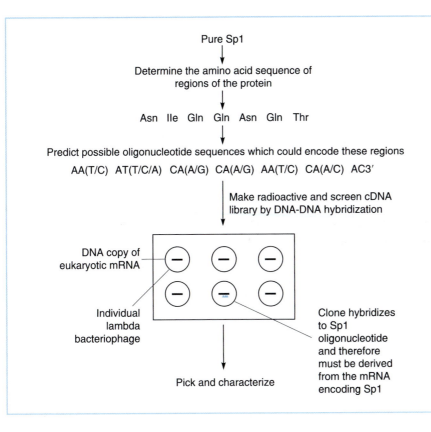

Pure Sp1

Determine the amino acid sequence of
regions of the protein

Asn Ile Gln Gln Asn Gln Thr

Predict possible oligonucleotide sequences which could encode these regions
AA(T/C) AT(T/C/A) CA(A/G) CA(A/G) AA(T/C) CA(A/C) AC3′

Make radioactive and screen cDNA
library by DNA-DNA hybridization

DNA copy of
eukaryotic mRNA

Individual
lambda
bacteriophage

Clone hybridizes
to Sp1
oligonucleotide
and therefore
must be derived
from the mRNA
encoding Sp1

Pick and characterize

Figure 2.12
Isolation of cDNA clones for the Sp1 transcription factor by screening with short oligonucleotides predicted from the protein sequence of Sp1. Because several different triplets of bases can code for any given amino acid, multiple oligonucleotides that contain every possible coding sequence are made. Positions at which these oligonucleotides differ from one another are indicated by the brackets containing more than one base.

Moreover, eventual determination of the partial amino acid sequence of the protein requires access to expensive protein sequencing apparatus.

To bypass these problems Singh *et al.* (1988) devised a procedure which is based on the fact that information is usually available about the specific DNA sequence to which a particular transcription factor binds. Hence a cDNA clone expressing the factor can be identified in a library by its ability to bind the appropriate DNA sequence. This method relies therefore on DNA–protein binding rather than DNA–DNA binding. Hence the library must be prepared in such a way that the cloned cDNA inserts are translated by the bacteria into their corresponding proteins. This is normally achieved by inserting the cDNA into the coding region of the bacteriophage lambda beta-galactosidase gene resulting in its translation as part of the bacteriophage protein. The resulting fusion protein binds DNA with the same sequence specificity as the original factor. Hence a cDNA clone encoding a particular factor can be identified in the library by screening with a radioactive oligo-nucleotide containing the binding site (Fig. 2.13).

This technique has been used to isolate cDNA clones encoding several transcription factors, such as the CCAAT box binding factor C/EBP

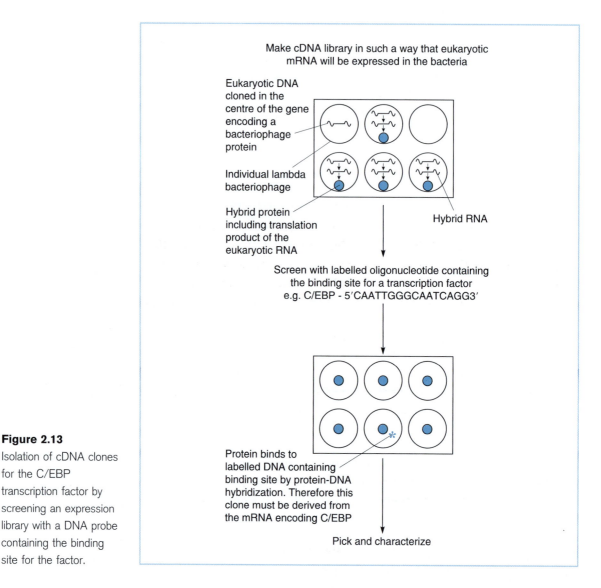

Make cDNA library in such a way that eukaryotic mRNA will be expressed in the bacteria

Eukaryotic DNA cloned in the centre of the gene encoding a bacteriophage protein

Individual lambda bacteriophage

Hybrid protein including translation product of the eukaryotic RNA

Hybrid RNA

Screen with labelled oligonucleotide containing the binding site for a transcription factor e.g. C/EBP - 5'CAATTGGGCAATCAGG3'

Protein binds to labelled DNA containing binding site by protein-DNA hybridization. Therefore this clone must be derived from the mRNA encoding C/EBP

Pick and characterize

Figure 2.13
Isolation of cDNA clones for the C/EBP transcription factor by screening an expression library with a DNA probe containing the binding site for the factor.

(Vinson *et al.*, 1988) and the octamer binding proteins Oct-1 (Sturm *et al.*, 1988) and Oct-2 (Staudt *et al.*, 1988) (for methodological details see Cowell and Hurst, 1999).

(c) Cloning of novel transcription factors by homology to known factors

The development of the two methods described above involving screening with oligonucleotides derived from the protein sequence or oligonucleotides derived from the binding site has therefore resulted in the isolation of cDNA clones corresponding to very many transcription factors.

More recently, however, novel transcription factors are increasingly being cloned on the basis of their relationship to previously characterized factors. In an early example of this approach He *et al.* (1989) identified short amino acid sequences which were highly conserved in the known members of the POU family of transcription factors (Fig. 2.14) (see Chapter 4, section 4.2.6 for a description of this family of proteins). They then prepared degenerate oligonucleotides which contained all the possible DNA sequences able to encode these sequences. Two of these degenerate oligonucleotides were then used in a polymerase chain reaction (PCR) to amplify cDNA prepared from the mRNA of different tissues. Evidently, cDNAs derived from mRNAs encoding novel POU proteins which contain these sequences will be amplified in the PCR procedure and can be isolated and characterized. Indeed, He *et al.* (1989) cloned several novel POU factors by this means and this approach has been applied by a number of others to both the POU family and other transcription factor families (for review and full description of the methods involved see Ashworth, 1999).

Of course, as more and more genomes, including the human genome, are fully sequenced, this approach can now be conducted *in silico* by using the DNA sequences of known transcription factors to search for related sequences in computer databases and this is now perhaps the most common means by which DNA sequences able to encode novel transcription factors are identified.

2.4 USE OF CLONED GENES

2.4.1 DOMAIN MAPPING OF TRANSCRIPTION FACTORS

The cloning of transcription factors by the means described above has, in turn, resulted in an explosion of information on these factors. Thus, once a

Figure 2.14

Cloning of novel members of the POU family of transcription factors on the basis of all family members having two conserved amino acid sequences, one in the POU-specific domain and one in the POU-homeodomain. Degenerate oligonucleotides containing all possible sequences able to encode these conserved sequences are used in a polymerase chain reaction with cDNA prepared from mRNA of a particular tissue. Novel POU factors expressed in this tissue will be amplified on the basis that they contain the conserved sequences and can then be characterized.

clone has been isolated, its DNA sequence can be obtained allowing prediction of the corresponding protein sequence and comparison with other factors. Similarly, the clone can be used to identify the mRNA encoding the protein and examine its expression in various tissues by Northern blotting, to study the structure of the gene itself within genomic DNA by Southern blotting and as a probe to search for related genes expressed in other tissues or other organisms.

Most importantly, however, the isolation of cDNA clones provides a means of obtaining large amounts of the corresponding protein for functional study. This can be achieved either by coupled *in vitro* transcription and translation (Fig. 2.15a: see for example Sturm *et al.*, 1988) or by expressing the gene in bacteria either in the original expression vector used in the screening procedure (see above section 2.3.2b) or more commonly by sub-cloning the cDNA into a plasmid expression vector (Fig. 2.15b: see for example Kadonaga *et al.*, 1987).

Figure 2.15

Methods of producing transcription factor protein from a cloned transcription factor cDNA. In the coupled *in vitro* transcription and translation method (a) the cDNA is cloned downstream of a promoter recognized by a bacteriophage polymerase and transcribed *in vitro* by addition of the appropriate polymerase. The resulting RNA is translated in an *in vitro* protein synthesis system to produce transcription factor protein. Alternatively, the cDNA can be cloned downstream of a prokaryotic promoter in a bacterial expression vector (b). Following introduction of this vector into bacteria, the bacteria will transcribe the cDNA into RNA and translate the RNA into protein which can be isolated from the bacteria.

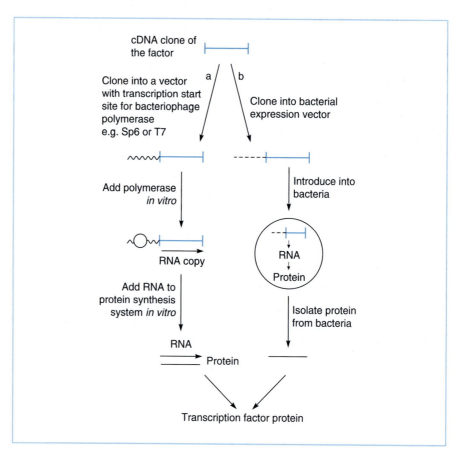

The protein produced in this way has similar activity to the natural protein, being capable of binding to DNA in footprinting or mobility shift assays (see for example, Kadonaga *et al.*, 1987) and of stimulating the transcription of appropriate DNAs containing its binding site when added to a cell free transcription system (see for example Mueller *et al.*, 1990).

Moreover, once a particular activity has been identified in a protein produced in this way, it is possible to analyse the features of the protein which produce this activity in a way that would not be possible using the factor purified from cells which normally express it. Thus, because the cDNA clone of the factor can be readily cut into fragments and each fragment expressed as a protein in isolation, particular features exhibited by the intact protein can readily be mapped to a particular region. Using the approach outlined in Figure 2.16 for example, it has proved possible to map the DNA binding abilities of specific transcription factors such as the octamer binding proteins Oct-1 (Sturm *et al.*, 1987) and Oct-2 (Clerc *et al.*, 1988) to a specific short region of the protein. Once this has been done, particular bases in the DNA encoding the DNA binding domain of the factor can then be mutated so as to alter its amino acid sequence and the effect of these mutations on DNA binding can be assessed as before by expressing the mutant protein and measuring its ability to bind to DNA.

Approaches of this type have proved particularly valuable in defining DNA binding motifs present in many factors and in analysing how differences in the protein sequence of related factors define which DNA sequence they bind. This is discussed in Chapter 4.

One other piece of information to emerge from these studies is that the binding to DNA of a small fragment of the factor does not normally result in the activation of transcription. Thus, a sixty amino acid region of the yeast transcription factor GCN4 can bind to DNA in a sequence specific manner but does not activate transcription of genes bearing its binding site (Hope and Struhl, 1986). Although DNA binding is necessary for transcription therefore, it is not sufficient. This indicates that transcription factors have a modular structure in which the DNA binding domain is distinct from another domain of the protein which mediates transcriptional activation.

The identification of the activation domain in a particular factor is complicated by the fact that DNA binding is necessary prior to activation. Hence the activation domain cannot be identified simply by expressing fragments of the protein and monitoring their activity. Rather the various regions of the cDNA encoding the factor must each be linked to the region encoding the DNA binding domain of another factor and the hybrid proteins produced. The ability of the hybrid factor to activate a target gene bearing the DNA binding site of the factor supplying the DNA binding domain is then assessed

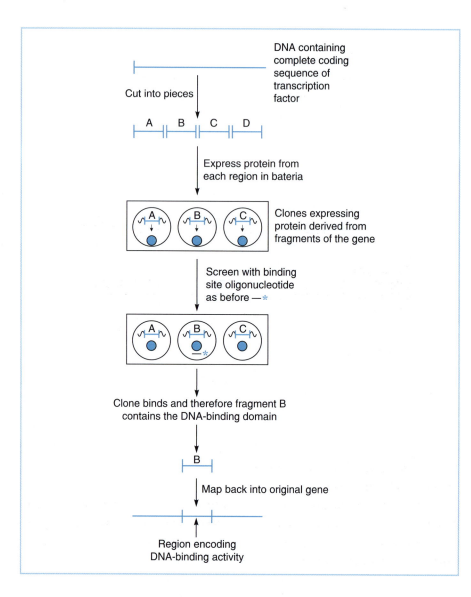

Figure 2.16

Mapping of the DNA-binding region of a transcription factor by testing the ability of different regions to bind to the appropriate DNA sequence when expressed in bacteria.

(Fig. 2.17). In these so called 'domain swap' experiments binding of the factor to the appropriate DNA binding site will be followed by gene activation only if the hybrid factor contains the region encoding the activation domain of the factor under test, allowing the activation domain to be identified.

Thus, if another sixty amino acid region of GCN4 distinct from the DNA binding domain is linked to the DNA binding domain of the bacterial Lex A protein, it can activate transcription in yeast from a gene containing a binding site for Lex A. This cannot be achieved by the Lex A DNA binding domain or this region of GCN4 alone indicating that this region of GCN4 contains the activation domain of the protein which can activate transcription following

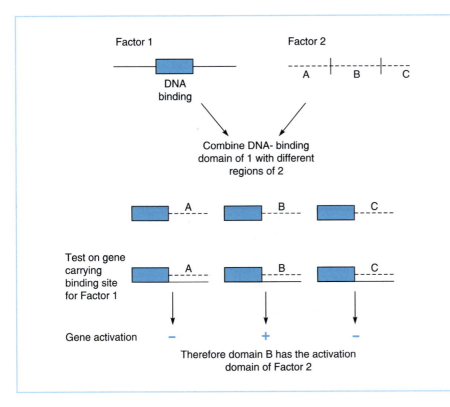

Figure 2.17
Domain swapping experiment in which the activation domain of factor 2 is mapped by combining different regions of factor 2 with the DNA-binding domain of factor 1 and assaying the hybrid proteins for the ability to activate transcription of a gene containing the DNA-binding site of factor 1

DNA binding and is distinct from the GCN4 protein DNA binding domain (Hope and Struhl, 1986).

As with DNA binding domains, the identification of activation domains and comparisons between the domains in different factors has provided considerable information on the nature of activation domains and the manner in which they function. This is discussed in Chapter 5.

2.4.2 DETERMINING THE DNA BINDING SPECIFICITY OF AN UNCHARACTERIZED FACTOR

As indicated above it is common for a transcription factor to be identified on the basis of its binding to a known DNA sequence and the gene encoding the factor then cloned. It is also possible, however, for a novel gene to be cloned on the basis, for example that its expression changes in response to a particular stimulus (see Chapter 7) or that it is mutated in a specific disease (see Chapter 9). On inspection of the DNA sequence and predicted protein sequence, it then appears that this gene encodes a transcription factor either because it is homologous to known transcription factors or because it contains regions with structures similar to those known to mediate DNA binding (see Chapter 4) or transcriptional activation (see Chapter 5). Alternatively, as described above (section 2.3.2c) the novel factor may have been identified by experimental or computer methods simply on the basis of its homology to known transcription factors.

Obviously all the techniques for analysing a cloned factor in section 2.4.1 above can be applied to analysing this factor examining for example, its expression pattern or determining whether regions within it mediate transcriptional activation when linked to the DNA binding domain of another factor. Unlike the situation for transcription factors which were identified on the basis of their DNA binding specificity, however, no information will be available on the DNA sequences to which this novel factor binds. It is evidently essential for the further study of this novel factor that such sequences are identified so allowing, for example, an analysis of the effect of the factor on artificial promoters carrying its binding site and the identification of its target genes.

To do this, Pollock and Treisman (1990) used a method in which oligonucleotides containing a randomized central twenty-six base pair sequence flanked by two defined twenty-five nucleotide sequences were prepared (Fig. 2.18). These sequences were then mixed with transcription factor protein. An antibody to the transcription factor was then used to immunoprecipitate the factor together with the oligonucleotides to which it had bound. This procedure should select from the pool of random oligonucleotides those which contain the binding site for the factor within their central twenty-six base pair sequence while removing those which contain all other sequences. However, after a single round of immunoprecipitation these oligonucleotides will be present in insufficient amounts and purity for further analysis. The immunoprecipitated sequences are therefore amplified by the polymerase chain reaction (PCR) using primers corresponding to the defined twenty-five base pair sequences at the ends of each oligonucleotide. Further cycles of transcription factor binding, immunoprecipitation and PCR are then carried out to purify further the binding sequences. Ultimately, the oligonucleotides which bind the factor are cloned and subjected to sequence analysis to identify the common sequence which they contain and which is therefore the binding site for the factor.

This method thus allows the identification of specific binding sites for the transcription factor and has been used, for example, to identify the DNA binding site for the Brn-3 POU family transcription factors (Gruber *et al.*, 1997) which were originally isolated on the basis of homology to other members of the POU family as described in section 2.3.2c (He *et al.*, 1989) (see Chapter 4, section 4.2.6 for further discussion of POU family transcription

Figure 2.18

Transcription factor binding sites can be cloned using oligonucleotides containing a random central sequence (NNNNN) flanked by defined sequences (solid lines). Repeated cycles of trancription factor binding (X), immunoprecipitation and PCR amplification with primers complementary to the defined end sequences (dotted lines) will eventually result in the purification of oligonucleotides containing the binding site for the factor (ACGAT in this case).

factors). Binding sites identified in this way can then, for example, be linked to a gene promoter and introduced into cells with an expression vector encoding the transcription factor itself to determine whether the factor acts as an activator or repressor of gene expression. Similarly, by inspecting the sequences of promoter or enhancer elements of known genes, it may be possible to identify putative target genes for the factor.

2.4.3 IDENTIFICATION OF TARGET GENES FOR TRANSCRIPTION FACTORS

(a) In vitro analysis of transcription factor binding to genomic DNA fragments

Although the approach described above can identify binding sites for transcription factors, it does not directly identify their target genes. A direct approach to identify such target genes for a previously uncharacterized factor was devised by Kinzler and Vogelstein (1989). This method is essentially the same as that of Pollock and Treisman (1990) except that the starting material is not random oligonucleotides but total genomic DNA. This DNA is digested with a restriction enzyme and small defined DNA sequences are added to the ends of the fragments. The transcription factor binding and immunoprecipitation steps are carried out as before, resulting in the purification of pieces of genomic DNA containing the binding site for the transcription factor. These are then PCR amplified as before using primers corresponding to the defined DNA sequences which were added at the fragment ends and are then cloned.

Although this method is more technically difficult than the use of oligonucleotides due to the complexity of genomic DNA, it has the great advantage that the DNA binding sites are obtained linked to the sequences to which they are normally joined in the genome rather than in isolation (Fig. 2.19). Hence these linked sequences can immediately be characterized and used to identify a target gene for the factor. This method has thus been used for example to identify novel target genes for members of the nuclear receptor transcription factor family discussed in Chapter 4 (section 4.4) such as the oestrogen receptor (Inoue *et al.*, 1993) and the thyroid hormone receptor (Caubin *et al.*, 1994).

(b) Chromatin immunoprecipitation (ChIP)

The above method using genomic fragments thus represents an advance over the oligonucleotide method in the identification of potential target genes for a specific factor. However, since the genome DNA fragments and the transcription factor are mixed in the test tube, it indicates which genomic fragments can bind the factor of interest *in vitro* rather than identifying those genes to which it actually binds in the cell.

In a further advance, the chromatin immunoprecipitation method (ChIP) actually involves the direct identification of target genes for known or unknown factors in the intact cell. In this method (for review see Orlando, 2000), living cells are first fixed with formaldehyde. This has the effect of stably cross-linking transcription factors to the DNA sequences to which they are bound in the cell (Fig. 2.20). The chromatin in the cell is then broken up into small pieces and isolated. An antibody to the transcription factor is

Figure 2.19

While the purification of transcription factor binding sites using random oligonucleotides (as in Fig. 2.18) simply isolates the binding site (a), the use of genomic DNA sequences (dotted lines) in the purification results in the isolation of the binding site linked to a fragment of its target gene (boxed) which can then be characterized (b).

(a) Random oligonucleotide

(b) Genomic DNA fragments

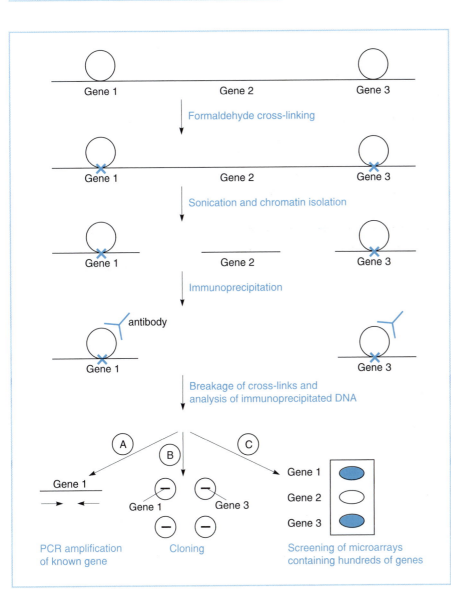

Figure 2.20

In the chromatin immunoprecipitation (ChIP) assay, transcription factors (circle) are cross-linked to their target DNA in the intact cell. The chromatin is then fragmented by sonication and immunoprecipitated with an antibody to the transcription factor of interest. The resulting immunoprecipitated DNA is then analysed by either (A) PCR to detect known potential target genes, (B) cloning and DNA sequencing, or (C) hybridization to a microarray that contains a very large number of genes. Note that in this case, gene 1 and gene 3 bind the factor and gene 2 does not. Methods (B) and (C) detect both genes 1 and 3, whereas method (A) with primers for gene 1 evidently detects only that gene.

then added to immunoprecipitate it, together with the target DNA to which it is cross-linked.

Following breakage of the cross-links and release of the immunoprecipitated DNA from the transcription factor protein, the DNA can be analysed in a number of ways. In the simplest method (Fig. 2.20A), one can test whether a particular gene has been immunoprecipitated by carrying out a PCR amplification with primers for that gene. This will test whether a particular transcription factor binds to a specific gene in intact cells as well as *in vitro*. Similarly, by carrying out the ChIP assay in cells incubated under different conditions or in different cell types, one can detect the changes in such binding which occur in these situations.

In addition, however, methods exist to identify all the genes immuno-precipitated by the ChIP assay rather than testing for the presence of indivi-dual genes (for review see Weinmann and Farnham, 2002). Thus, the immunoprecipitated DNA can be cloned and subjected to DNA sequence analysis to identify all the different DNA fragments (Fig. 2.20B). Alternatively, it is now possible to prepare microarrays containing thousands of DNA sequences representing the entire genome of an organism. The immunoprecipitated DNA can be labelled and hybridized to such an array, allowing all the genes to which the protein is bound in the cell to be identified (Fig. 2.20C). This extraordinarily powerful method of using the ChIP assay has been used, for example, to identify all the genes bound by specific transcrip-tion factors in yeast under different conditions so as to define global transcriptional regulatory networks (Ren *et al.*, 2000; Lee *et al.*, 2002).

2.5 CONCLUSIONS

This chapter has described a number of methods which allow the investigation of the interaction of a transcription factor with DNA, its purification, gene cloning and dissection of its functional domains as well as the identification of its DNA binding site and its target genes. The information obtained by the application of these procedures to particular factors is discussed in subsequent chapters.

REFERENCES

Ares, M. Jr, Chung, J-S., Giglio, L. and Weiner, A.M. (1987) Distinct factors with Sp1 and NF-A specificities bind to adjacent functional elements of the human U2 snRNA gene enhancer. Genes and Development 1, 808–817.

Ashworth, A. (1999) Cloning transcription factors by sequence homology. In: Transcription Factors: a practical approach Second Edition, Latchman, D.S. (ed) Oxford University Press. pp.145–164.

Baumruker, T., Sturm, R. and Herr, W. (1988) OBP 100 binds remarkably degenerate octamer motifs through specific interactions with flanking sequences. Genes and Development 2, 1400–1413.

Caubin, J., Iglesias, T., Bernal, J. *et al.* (1994) Isolation of genomic DNA fragments corre-sponding to genes modulated *in vivo* by a transcription factor. Nucleic Acids Research 22, 4132–4138.

Clerc, R.G., Corcoran, L.M., LeBowitz, J.H. *et al.* (1988) The B-cell specific Oct-2 protein contains POU box and homeo box type domains. Genes and Development 2, 1570–1581.

Cowell, I.G. and Hurst, H.C. (1999) Cloning transcription factors from a cDNA expression library. In: Transcription Factors: a practical approach, Second Edition, Latchman, D.S.(ed) Oxford University Press. pp. 123–143.

Dignam, J.D., Lebovitz, R.M. and Roeder, R.G. (1983) Accurate transcription initiation by RNA polymerase II in a soluble extract from isolated mammalian nuclei. Nucleic Acids Research 11, 1475–1489.

Dynan, W.S. and Tjian, R. (1983) The promoter specific transcription factor Sp1 binds to upstream sequences in the SV40 promoter. Cell 35, 79–87.

Fried, M. and Crothers, D.M. (1981) Equilibria and kinetics of lac repressor-operator interactions by polyacrylamide gel electrophoresis. Nucleic Acids Research 9, 6505–6525.

Galas, D. and Schmitz, A. (1978) DNAse footprinting: A simple method for the detection of protein-DNA binding specificity. Nucleic Acids Research 5, 3157–3170.

Garner, M.M. and Revzin, A. (1981) A gel electrophoresis method for quantifying the binding of proteins to specific DNA regions: application to components of the *Escherichia coli* lactose operon regulatory system. Nucleic Acids Research 9, 3047–3060.

Gruber, C.A., Rhee, J.M., Gleiberman, A. and Turner, E.E. (1997) POU domain factors of the Brn-3 class recognize functional DNA elements which are distinctive, symmetrical and highly conserved in evolution. Molecular and Cellular Biology 17, 2391–2400.

He, X., Treacey, M.N., Simmonds, D.M. *et al.* (1989) Expression of a large family of POU-domain genes in mammalian brain development. Nature 340, 35–42.

Herrera, R.E., Shaw, P.E. and Nordheim, A. (1989) Occupation of the *c-fos* serum response element *in vivo* by a multi-protein complex is unaltered by growth factor induction. Nature 340, 68–71.

Hope, I.A. and Struhl, K. (1986) Functional dissection of a eukaryotic transcriptional activator GCN4 of yeast. Cell 46, 885–894.

Inoue, S., Orimo, A., Hosoi, T. *et al.* (1993) Genomic binding site cloning reveals an estrogen responsive gene that encodes a RING finger protein. Proceedings of the National Academy of Sciences, USA 90, 11117–11121.

Kadonaga, J.T. and Tjian, R. (1986) Affinity purification of sequence-specific DNA binding proteins. Proceedings of the National Academy of Sciences, USA 83, 5889–5893.

Kadonaga, J.T., Carner, K.R., Masiarz, F.R. and Tjian, R. (1987) Isolation of cDNA encoding the transcription factor Sp1 and functional analysis of the DNA binding domain. Cell 51, 1079–1090.

Kinzler, K.W. and Vogelstein, B. (1989) Whole genome PCR: application to the identification of sequences bound by gene regulatory proteins. Nucleic Acids Research 17, 3645–3653.

Kreale, G.G. (ed) (1994) DNA-protein interactions. Humana Press.

Latchman, D.S. (ed) (1999) Transcription factors, a practical approach. Second Edition, Oxford University Press. p. 303.

Lee, T. I., Rinaldi, N. J., Robert, F. *et al.* (2002) Transcriptional regulatory networks in *Saccharomyces* cerevisiae. Science 298, 799–804.

Manley, J.L., Fire, A., Cano, A. *et al.* (1980) DNA-dependent transcription of adenovirus genes in a soluble whole-cell extract. Proceedings of the National Academy of Sciences, USA 77, 3855–3859.

Maxam, A.M. and Gilbert, W. (1980) Sequencing end labelled DNA with base-specific chemical cleavages. Methods in Enzymology 65, Part 1, 499–560.

Mueller, C.R., Macre, P. and Schibler, U. (1990) DBP a liver-enriched transcriptional activator is expressed late in ontogeny and its tissue specificity is determined post-transcriptionally. Cell 61, 279–291.

Mueller, P.R. and Wold, B. (1989) *In vivo* footprinting of a muscle specific enhancer by ligation mediated PCR. Science 246, 780–786.

Nicolas, R.H., Hynes, G. and Goodwin G.H. (1999) Purification and cloning of DNA binding transcription factors. In: Transcription Factors: a practical approach. Second Edition. Latchman, D.S. (ed), Oxford University Press. pp. 97–122.

Norman, C., Runswick, M., Pollock, R. and Treisman, R. (1988) Isolation and properties of cDNA clones encoding SRF, a transcription factor that binds to the c-*fos* serum response element. Cell 55, 989–1003.

Orchard, K., Perkins, N.D., Chapman, C. *et al.* (1990) A novel T cell protein recognizes a palindromic element in the negative regulatory element of the HIV-1 LTR. Journal of Virology 64, 3234–3239.

Orlando, V. (2000) Mapping chromosomal proteins *in vivo* by formaldehyde-crosslinked-chromatin immunoprecipitation. Trends in Biochemical Sciences 25, 99–104.

Papavassilou, A.G. (1995) Chemical nucleases as probes for studying DNA-protein interactions. Biochemical Journal 305, 345–357.

Pollock, R. and Treisman, R. (1990) A sensitive method for the determination of protein DNA binding specificities. Nucleic Acids Research 18, 6197–6204.

Ren, B., Robert, F., Wyrick, J. J. *et al.* (2000) Genome-wide location and function of DNA binding proteins. Science 290, 2306–2309.

Rosenfeld, P.J. and Kelley, T.J. (1986) Purification of nuclear factor 1 by DNA recognition site affinity chromatography. Journal of Biological Chemistry 261, 1398–1408.

Santoro, C., Mermod, N., Andrews, P.C. and Tjian, R. (1988) A family of human CCAAT box binding proteins active in transcription and DNA replication: cloning and expression of multiple cDNAs. Nature 334, 218–224.

Siebenlist, U. and Gilbert, W. (1980) Contacts between the RNA polymerase and an early promoter of phage T7. Proceedings of the National Academy of Sciences, USA 77, 122–126.

Singh, H., Le Bowitz, J.H., Baldwin, A.S. and Sharp, P.A. (1988) Molecular cloning of an enhancer binding protein: isolation by screening of an expression library with a recognition site DNA. Cell 52, 415–429.

Smith, M.D., Dent, C.L. and Latchman, D.S. (1999) The DNA mobility shift assay. In: Transcription Factors: a practical approach, Second Edition, Latchman, D.S. (ed) Oxford University Press. pp. 1–25.

Spiro, C. and McMurray, C.T. (1999) Footprint analysis of DNA-protein complexes *in vitro* and *in vivo*. In: Transcription factors: a practical approach. Second Edition, Latchman, D.S. (ed). Oxford University Press. pp. 27–62.

Staudt, L.M., Clerc, R.G., Singh, H. *et al*. (1988) Cloning of a lymphoid-specific cDNA encoding a protein binding the regulatory octamer DNA motif. Science 241, 577–580.

Sturm, R., Baumruker, T., Franza, R. Jr and Herr, W. (1987) A 100-kD HeLa cell octamer binding protein (OBP 100) interacts differently with two separate octamer-related sequences within the SV40 enhancer. Genes and Development 1, 1147–1160.

Sturm, R.A., Das, G. and Herr, W. (1988) The ubiquitous octamer-binding protein Oct-1 contains a POU domain with a homeobox subdomain. Genes and Development 2, 1582–1599.

Treisman, R. (1987) Identification and purification of a polypeptide that binds to the c-fos serum response element. The EMBO Journal 6, 2711–2717.

Vinson, C.R., La Marco, K.L., Johnson, P.F. *et al*. (1988) In situ detection of sequence-specific DNA binding activity specified by a recombinant bacteriophage. Genes and Development 2, 801–806.

Weinmann, A. S. and Farnham, P. J. (2002) Identification of unknown target genes of human transcription factors using chromatin immunoprecipitation. Methods 26, 37–47.

RNA POLYMERASES AND THE BASAL TRANSCRIPTIONAL COMPLEX

3.1 RNA POLYMERASES

Transcription involves the polymerization of ribonucleotide precursors into an RNA molecule using a DNA template. The enzymes that carry out this reaction are known as RNA polymerases. In eukaryotes three different enzymes of this type exist which are active on different sets of genes and can be distinguished on the basis of their different sensitivities to the fungal toxin alpha-amanitin (Table 3.1, for review see Sentenac, 1985). All the genes that code for proteins, as well as those encoding some of the small nuclear RNAs involved in splicing, are transcribed by RNA polymerase II. Because of the very wide variety of regulatory processes that these genes exhibit, much of this book is concerned with the interaction of different transcription factors with RNA polymerase II. Information is also available, however, on the interaction of such factors with RNA polymerase I which transcribes the genes encoding the 28S, 18S and 5.8S ribosomal RNAs (Sommerville, 1984) and with RNA polymerase III which transcribes the transfer RNA and 5S ribosomal RNA genes (Cilberto *et al.*, 1983). These interactions are therefore discussed where appropriate.

Table 3.1

Eukaryotic RNA polymerases

Genes transcribed	Sensitivity to α-amanitin
I Ribosomal RNA (45S precursor of 28S, 18S and 5.8S rRNA)	Insensitive
II All protein-coding genes, small nuclear RNAs U1, U2, U3, etc.	Very sensitive (inhibited 1 μg/ml)
III Transfer RNA, 5S ribosomal RNA, small nuclear RNA U6, repeated DNA sequences: Alu, B1, B2 etc., 7SK, 7SL RNA	Moderately sensitive (inhibited 10 μg/ml)

All three RNA polymerases are large multi-subunit enzymes, RNA polymerase II for example having 10–14 subunits with sizes ranging from 220 to 10 kilo-daltons (Sentenac, 1985; Saltzman and Weinmann, 1989), which interact with one another to form a highly complex multimeric molecule that has recently been crystallized allowing structural analysis (for review see Klug, 2001; Landick, 2001). Interestingly, the cloning of the genes encoding the largest subunits of each of the three polymerases has revealed that they show homology to one another (Memet *et al.*, 1988). Similarly chemical labelling experiments have indicated that the second largest subunit of each polymerase contains the active site of the enzyme (Riva *et al.*, 1987) while at least three smaller, non-catalytic subunits are shared by the three yeast polymerases (Woychik *et al.*, 1990). Such relationships evidently indicate a basic functional similarity between the three eukaryotic RNA polymerases and may also be indicative of a common evolutionary origin.

In addition to the conservation of function between the three eukaryotic enzymes, each individual enzyme exhibits a strong conservation between different organisms. Thus the largest subunit of the mammalian RNA polymerase II enzyme is 75% homologous to that of the fruit fly *Drosophila* (Saltzman and Weinmann, 1989) and also shows homology to the equivalent enzymes in yeast (Memet *et al.*, 1988) and even *E. coli* (Ahearn *et al.*, 1987). All the eukaryotic RNA polymerase II enzymes contain a repeated region at the carboxyl end of the largest subunit which contains multiple copies of the sequence Tyr-Ser-Pro-Thr-Ser-Pro-Ser. This sequence is unique to the largest subunit of RNA polymerase II and is present in multiple copies being repeated fifty-two times in the mouse protein and twenty-six times in the yeast protein. This repeated region is highly evolutionarily conserved (for review see Stiller and Hall, 2002) and, as expected from this, is essential for the proper functioning of the enzyme and hence for cell viability, although its size can be reduced to some extent without affecting the activity of the enzyme (for review see Young, 1991).

Interestingly, this repeated region serves as a site for phosphorylation and it is likely that such phosphorylation is critical for functioning of the polymerase (for review see Drapkin *et al.*, 1993). Thus it appears that the dephosphorylated form of RNA polymerase II is the form which enters the basal transcriptional complex (see section 3.5.1), while its phosphorylation triggers the start of transcriptional elongation to produce the RNA product. Such phosphorylation appears to be a means of regulating the rate of transcription with specific stimuli, such as growth factors resulting in enhanced phosphorylation of the polymerase (Dubois *et al.*, 1994).

In addition, as will be discussed in Chapter 5 (section 5.3.3), this region may be a target for transcriptional activators either directly or, more probably,

indirectly via intermediate proteins. Moreover, recent studies have indicated that factors involved in post-transcriptional processes such as RNA splicing, associate with this region of the polymerase so that the nascent RNA transcript produced by the polymerase can actually be spliced by factors which are bound to the polymerase itself (for review see Hirose and Manley, 2000; Proudfoot, *et al.*, 2002). Hence this region appears to represent a critical target for cellular transcriptional and post-transcriptional regulatory processes.

Whether this is the case or not, it is clear that while the RNA polymerases possess the enzymatic activity necessary for transcription, they cannot function independently. Rather transcription involves numerous transcription factors which must interact with the polymerase and with each other if transcription is to occur. The role of these factors is to organize a stable transcriptional complex containing the RNA polymerase and which is capable of repeated rounds of transcription.

3.2 THE STABLE TRANSCRIPTIONAL COMPLEX

For all three eukaryotic polymerases, the initiation of transcription requires a multi-component complex containing the RNA polymerase and transcription factors. This complex has several characteristics which have led to it being referred to as a stable transcriptional complex (Brown, 1984). These are:

1   The assembled complex is stable to treatment with low concentrations of specific detergents or to the presence of a competing DNA template, both of which would prevent its assembly.
2   The complex contains factors which are necessary for its assembly but not for transcription itself. These factors can therefore be dissociated once the complex has formed without affecting transcription.
3   The complex of RNA polymerase and other factors necessary for transcription is stable through many rounds of transcription, resulting in the production of many RNA copies from the gene.

These characteristics are illustrated in Figure 3.1.

Much of the information on these complexes has been obtained by studying the relatively simple systems of RNA polymerases I and III and applying the information obtained to the RNA polymerase II situation. The stable complex formed by each of these enzymes will therefore be discussed in turn.

Figure 3.1

Stages in the formation of the stable transcriptional complex. The initial binding of the transcription factor (A) and the assembly factor (B) results in a metastable complex which can be dissociated by low levels of detergent or competing DNA. Following RNA polymerase binding, however, a stable complex is formed. This complex cannot be dissociated by low levels of detergent or competing DNA, is stable through multiple rounds of transcription and retains activity if the assembly factor (B) is removed.

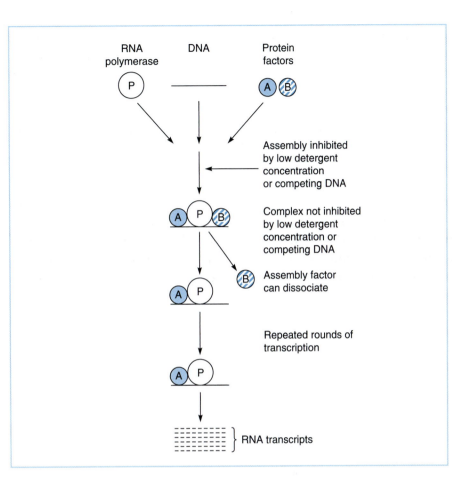

3.3 RNA POLYMERASE I

The simplest complex known for the transcription of the ribosomal RNA genes by RNA polymerase I is found in *Acanthamoeba* (for review see Paule, 1990; Paule and White, 2000). In this organism only one transcription factor, known as TIF-1, is required for transcription by the polymerase. This factor binds to the ribosomal RNA promoter protecting a region from twelve to seventy bases upstream of the transcriptional start site from DNAseI digestion. Subsequently, the polymerase itself binds to the DNA just downstream of TIF-1 protecting a region between eighteen and fifty-two bases upstream of the start site. Interestingly, binding of the polymerase is not dependent on the specific DNA sequence within this region since it can be replaced with a completely random sequence without affecting binding of the polymerase. Hence RNA polymerase is positioned on the promoter by protein–protein

interaction with TIF-1 which has previously bound in a sequence specific manner (Fig. 3.2). When the RNA polymerase moves along the DNA transcribing the gene, TIF-1 remains bound at the promoter allowing subsequent rounds of transcription to occur following binding of another polymerase molecule.

This system therefore represents a simple one in which one single factor is necessary for transcription and is active through multiple rounds of transcription. In vertebrate rRNA gene transcription, the situation is more complex, however, with an additional factor UBF (upstream binding factor) also being involved (for review see Jacob, 1995). UBF binds specifically to the promoter and upstream elements of the ribosomal RNA genes and stimulates transcription. This is achieved, however, by interaction with the vertebrate TIF-1 homologue, known as SL1. Thus, although a low basal rate of transcription is observed in the absence of UBF, no transcription is detectable unless SL1 is

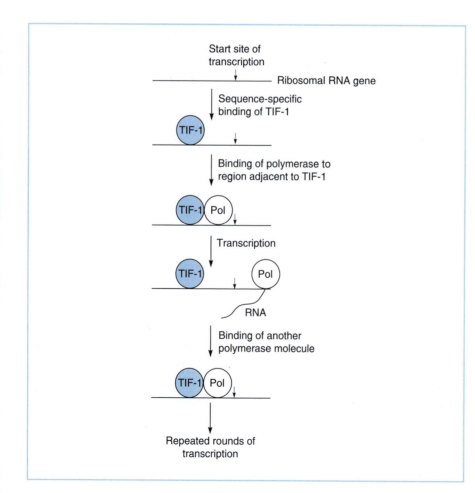

Figure 3.2

Transcription by *Acanthamoeba* RNA polymerase I involves the binding of transcription factor TIF-1 to a specific DNA sequence followed by binding of the polymerase in a non-sequence specific manner to the DNA region adjacent to TIF-1. When the RNA polymerase moves away as it transcribes the gene, TIF-1 remains bound at the promoter allowing another RNA polymerase molecule to bind and initiate a new round of transcription.

present. Unlike TIF-1, SL1 does not exhibit sequence specific binding to the ribosomal RNA promoter. Hence UBF acts by binding to the DNA in a sequence specific manner and facilitating the binding of SL1. Thus, while both SL1 and its homologue TIF-1 act as transcription factors necessary for polymerase I binding, UBF is an additional assembly factor required for binding of SL1 in vertebrates but not of TIF-1 in *Acanthamoeba*. This example therefore illustrates the distinction between factors required only for assembly of the complex or for binding of the polymerase and transcription itself (Fig. 3.3).

Figure 3.3

Comparison of ribosomal RNA gene transcription in *Acanthamoeba* and vertebrates. In vertebrates, transcription requires both the TIF-1 homologue SL1 and an additional assembly factor UBF whose prior binding is necessary for subsequent binding of SL1.

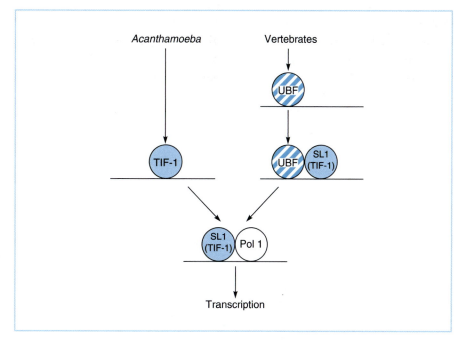

3.4 RNA POLYMERASE III

The different roles of transcription factors and assembly factors are also well illustrated by the RNA polymerase III system (for reviews see Geiduschek and Kassavetis, 2001; Paule and White, 2000; Schramm and Hernandez, 2002). Thus three different classes (I–III) of RNA polymerase III transcription unit exist, all of which require the essential factor TFIIIB for transcription (for review see Hernandez, 1993).

In the case of class I transcription units encoding the 5S ribosomal RNAs, transcription by RNA polymerase III requires the binding of three additional

factors TFIIIA, TFIIIB and TFIIIC. Although both TFIIIA and TFIIIC exhibit the ability to bind to 5S DNA in a sequence specific manner, TFIIIB like SL1 cannot do so unless TFIIIC has already bound. Once the complex of all these factors has formed and the RNA polymerase has bound, TFIIIA and TFIIIC can be removed and transcription continues with only TFIIIB and the polymerase bound to the DNA. Hence like UBF, TFIIIA and TFIIIC are assembly factors which are required for the binding of the transcription factor TFIIIB. In turn, bound TFIIIB is recognized by the polymerase itself and transcription begins (Fig. 3.4). As with RNA polymerase I, RNA polymerase III binds to the region of DNA adjacent to that which has bound the transcription factor, binding of the polymerase being independent of the DNA sequence in this region.

Although the transcription of the class II RNA polymerase III transcription units, such as those encoding the tRNAs, is similar to that described for the 5S RNA genes, TFIIIA is not required. Rather transcription is dependent only upon TFIIIB and TFIIIC with binding of TFIIIC being sufficient for subsequent binding of TFIIIB and the polymerase. Similarly, the class III RNA polymerase III transcription units, which have a TATA box in the promoter (for review see Sollner-Webb, 1988) that resembles that found in RNA poly-

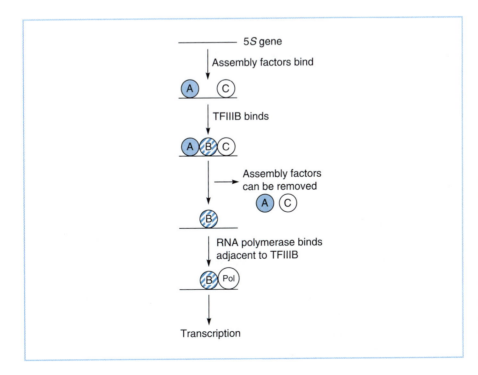

Figure 3.4

Binding of factors to the 5S RNA gene. Transcription requires the initial binding of the assembly factors TFIIIA and TFIIIC with subsequent binding of the transcription factor TFIIIB and of RNA polymerase III itself.

merase II promoters (see Chapter 1, section 1.3.2) also require TFIIIB for transcription together with other accessory factors (for discussion see Hernandez, 1993).

The process of transcription by RNA polymerases I and III therefore involves the binding of a single transcription factor to the promoter, allowing subsequent binding of the RNA polymerase to an adjacent region of DNA. The transcription factor remains bound at the promoter as the polymerase moves down the DNA allowing repeated binding of polymerase molecules and hence repeated rounds of transcription. Binding of the polymerase to the promoter requires prior binding of the transcription factor since the polymerase does not recognize a specific sequence in the promoter but rather makes protein–protein contact with the transcription factor and binds to the adjacent region of the DNA.

In different systems, however, different requirements exist for the binding of the transcription factor itself. Thus in the *Acanthamoeba* system, TIF-1 can bind to DNA in a sequence specific manner and hence is the only factor required. In most other systems, this is not the case and the transcription factors do not bind to the DNA unless other assembly factors, which exhibit sequence specific DNA binding, are present. Once the transcription factor has bound, these assembly factors can be removed, for example, by detergent treatment without affecting subsequent transcription. It is unclear, however, whether these factors do actually dissociate from the complex under normal conditions *in vivo* once the transcription factor has bound (for discussion see Paule, 1990). Whatever the case, the transcription factor itself remains bound at the promoter even after the polymerase has moved down the gene, allowing repeated binding of polymerase molecules and hence repeated rounds of transcription.

Although assembly factors play only an accessory role in transcription itself, they are essential if the complex is to assemble. Hence both assembly factors and transcription factors can be the target for processes which regulate the rate of transcription (for review see Brown *et al.*, 2000). Thus, while the high rate of polymerase III transcription in embryonal carcinoma cells is dependent on a high level of transcription factor TFIIIB, the increase in transcription by this polymerase following adenovirus infection is due to an increase in the activity of the assembly factor TFIIIC. Similarly, alterations in the level of TFIIIA during *Xenopus* development control the nature of the 5S rRNA genes that are transcribed at different developmental stages. In addition, as will be discussed in Chapter 9 (section 9.4.3) the retinoblastoma anti-oncoprotein inhibits cellular growth by interacting with UBF to inhibit RNA polymerase I activity and with TFIIIB to inhibit RNA polymerase III activity.

3.5 RNA POLYMERASE II

3.5.1 STEPWISE ASSEMBLY OF THE RNA POLYMERASE II BASAL TRANSCRIPTIONAL COMPLEX

Although some regulation of RNA polymerase I and III activity does occur therefore, this is much less extensive compared to the very wide variety of regulatory events affecting the activity of genes transcribed by RNA polymerase II. As discussed above, this results in a bewildering array of transcription factors interacting with this enzyme and conferring particular patterns of regulation. Interestingly, however, even the basal transcriptional complex, which is essential for any transcription by this enzyme, contains far more components than is the case for the other RNA polymerases (for reviews see Orphanides *et al.*, 1996; Roeder, 1996; Woychick and Hampsey, 2002).

One component of this complex which has been intensively studied and plays an essential role in RNA polymerase II mediated transcription is TFIID (for review see Burley and Roeder, 1996). In promoters containing a TATA box (see Chapter 1, section 1.3.2), TFIID binds to this element, protecting a region from thirty-five bases to nineteen bases upstream of the start site of transcription in the human hsp70 promoter, for example. The binding of TFIID to the TATA box or equivalent region is the earliest step in the formation of the stable transcriptional complex, such binding being facilitated by another factor TFIIA (Fig. 3.5a).

Interestingly, as TFIID is progressively purified, its requirement for TFIIA to aid its activity decreases. This is because in less purified preparations and in the intact cell, TFIID is associated with a number of inhibitory factors such as Dr1 and Dr2 (for review see Drapkin *et al.*, 1993) which act by preventing its binding to the DNA and/or its interaction with other components of the basal complex such as TFIIB (see below) (for further discussion of the role of Dr1, see Chapter 6, section 6.3.3). One role of TFIIA appears to be to bind to TFIID and overcome this inhibition, thereby stimulating the activity of TFIID. Hence the need for TFIIA decreases as TFIID is purified away from these inhibitory factors, although it is likely to play a critical role in the intact cell. In addition, TFIIA may also play a role in the response to transcriptional activators acting as a co-activator molecule linking DNA-bound activators and the basal transcriptional complex.

Hence rather than acting as a basal transcription factor essential for all transcription, TFIIA appears to play a key role in the response of the complex to activating and inhibiting molecules. Such a role is of particular importance since the antagonism between positively and negatively acting factors in the assembly of the basal transcriptional complex may play a critical role in reg-

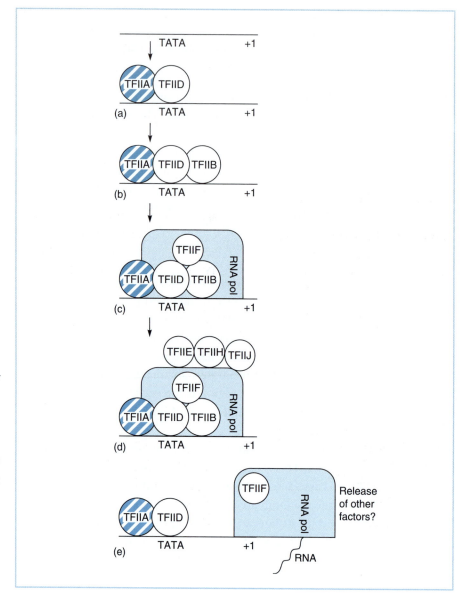

Figure 3.5

Stages in the assembly of the stable transcriptional complex for RNA polymerase II transcription. As the polymerase moves away from the promoter to transcribe the gene, TFIIF remains associated with it while TFIIA and TFIID remain bound at the TATA box allowing the formation of a new stable complex and further rounds of transcription.

ulating the rate of transcription, representing a major target for activators and repressors of transcription (see Chapters 5 and 6, for a further discussion of the mechanisms by which specific factors activate or inhibit transcription).

Once TFIID has bound to the DNA, another transcription factor TFIIB, joins the complex by binding to TFIID (Fig. 3.5b). This binding of TFIIB is an essential step in initiation complex formation since, as well as binding to TFIID, TFIIB can also bind to the RNA polymerase itself. Hence it acts as a bridging factor allowing the recruitment of RNA polymerase to the complex in association with another factor TFIIF (Fig. 3.5c). Following polymerase binding, three other transcription factors TFIIE, TFIIH and TFIIJ, rapidly associate with the complex (Fig. 3.5d). At this point, TFIIH, which has a DNA helicase activity, unwinds the double-stranded DNA so allowing it to be copied into RNA. Subsequently, the kinase activity of TFIIH, which allows it to phosphorylate other proteins, phosphorylates the C-terminal domain of RNA polymerase (for review see Orphanides *et al.*, 1996). This converts it from the non-phosphorylated form which joins the complex to the phosphory-lated form which is capable of transcriptional elongation to produce the RNA product (Fig. 3.6) (see section 3.1).

Hence TFIIH, via its kinase and helicase activities, plays a critical role in allowing the basal transcriptional complex to initiate transcription. Moreover, TFIIH also plays a critical role in the repair of damaged DNA, providing a possible link between the processes of DNA repair and transcription (for reviews of TFIIH see Hoeijmakers *et al.*, 1996; Svejstrup *et al.*, 1996). Interestingly, it has recently been shown that the kinase activity associated with TFIIH can also phosphorylate the retinoic acid receptor which is a mem-

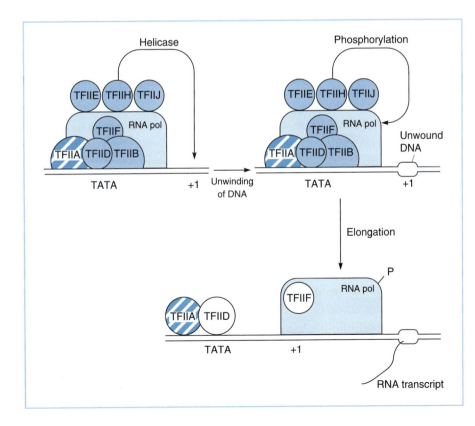

Figure 3.6

TFIIH has a helicase activity which unwinds the DNA allowing its transcription into RNA and a kinase activity that phosphorylates the C-terminal region of RNA polymerase which allows it to begin transcription.

ber of the nuclear receptor transcription factor family discussed in Chapter 4 (section 4.4). This phosphorylation stimulates the ability of the retinoic acid receptor to activate transcription (Rochette-Egly *et al.*, 1997) indicating that TFIIH may play a role in the regulation of transcription factor activity by phosphorylation (see Chapter 8, section 8.4.2).

The complex of the seven factors (TFIIA, B, D, E, F, H and J) and the polymerase is thus sufficient for transcription to occur. As the polymerase moves down the gene during this process TFIIF remains associated with it while TFIIA and TFIID remain bound at the promoter and are capable of binding another molecule of polymerase allowing repeated rounds of transcription as with the other polymerases (see Fig. 3.5e).

3.5.2 THE RNA POLYMERASE HOLOENZYME

Although the step-by-step pathway of assembling the basal transcriptional complex described above was proposed on the basis of a number of studies, an alternative pathway has also been identified based on the finding that some RNA polymerase is found in solution already associated with TFIIB, TFIIF and TFIIH in the absence of DNA. This so-called RNA polymerase holoenzyme has now been observed in a wide range of organisms ranging from yeast to man. It is clear therefore that, in some cases, following binding of TFIIA and TFIID to the promoter, this complex of RNA polymerase and associated factors may bind resulting in a reduced number of steps being required for complex formation (Fig. 3.7) (for discussion see Pugh, 1996; Greenblatt, 1997; Myer and Young, 1998).

Interestingly, the RNA polymerase holoenzyme also contains a number of other components apart from RNA polymerase itself and the basal transcription factors. Thus it includes a complex of proteins, known as the mediator complex, which appears to be required, at least in yeast, for the response to transcriptional activators (see Chapter 5, section 5.4.1). Hence the mediator may serve as a link between these activators and the components of the basal transcriptional complex whose activity they stimulate. In addition, the holoenzyme can also associate with the SWI/SNF complex discussed in Chapter 1 (section 1.2.2) whose role is to remodel the chromatin into a form which allows the binding of transcriptional activators and transcription itself. Hence, at least in some cases, this remodelling complex can be recruited to DNA together with the RNA polymerase and its associated proteins.

The RNA polymerase holoenzyme is thus a highly complex structure which, as well as RNA polymerase itself and basal transcription factors, also contains factors involved in the response to transcriptional activators and others which remodel chromatin structure. Although this holoenzyme represents only one

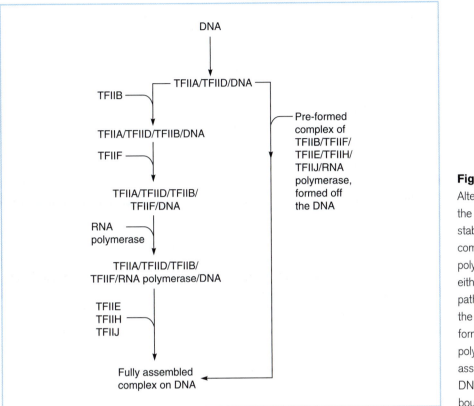

Figure 3.7

Alternative pathways in the assembly of the stable transcriptional complex for RNA polymerase II involving either the step by step pathway (see Fig. 3.5) or the binding of a pre-formed complex of RNA polymerase and its associated factors to DNA which has already bound TFIIA and TFIID.

of the two possible methods by which the basal transcription complex assembles on the DNA, it is clear that regardless of its method of assembly the basic stable transcriptional complex for RNA polymerase II requires a number of factors in addition to the polymerase itself and is therefore much more complex than that of RNA polymerase I or III.

3.6 TBP, THE UNIVERSAL TRANSCRIPTION FACTOR?

Most of the transcription factors described in the previous sections were isolated by the biochemical fractionation of cellular extracts and were then shown to have a particular functional activity in modulating the rate of transcription when mixed with RNA polymerase and other subcellular fractions. When these factors were characterized in more detail by further fractionation and subsequent cloning, however, many of them were shown to consist of several different proteins which together are responsible for the properties ascribed to the original factor. Thus, although these factors have been dealt with for simplicity in the previous sections as single factors, most of them are in fact complexes of several different proteins, for example, TFIIE and TFIIF both contain two distinct proteins. Similarly, TFIIH is a multi-protein complex whose structure has been determined (Chang and Kornbeg, 2000; Schultz *et al.*, 2000) with one of the component proteins having the kinase activity which results in phosphorylation of the RNA polymerase while another has the helicase activity which unwinds the DNA (see section 3.5.1) (for review see Hoeijmakers *et al.*, 1996; Svejstrup *et al.*, 1996).

This responsibility of one component of the complex for an activity formerly ascribed to the whole complex is seen most clearly in TFIID. Thus, TFIID is a multi-protein complex in which only one protein known as TBP (TATA-binding protein) directs the binding to the TATA box while the other components of the complex, known as TAFs (TBP-associated factors), do not bind directly to the TATA box and appear to allow TFIID to respond to stimulation by transcriptional activators (see Chapter 5, section 5.4.2) (for review see Hahn, 1998; Green, 2000). They thus represent co-activator molecules, linking transcriptional activators and the basal transcriptional complex.

Hence TBP plays a critical role in the transcription of TATA box-containing RNA polymerase II promoters by binding to the TATA box as the first step in assembly of the basal transcriptional complex. In view of this critical role, it is not surprising that TBP is one of the most highly conserved eukaryotic proteins. The structure of this protein has been defined by X-ray crystallography and shown to have a saddle structure in which the concave underside binds to DNA and the convex outer surface is accessible for interactions with other factors. Most interestingly, binding of TBP to the DNA deforms the DNA so that it follows the concave curve of the saddle (Fig. 3.8). Moreover, structural studies of the TFIID complex (consisting of TBP and the TAFs) bound to DNA have indicated that it resembles the complex of the eight histone molecules around which DNA is wound in the nucleosome to form the normal chromatin structure (see Chapter 1, section 1.2.1). Hence the DNA may bend around TFIID at the promoter in a similar manner to the folding of the rest of DNA in the basic nucleosome structure of chromatin (for reviews see Hoffmann *et al.*, 1997; Gangloff *et al.*, 2001). This role for TFIID in altering nucleosome structure at the promoter is also supported by the finding that TAFII$_{250}$, one of the subunits of TFIID, has histone acetyltransferase activity (Mizzen *et al.*, 1996), since acetylation of histones appears to play a key role in modulating chromatin structure (see Chapter 1, section 1.2.3).

The bent DNA with TFIID bound to it serves as the central platform on which the basal transcriptional complex assembles. Thus, structural studies have shown that TFIIA binds to the amino terminal stirrup of the TBP saddle

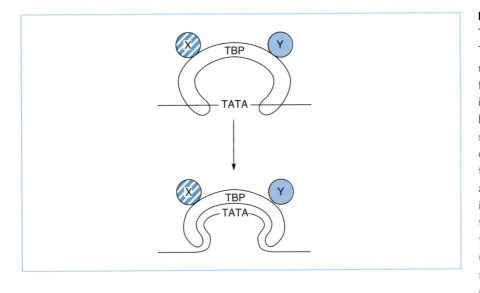

Figure 3.8
The saddle structure of TBP as determined by X-ray crystallography allows the concave surface to interact with the TATA box while the convex surface associates with other accessory transcription factors (X and Y). The initial binding induces the bending of the DNA so that it follows the concave under surface of the saddle. See also colour plate 1.

and interacts only with the DNA upstream of the TATA box. This allows it to fulfil its role of protecting TFIID from inhibition by transcriptional repressors and allowing it to respond to activators bound to upstream DNA sequences (see section 3.5.1). In contrast, TFIIB binds to the carboxyl-terminal stirrup of the TBP saddle and binds to the DNA downstream (as well as upstream) of the TATA box (Andel *et al.*, 1999). This allows it to fulfill its role of acting as a bridge between TBP and RNA polymerase II so positioning the start site of transcription by the polymerase relative to the TATA box (see Plate 1; Geiger *et al.*, 1996) (for reviews see Roeder, 1996; Nikolov and Burley, 1997; Woychick and Hampsey, 2002). Interestingly, a recent study indicates that binding of TFIIB promotes bending of the DNA by TBP, indicating that TFIIB acts by interacting with the partially assembled complex as well as by recruiting new factors to the complex (Zhao and Herr, 2002).

Paradoxically, in view of its TATA box binding ability, TBP also plays a critical role in the transcription of the subset of RNA polymerase II genes which do not contain a TATA box (see Chapter 1, section 1.3.2). In this case, however, TBP does not bind to the DNA but is recruited to the promoter by another DNA binding protein which binds to the initiator element overlapping the transcriptional start site. TBP then binds to this initiator binding protein allowing the recruitment of TFIIB and the RNA polymerase itself as for promoters containing a TATA box. Hence TBP plays a critical role in the assembly of the transcription complex for RNA polymerase II, although it joins the complex by binding to DNA in the case of TATA-box-containing promoters (Fig. 3.9a) and is recruited by protein–protein interactions in the case of promoters which lack a TATA-box (Fig. 3.9b).

Figure 3.9

Transcription of promoters by RNA polymerase II involves the recruitment of TBP (and associated factors (X) forming the TFIID complex) to the promoter. This may be achieved by direct DNA binding to the TATA box where this is present (panel a) or by protein–protein interaction with a factor (IN) bound to the initiator element where the TATA box is absent (panel b).

These findings have led to the suggestion that TBP represents the basic transcription factor for RNA polymerase II, paralleling the role of SL1 for RNA polymerase I and TFIIIB for RNA polymerase III. This idea was supported by the amazing finding that TBP is actually also a component of both SL1 and TFIIIB (for review see White and Jackson, 1992). Thus the SL1 factor is actually a complex of four factors, one of which is TBP. Hence when SL1 is recruited to the promoter by UBF (see section 3.3) TBP is delivered to the DNA exactly as in the non-TATA-box containing RNA polymerase II promoters where TBP is recruited by the prior binding of another protein to the initiator element.

Similarly, in the case of RNA polymerase III transcription where TBP is part of the multi-component TFIIIB complex (for review see Rigby, 1993), TBP is delivered to class I polymerase III promoters by protein–protein interaction following the prior binding of TFIIIA and TFIIIC and is delivered to class II polymerase III promoters by the prior binding of TFIIIC (section 3.4) (Fig. 3.10a). Interestingly, however, as noted in section 3.4, the class III group of RNA polymerase III promoters contains a TATA box and hence in this case TBP can bind directly (Fig. 3.10b). As in RNA polymerase II promoters, distinct mechanisms therefore ensure the recruitment of TBP to all RNA polymerase III promoters.

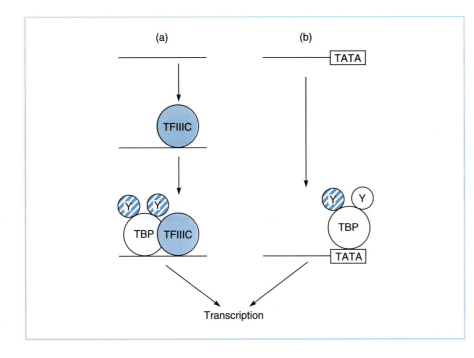

Figure 3.10
Transcription of promoters by RNA polymerase III involves the recruitment of TBP (and associated factors (Y) forming the TFIIIB complex) to the promoter. This may be achieved by protein–protein interactions with TFIIIA and TFIIIC in the case of class I promoters, with TFIIIC alone in the case of class II promoters (panel a) or by direct DNA binding to the TATA box in class III promoters where the TATA box is present (panel b).

The similarities between the three RNA polymerases discussed in section 3.1 are therefore paralleled by the involvement of a common factor, TBP, in transcription by all three RNA polymerases (Fig. 3.11). Interestingly, this relationship has been extended further by the recent finding that TFIIH also plays an essential role in transcription by RNA polymerase I as well as that mediated by RNA polymerase II (Iben *et al.*, 2002).

In all three RNA polymerase complexes, TBP forms a part of the multi-protein complexes which have been shown to be essential for transcription itself, binding via the TATA box or by protein–protein interactions with

Figure 3.11
TBP is involved in transcription by all three RNA polymerases.

assembly factors (for review see Struhl, 1994). In view of this it has been suggested that TBP represents an evolutionarily ancient transcription factor preceding the division of the three RNA polymerases and having a universal and essential role in eukaryotic transcription (for review see Hernandez, 1993). Indeed, a TBP homologue is also found in the archaebacteria, which constitutes a separate kingdom distinct from the eukaryotes and the eubacteria. Hence the existence of TBP appears to predate, not only the divergence of the three RNA polymerases but also the divergence of the eukaryotic and archaebacterial kingdoms (for review of archaebacterial transcription see Reeve *et al.*, 1997).

It was initially believed that each organism would have only one form of TBP encoded by a single gene. However, studies of the genomes of multicellular organisms have identified genes encoding other TBP-related proteins with humans, for example, having one such TBP-like factor (TLF) whereas *Drosophila* has two (for reviews see Berk, 2000; Veenstra and Wolfe, 2001).

It has been shown that, in some cases, the basal transcriptional complex contains a TLF rather than a TBP. For example, specific stages of development in the amphibian *Xenopus* require a TLF activity which cannot be substituted by TBP (Veenstra *et al.*, 2000). Similarly, the *Drosophila* PCNA gene has two promoters, one of which is recognized by a basal transcriptional complex containing TBP and the other by a complex containing TRF2 which is a TLF (Hochheimer *et al.*, 2002). Hence, it is clear that some specific transcription complexes contain a TLF rather than TBP and that this is required for their proper functioning (Fig. 3.12b). The existence of the TLFs thus offers a further means of regulating gene transcription in specific situations.

Interestingly, as well as basal transcriptional complexes containing either TBP or a TLF, it has also been shown that RNA polymerase II transcription can be driven by a complex which does not contain TBP or a TLF (Wieczorek *et al.*, 1998). This suggests that, under some circumstances, neither TBP or TLF is required for transcription. In agreement with this, some RNA polymerase II transcription occurs in early stage embryos from knock out mice lacking functional TBP, although transcription by RNA polymerases I and III does not occur. As early stage mouse embryos do not express a TLF, this indicates the existence of TBP/TLF independent transcription, at least for RNA polymerase II (Martianov *et al.*, 2002) (Fig. 3.12c).

Hence, TBP is a highly ancient transcription factor which is involved in transcription by all three RNA polymerases (see Fig. 3.11). However, in some situations transcription can occur in a TBP-independent manner involving either a TBP-like factor or a complex which lacks TBP or a TLF (Fig. 3.12).

Figure 3.12
On different target promoters, the basal transcriptional complex (BTC) for RNA polymerase II can contain TBP (panel a), a TBP-like factor (TLF) (panel b) or lack either TBP or TLF which are presumably replaced by an unrelated factor (X) (panel c).

3.7 CONCLUSIONS

The binding of each of the three eukaryotic RNA polymerases to appropriate gene promoters and subsequent transcription is dependent on the prior binding of a specific transcription factor to the promoter. Binding of the polymerase to the DNA adjacent to this factor occurs by recognition of the bound protein rather than by recognition of the specific DNA sequence in this region. In most cases, the binding of the transcription factor itself requires the prior binding of other factors to the DNA. These assembly factors therefore play a critical role in the formation of the stable transcriptional complex, but can be dissociated once the complex has formed without affecting its activity. In the case of RNA polymerase II transcription, either the stability of the complex or its activity is greatly affected by the binding of other proteins to sequences upstream of the promoter. The roles of these transcription factors and the mechanisms by which they function are described in the remainder of this book.

REFERENCES

Ahearn, J.M. Jr, Bartolomei, M.S., West, M.L. *et al.* (1987) Cloning and sequence analysis of the mouse genomic locus encoding the large subunit of RNA polymerase II. Journal of Biological Chemistry 262, 10695–10705.

Andel, F., Ladurner, A. G., Inouye, C. *et al.* (1999) Three-dimensional structure of the human TFIID-IIa-IIB complex. Science 286, 2153–2156.

Berk, A. J. (2000) TBP-like factors come into focus. Cell 103, 5–8.

Brown, D.D. (1984) The role of stable complexes that repress and activate eukaryotic genes. Cell 37, 359–365.

Brown, T. R., Scott, P. H., Stein, T. *et al.* (2000) RNA polymerase III transcription: Its control by tumour suppressors and its deregulation by transforming agents. Gene Expression 9, 15–28.

Burley, S.K. and Roeder, R.G. (1996) Biochemistry and structural biology of transcription factor IID (TFIID). Annual Review of Biochemistry 65, 769–799.

Chang, W-H. and Kornberg, R. D. (2000) Electron crystal structure of the transcription factor and DNA repair complex, core TFIIH. Cell 102, 609–613.

Cilberto, G., Castagnoli, L. and Cortese, R. (1983) Transcription by RNA polymerase III. Current Topics in Developmental Biology 18, 59–88.

Drapkin, R., Merino, A. and Reinberg, D. (1993) Regulation of RNA polymerase II transcription. Current Opinion in Cell Biology 5, 469–476.

Dubois, M.F., Nguyer, V.T., Dahmus, M.E. *et al.* (1994) Enhanced phosphorylation of the C-terminal domain of RNA polymerase II upon serum stimulation of quiescent cells: possible involvement of MAP kinases. EMBO Journal 13, 4787–4797.

Gangloff, Y-G., Romier, C., Thuault, S. *et al.* (2001) The histone fold is a key structural motif of transcription factor TFIID. Trends in Biochemical Sciences 26, 250–256.

Geiduschek, E. P. and Kassavetis, G. A. (2001) The RNA polymerase III transcription apparatus. Journal of Molecular Biology 310, 1–26.

Geiger, J.H., Hahn, S., Lee, S. and Sigler, P.B. (1996) Crystal structure of the yeast TFIIA/ TBP/DNA complex. Science 272, 830–836.

Green, M. R. (2000) TBP-associated factors (TAF$_{IIs}$): multiple, selective transcriptional mediators in common complexes. Trends in Biochemical Sciences 25, 59–63.

Greenblatt, J. (1997) RNA polymerase II holoenzyme and transcriptional regulation. Current Opinion in Cell Biology 9, 310–319.

Hahn, S. (1998) The role of TAFs in RNA polymerase II transcription. Cell 95, 579–582.

Hernandez, N. (1993) TBP, a universal eukaryotic transcription factor. Genes and Development 7, 1291–1308.

Hirose, Y. and Manley, J. (2000) RNA polymerase II and the integration of nuclear events. Genes and Development 14, 1415–1429.

Hochheimer, A., Zhou, S., Zheng, S. *et al.* (2002) TRF2 associates with DREF and directs promoter-selective gene expression in *Drosophila*. Nature 420, 439–445.

Hoeijmakers, J.H.J., Egly, J.M. and Vermueler, W. (1996) TFIIH: a key component in multiple DNA transactions. Current Opinion in Genetics and Development 6, 26–33.

Hoffman, A., Oelgeschlage, T. and Roeder, R. (1997) Considerations of transcriptional control mechanisms: Do TFIID-core promoter complexes recapitulate nucleosome-like functions. Proceedings of the National Academy of Sciences, USA 94, 8923–8925.

Iben, S., Tschochner, H., Bier, M. *et al.* (2002) TFIIH plays an essential role in RNA polymerase I transcription. Cell 109, 297–306.

Jacob, S.T. (1995) Regulation of ribosomal gene transcription. Biochemical Journal 306, 617–626.

Klug, A. (2001) A marvellous machine for making messages. Science 292, 1844–1846.

Landick, R. (2001) RNA polymerase clamps down. Cell 105, 567–570.

Martianov, I., Viville, S. and Davidson, I. (2002) RNA polymerase II transcription in murine cells lacking the TATA binding protein. Science 298, 1036–1039.

Memet, S., Saurn, W. and Sentenac, A. (1988) RNA polymerases B and C are more closely related to each other than to RNA polymerase A. Journal of Biological Chemistry 263, 10048–10051.

Mizzen, C.A., Yang, X-J., Kokubo, T. *et al.* (1996) The TAFII 250 subunit of TFIID has histone acetyltransferase activity. Cell 87, 1261–1270.

Myer, V. E. and Young, R. A. (1998) RNA Polymerase II holoenzymes and subcomplexes. Journal of Biological Chemistry 273, 27757–27760.

Nikolov, D. B. and Burley, S. K. (1997) RNA polymerase II transcription initiation: a structural view. Proceedings of the National Academy of Sciences USA 94, 15–22.

Orphanides, G., Lagrange, T. and Reinberg, D. (1996) The general transcription factors of RNA polymerase II. Genes and Development 10, 2657–2683.

Paule, M.R. (1990) In search of the single factor. Nature 344, 819–820.

Paule, M. R. and White, R. J. (2000) Transcription by RNA polymerases I and III. Nucleic Acids Research 28, 1283–1298.

Proudfoot, N. J., Furger, A. and Dye, M. J. (2002) Integrating mRNA processing with transcription. Cell 108, 501–512.

Pugh, B.F. (1996) Mechanisms of transcription complex assembly. Current Opinion in Cell Biology 8, 303–311.

Reeve, J.N., Sandman, K. and Daniels, C.J. (1997) Archaeal histones, nucleosomes and transcriptional initiation. Cell 89, 999–1002.

Rigby, P.W.J. (1993) Three in one and one in three: it all depends on TBP. Cell 72, 7–10.

Riva, M., Schaffner, A.R., Sentenac, A. *et al.* (1987) Active site labelling of the RNA polymerases A, B and C from yeast. Journal of Biological Chemistry 262, 14377–14380.

Rochette-Egly, C., Adam, S., Rossignol, M. *et al.* (1997) Stimulation of RARαactivation function AF-1 through binding to the general transcription factor TFIIH and phosphorylation by CDK7. Cell 90, 97–107.

Roeder, R.G. (1996) The role of general initiation factors in transcription by RNA polymerase II. Trends in Biochemical Sciences 21, 327–334.

Saltzman, A.G. and Weinmann, R. (1989) Promoter specificity and modulation of RNA polymerase II transcription. FASEB Journal 3, 1723–1733.

Schramm, L. and Hernandez, N. (2002) Recruitment of RNA polymerase III to its target promoters. Genes and Development 16, 2593–2620.

Schultz, P., Fribourg, S., Poterszman, A. *et al.* (2000) Molecular structure of human TFIIH. Cell 102, 599–607.

Sentenac, A. (1985) Eukaryotic RNA polymerases. CRC Critical Reviews in Biochemistry 1, 31–90.

Sollner-Webb, B. (1988) Surprises in RNA polymerase III transcription Cell 52,153–154.

Somerville, J. (1984) RNA polymerase I promoters and cellular transcription factors. Nature 310, 189–190.

Stiller, J. W. and Hall, B. D. (2002) Evolution of the RNA polymerase II C-terminal domain. Proceedings of the National Academy of Sciences USA 99, 6091–6096.

Struhl, K. (1994) Duality of TBP: The universal transcription factor. Science 263, 1103–1104.

Svejstrup, J.Q., Vichi, P. and Egly, J-M. (1996) The multiple roles of transcription/repair factor TFIIH. Trends in Biochemical Sciences 21, 346–350.

Veenstra, G. J. C. and Wolffe, A. P. (2001) Gene-selective developmental roles of general transcription factors. Trends in Biochemical Sciences 26, 665–671.

Veenstra, G. J. C., Weeks, D. L. and Wolffe, A. P. (2000) Distinct roles for TBP and TBP-like factor in early embryonic gene transcription in *Xenopus*. Science 290, 2312–2315.

White, R.J. and Jackson, S.P. (1992) The TATA binding protein: a central role in transcription by RNA polymerases I and III. Trends in Genetics 8, 284–288.

Wieczorek, E., Brand, M., Jacq, X. and Tora, L. (1998) Function of TAF$_{II}$-containing complex without TBP in transcription by RNA polymerase II. Nature 393, 187–191.

Woychik, N. A. and Hampsey, M. (2002) The RNA polymerase II machinery: Structure illuminates function. Cell 108, 454–463.

Woychik, N.A., Liao, S-M., Koldrieg, P.A. and Young, R.A. (1990) Subunits shared by eukaryotic RNA polymerases. Genes and Development 4, 313–323.

Young, R. (1991) RNA polymerase II. Annual Review of Biochemistry 60, 689–715.

Zhao, X. and Herr, W. (2002) A regulated two-step mechanism of TBP binding to DNA: a solvent-exposed surface of TBP inhibits TATA box recognition. Cell 108, 615–627.

CHAPTER 4

FAMILIES OF DNA BINDING TRANSCRIPTION FACTORS

4.1 INTRODUCTION

In previous chapters we have considered the role of chromatin structure, DNA sequences and RNA polymerase in the process of transcription and its regulation. The remaining aspect of this process, and the major subject of this book, are the transcription factors which bind to specific DNA sequences that have been exposed by changes in chromatin structure and then alter transcription by interacting directly or indirectly with RNA polymerase. To fulfil this role, transcription factors must possess certain features allowing them to modulate gene expression.

Clearly the first feature that many of these factors require is the ability to bind to DNA in a sequence specific manner and this is discussed in this chapter. Following binding, the factor must interact with other factors or with the RNA polymerase itself in order to influence transcription either positively or negatively and these aspects are discussed respectively in Chapters 5 and 6. Finally, in the case of factors modulating inducible, tissue-specific or developmentally regulated gene expression, some means must exist to regulate the synthesis or activity of the factor so that it is active only in a particular situation. This regulation of factor synthesis or activity is discussed respectively in Chapters 7 and 8.

Following the cloning of many different eukaryotic transcription factors, the domain mapping experiments described in Chapter 2 (section 2.4.1) have led to the identification of several distinct structural elements in different factors which can mediate DNA binding. These motifs have been used to classify transcription factors into families. These families and the DNA motifs that define them will be discussed in turn using transcription factors which contain them to illustrate their properties (for reviews see Harrison, 1991; Pabo and Sauer, 1992; Travers, 1993; Garvie and Wolberger, 2001).

4.2 THE HOMEODOMAIN

4.2.1 TRANSCRIPTION FACTORS IN DROSOPHILA DEVELOPMENT

Detailed genetic studies in the fruit fly *Drosophila melanogaster* have led to the identification of a very large number of mutations which affect the development of this organism and their corresponding genes have been named on the basis of the observed phenotype of the mutant fly (for reviews see Ingham; 1988; Lawrence and Morata, 1994). Thus mutations in the so-called homeotic genes result in the transformation of one particular segment of the body into another; mutations in the Antennapedia gene for example causing the transformation of the segment which normally produces the antenna into one which produces a middle leg (Fig. 4.1). Similarly mutations in genes of the gap class result in the total absence of particular segments; mutations in the Knirps gene for example resulting in the absence of most of the abdominal segments although the head and thorax develop normally.

The products of genes of this type therefore play critical roles in *Drosophila* development. The products of the gap genes, for example, are necessary for the production of particular segments while the homeotic gene products

Figure 4.1

Effect of the homeotic mutation Antennapedia, which produces a middle leg (B) in the region that would contain the antenna of a normal fly (A), a1, all, alll: 1st, 2nd, and 3rd antennal segments; ar: arista; ta: tarsus; ti: tibia; fe: femur; ap: apical bristle.

specify the identity of these segments. Given that these processes are likely to require the activation of genes whose protein products are required in the particular segment, it is not surprising that many of these genes have been shown to encode transcription factors. Thus the Knirps gene product and that of another gap gene Kruppel, contain multiple zinc finger motifs characteristic of DNA binding transcription factors and can bind to DNA in a sequence specific manner (see section 4.3). Similarly, the tailless gene, whose product plays a key role in defining the anterior and posterior regions of the *Drosophila* embryo, has been shown to be a member of the nuclear receptor super gene family (see section 4.4).

It is clear therefore that the genes identified by mutation as playing a role in *Drosophila* development can encode several different types of transcription factors. However, of the first twenty-five such genes that were cloned allowing a study of their protein products, well over half (fifteen) contain a motif known as the homeobox or homeodomain (Gehring *et al.*, 1994a) which was originally identified in the homeotic genes of *Drosophila*. The features of these homeodomain proteins and the manner in which they mediate DNA binding and transcriptional regulation will be extensively discussed since they serve as a paradigm for the manner in which transcription factors function and can control highly complex processes such as development.

4.2.2 THE HOMEOBOX

When the first homeotic genes were cloned, it was found that they shared a region of homology, approximately one hundred and eighty base pairs long and therefore capable of encoding sixty amino acids, which was flanked on either side by regions which differed dramatically between the different genes. This region was named the homeobox or homeodomain (for review see Gehring *et al.*, 1994a). Subsequently the homeobox was shown to be present in many other *Drosophila* regulatory genes. These include the Fushi-tarazu gene (Ftz), which is a member of the pair rule class of regulatory loci whose mutation causes alternate segments to be absent, and the engrailed gene (eng), which is a member of the class of genes whose products regulate segment polarity. The close similarity of the homeoboxes encoded by the homeotic genes Antennapedia and Ultrabithorax and that encoded by the Ftz gene is shown in Figure 4.2.

The presence of this motif in a large number of different regulatory genes of different classes strongly suggested that it was of importance in their activity. The evidence that the homeobox-containing proteins are transcription factors whose DNA binding activity is mediated by the homeobox is discussed

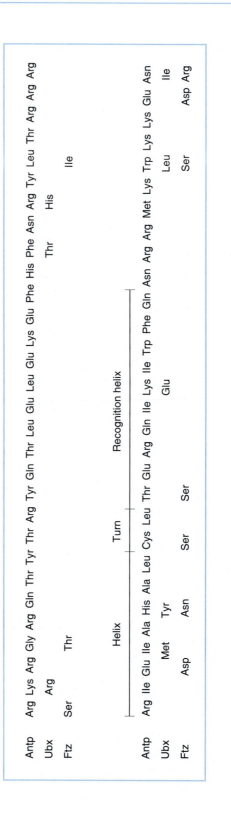

Figure 4.2

Amino acid sequences of several *Drosophila* homeodomains, showing the conserved helical motifs. Differences between the sequences of the Ubx and Ftz homeodomains from that of Antp are indicated; a blank denotes identity in the sequence. The helix-turn-helix region is indicated.

in the next section (for reviews see Hayashi and Scott, 1990; Gehring *et al.*, 1994).

4.2.3 DNA BINDING BY THE HELIX-TURN-HELIX MOTIF IN THE HOMEOBOX

The first indication that the homeobox proteins were indeed transcription factors came from the finding that the homeobox was also present in the yeast mating type a and α gene products which are known to be transcription factors that regulate the activity of a and α-specific genes (for review see Dolan and Fields, 1991), hence suggesting, by analogy, that the *Drosophila* proteins also fulfilled such a role.

Direct evidence that this is the case is available from a number of different approaches. Thus it has been shown that many of these proteins bind to DNA in a sequence specific manner as expected for transcription factors (Hoey and Levine, 1988). Moreover, binding of a specific homeobox protein to the promoter of a particular gene correlates with the genetic evidence that the protein regulates expression of that particular gene. For example, the Ultrabithorax (Ubx) protein has been shown to bind to specific DNA sequences within its own promoter and in the promoter of the Antennapedia gene, in agreement with the genetic evidence that Ubx represses Antennapedia expression (Fig. 4.3).

The ability of the homeobox-containing proteins to bind to DNA is directly mediated by the homeobox itself. Thus if the homeobox of the Antennapedia protein is synthesized in isolation either in bacteria or by chemical synthesis, it is capable of binding to DNA in the identical sequence specific manner characteristic of the intact protein.

This ability to define the sixty amino acid homeodomain as the region binding to DNA has led to intensive study of its structure in the hope of elucidating how the protein binds to DNA in a sequence specific manner (for reviews see Kornberg, 1993; Gehring *et al.*, 1994b). In particular, the crystal structure of the Antennapedia (Antp) homeodomain bound to DNA has been determined by nuclear magnetic resonance spectroscopy (NMR) while similar structural studies of the engrailed (eng) and the yeast MATα2 homeodomains bound to DNA have been carried out by X-ray crystallography.

By this means the Antp homeodomain was shown to contain a short N-terminal arm of six residues followed by four alpha helical regions (Fig. 4.4). The first two helices are virtually anti-parallel to each other with the other two helices arranged at right angles to the first. Most interestingly, helices II and III are separated by a beta turn forming a helix-turn-helix motif (Fig. 4.5). The

Figure 4.3

Assay of protein binding to a DNA fragment from the Antennapedia gene promoter (Antp) or a control fragment of plasmid DNA (pUC) using protein extracts from *E. coli* which have been genetically engineered to express the *Drosophila* Ubx protein (a) or protein extracts from control *E. coli* not expressing Ubx (b). Note the specific binding of Ubx protein to the Antennapedia DNA fragment.

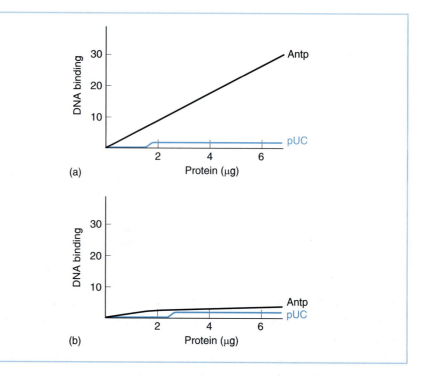

eng and MATα2 homeodomains also have a similar structure with an N-terminal arm and a subsequent helix-turn-helix motif. In this case, however, the third and fourth helices observed in Antp form a single helical region. Interestingly, the helix-turn-helix structure typical of the homeodomain is very similar to the DNA binding motif of several bacteriophage regulatory proteins such as the lambda cro protein or the phage 434 repressor which have also been crystallized and subjected to intensive structural study.

Figure 4.4

Structure of the Antennapedia homeodomain as determined by nuclear magnetic resonance spectroscopy. Note the four alpha-helical regions (I–IV) represented as cylinders with the amino acids at their ends indicated by numbers and the one letter amino acid code.

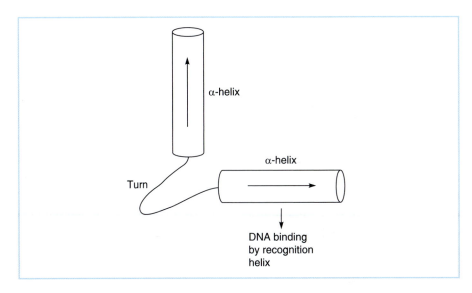

α-helix

α-helix

Turn

DNA binding
by recognition
helix

Figure 4.5
The helix-turn-helix motif.

In these bacteriophage proteins X-ray crystallographic studies have shown that the helix-turn-helix motif does indeed contact DNA. One of the two helices lies across the major groove of the DNA while the other lies partly within the major groove where it can make sequence specific contacts with the bases of DNA. It is this second helix (known as the recognition helix) that therefore controls the sequence specific DNA binding activity of these proteins (Fig. 4.6).

Figure 4.6
Binding of the helix-turn-helix motif to DNA with the recognition helix in the major groove of the DNA.

The similarity in structure of helices II and III in the eukaryotic homeodomains to the two helices of the bacteriophage proteins led to the suggestion that these two helices in the homeodomain are similarly aligned relative to the DNA with helix III constituting the recognition helix responsible for sequence specific DNA binding. Hence the precise amino acid sequence in the recognition helix in different homeodomain proteins would determine which DNA sequence they bound (for review see Treisman *et al.*, 1992).

In agreement with this idea, exchanging the recognition helix in the Bicoid (Bcd) homeodomain for that of Antp resulted in a protein with the DNA binding specificity of Antp and not that of Bicoid. Most interestingly a Bcd protein with the DNA binding specificity of Antp could also be obtained by exchanging only the ninth amino acid in the recognition helix, replacing the lysine residue in Bcd with the glutamine residue found in the Antp protein (Fig. 4.7), whereas the exchange of other residues which differ between the two proteins has no effect on the DNA binding specificity. Hence the ninth amino acid within the recognition helix of the homeodomain plays a critical role in determining DNA binding specificity.

Figure 4.7

Effect of changing the amino acid sequence in the recognition helix of the Bicoid protein on its binding to its normal recognition site and that of the Antennapedia (Antp) protein. Note the critical effect of changing the ninth amino acid in the helix which completely changes the specificity of the Bicoid protein.

					Recognition helix						Binding to Bicoid site TCTAATCCC	Binding to Antp site TCAATTAAAT
Bicoid	T	A	Q	V	K	I	W	F	K	N	+	–
	A	–	–	–	–	–	–	–	–	–	+	–
	–	–	–	A	–	–	–	–	–	–	+	–
	–	–	–	–	–	–	–	–	A	–	–	–
	A	–	–	–	A	A	–	–	A	–	–	–
	E	R	–	–	–	–	–	–	Q	–	–	+
	E	R	–	–	–	–	–	–	–	–	+	–
	–	–	–	–	–	–	–	–	Q	–	–	+
Antp	E	R	Q	I	K	I	W	F	Q	N	–	+

It is likely that the amino group of lysine found at the ninth position in the Bcd protein makes hydrogen bonds with the N6 and N7 positions of a guanine residue in the Bcd-specific DNA binding site whereas the amide group of glutamine found at the corresponding position in the Antp recognition helix forms hydrogen bonds with the N6 and N7 positions of an adenine residue at the equivalent position within the Antp-specific DNA binding site. Hence the replacement of lysine with glutamine results in the loss of two potential hydrogen bonds to a Bcd site and the gain of two potential hydrogen bonds to an Antp site explaining the observed change in DNA binding specificity (Fig. 4.8).

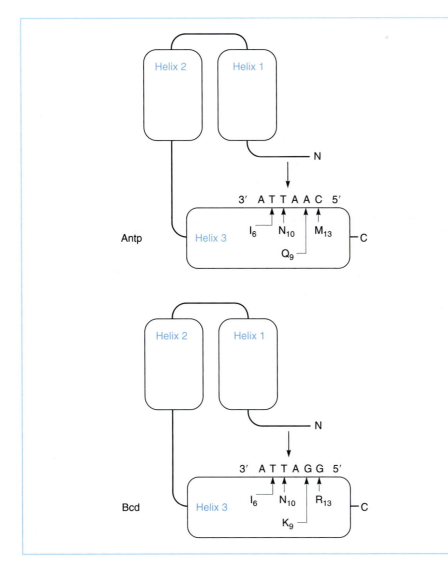

Figure 4.8
Contacts between DNA and the Antp or Bcd homeodomains. Note that the change in the ninth amino acid of the recognition helix (helix 3) alters the base that is preferentially bound from a G for Bcd to an A for Antp, as discussed in the text, while the N-terminal arm of the homeodomain contacts the ATTA sequence common to the recognition site of both proteins.

A similar critical role for the ninth amino acid in determining the precise DNA sequence which is recognized is also seen in other homeobox-containing proteins, replacement of the serine found at this position in the paired protein with the lysine found in Bicoid or the glutamine found in Antp, allowing the paired protein to recognize respectively Bcd or Antp-specific DNA sequences. Hence the DNA sequence recognized by a homeobox-containing protein appears to be primarily determined by the ninth amino acid in the recognition helix, proteins with different amino acids at this position recognizing different DNA sequences whereas proteins such as Antp and fushi-tarazu which have the same amino acid at this position recognize the same DNA sequence.

This critical role of the ninth amino acid is in contrast to the situation in the bacteriophage proteins in which the helix-turn-helix motif was originally defined. In these proteins, the most N terminal residues (1–3) in the recognition helix play a critical role in determining DNA binding specificity (for review see Pabo and Sauer, 1992). As shown in Figure 4.7, however, these amino acids appear to play little or no role in determining the DNA binding specificity of eukaryotic helix-turn-helix proteins suggesting, therefore, that the recognition helix of these proteins is oriented differently in the major groove of the DNA.

This idea is in agreement with the structural studies of the eukaryotic homeodomains bound to DNA which have identified the actual protein–DNA contacts. These studies have shown that as in the bacteriophage proteins, the recognition helix directly contacts the bases of DNA in the major groove. However, in the eukaryotic homeobox proteins this helix is oriented within the major groove somewhat differently such that the critical base-specific contacts are, as predicted, made by the C terminal end of the helix which contains residue nine (see Fig. 4.8).

It is clear therefore that the helix-turn-helix motif in the homeobox mediates both the DNA binding of the protein and also, via the recognition helix, controls the precise DNA sequence that is recognized. Interestingly, however, the short N-terminal arm of the homeodomain also contacts the bases of the DNA, although it makes contact in the minor groove rather than the major groove. Removal of this short N-terminal arm dramatically reduces the DNA binding affinity of the homeodomain indicating that this region contributes significantly to the DNA binding ability of the homeodomain probably by contacting the ATTA bases common to the DNA binding sites of several homeodomain proteins (see Fig. 4.8).

Although DNA binding is important for the modulation of transcription, it is necessary to demonstrate that the homeobox proteins do actually affect transcription following such binding. In the case of the Ubx protein, this was achieved by showing that co-transfection of a plasmid expressing Ubx with a plasmid in which the Antennapedia promoter drives a marker gene resulted in the repression of gene expression driven by the Antennapedia promoter (Fig. 4.9). Hence the observed binding of Ubx to the Antp promoter (see above) results in down regulation of its activity in agreement with the results of genetic experiments.

Most interestingly, the Ubx expression plasmid was able to up regulate activity of its own promoter in co-transfection experiments, this ability being dependent on the previously defined binding sites for Ubx within its

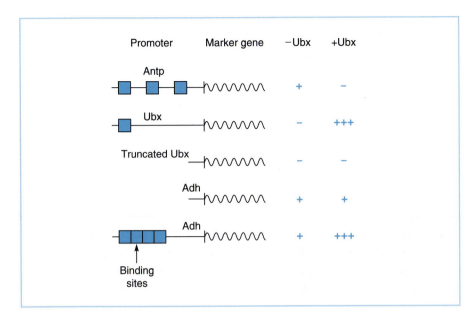

Figure 4.9

Effect of Ubx on various marker genes with or without binding sites (hatched boxes) for the Ubx protein. Note that Ubx can stimulate its own promoter which contains a Ubx binding site and this effect is abolished by deleting the Ubx binding site. Similarly, the alcohol dehydrogenase (Adh) gene which is normally unaffected by Ubx, is rendered responsive to Ubx stimulation by addition of Ubx binding sites. In contrast the Antennapedia promoter, which also contains Ubx binding sites, is repressed by Ubx. Hence binding of Ubx can activate or repress different promoters.

own promoter. Similarly, although Ubx normally has no effect on expression of the alcohol dehydrogenase (Adh) gene it can stimulate the Adh promotor following linkage of the promoter to a DNA sequence containing multiple binding sites for Ubx. Hence a homeobox protein can produce distinct effects following binding, Ubx activating its own promoter and a hybrid promoter containing Ubx binding sites but repressing the activity of the Antp promoter (Fig. 4.9).

A similar transcriptional activation effect of DNA binding has been demonstrated for the Fushi-tarazu (Ftz) protein. This protein binds specifically to the sequence TCAATTAAATGA. As with Ubx, linkage of this sequence to a marker gene confers responsivity to activation by Ftz, such activation being dependent upon binding of Ftz to its target sequence, a one base pair change which abolishes binding also abolishing the induction of transcription (Fig. 4.10).

Interestingly, the ability of Ubx to induce its own transcription provides a mechanism for the long-term maintenance of Ubx gene expression during development since once expression has been switched on and some Ubx protein made, it will induce further transcription of the gene via a simple positive feedback loop even if the factors which originally stimulated its expression are no longer present (Fig. 4.11). This long-term maintenance of Ubx expression is essential since, if the Ubx gene is mutated within the larval imaginal disc cells, which eventually produce the adult fly, the cells that would normally produce the haltere (balancer) will produce a wing instead. Thus, although these cells are known to already be committed to form the adult

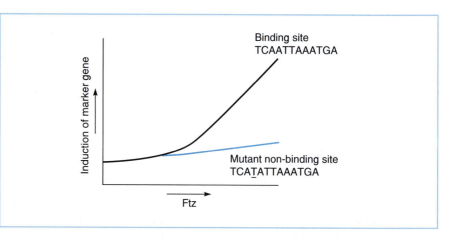

Figure 4.10
Effect of expression of the Ftz protein on the expression of a gene containing its binding site, or a mutated binding site containing a single base pair change which abolishes binding of Ftz.

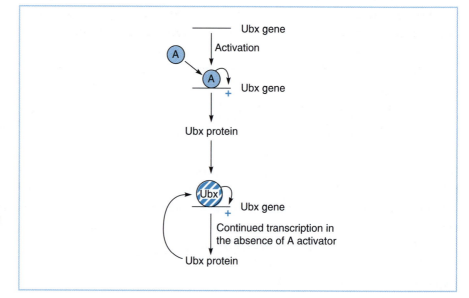

Figure 4.11
The stimulatory effect of the Ubx protein on the transcription of its own gene, ensures that once Ubx gene transcription is initially switched on by an activator protein (A), transcription will continue even if the activator protein is removed.

haltere at the larval stage, the continued expression of the Ubx gene is essential to maintain this commitment and allow eventual overt differentiation (see Hadorn, 1968 for a review of imaginal discs and their role in *Drosophila* development).

4.2.4 REGULATION OF DNA BINDING SPECIFICITY BY INTERACTIONS BETWEEN DIFFERENT HOMEOBOX PROTEINS

Although we have previously described the DNA binding specificity of individual homeobox proteins, it is possible for the DNA binding specificity of one factor to be altered in the presence of another factor. Thus several homeobox proteins, such as Ubx and Antp, bind to the same DNA sequences when tested in isolation *in vitro* (Hoey and Levine, 1988) yet, paradoxically, the effects of mutations which inactivate the genes encoding each of these proteins are different indicating that they cannot substitute for one another. Similarly, *in vivo* Ubx can bind to a site in the promoter of the decapentaplegic (dpp) gene and activate its expression whereas Antp cannot do so.

This paradox is explained by the presence in the dpp promoter of a binding site for another homeobox protein extradenticle (Exd) which lies adjacent to the site to which Ubx binds. The Exd protein interacts with the Ubx protein and both enhances its DNA binding affinity and modifies its DNA binding specificity so it can bind strongly to the dpp gene promoter and activate its expression (Fig. 4.12) (for review see Mann and Chan, 1996). As Antp does not interact with Exd, its specificity is not modified in this way. Hence, it does not bind to the dpp gene promoter and therefore cannot activate this promoter. Interestingly, structural studies have shown that Ubx and Exd bind to opposite sides of the DNA and that a short region of Ubx N-terminal to the homeodomain extends round the DNA and inserts into a cleft in the Exd homeodomain resulting in interaction of the proteins and enhanced DNA binding by the complex (Passner *et al.*, 1999; for review see Scott, 1999).

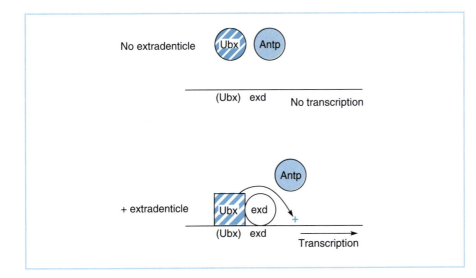

Figure 4.12
The Exd protein interacts with the Ubx protein to allow it to bind with high affinity to its potential binding site in the dpp promoter (indicated as (UBX)) and activate its expression. In contrast, the Antp protein cannot interact with Exd and so does not bind to the dpp promoter.

A similar interaction is observed in the case of the yeast homeodomain proteins a1 and α2, which control the mating type in this organism (for review see Dolan and Fields, 1991). Thus, in the absence of a1, the α2 protein has a weak DNA binding ability. However, in the presence of a1, an a1/α2 heterodimer forms and binds to specific gene promoters. As the α2 protein is a strong transcriptional repressor, this results in the repression of the genes which bind the a1/α2 heterodimer. In this case, however, unlike the Ubx/Exd case, the interaction is mediated by the C-terminal region of the α2 homeodomain which forms an additional α-helix and interacts with the homeodomain of the a1 protein (Andrews and Donoviel, 1995; Li *et al.*, 1995) (Plate 2).

Interestingly, α2 can also interact with the non-homeodomain protein MCM1 to form a heterodimer which has a different DNA binding specificity to that of the a1/α2 heterodimer and which therefore binds to and represses a different set of genes (Fig. 4.13). Hence, α2 is a repressor protein with a weak DNA binding specificity which is guided to different sets of target genes depending on whether it interacts with a1 or MCM1 to form heterodimers with different DNA binding specificities.

Hence, the DNA binding specificity of homeodomain proteins can be altered by interactions with other homeodomain and non-homeodomain-containing proteins with different regions within or adjacent to the homeodomain mediating this interaction in different cases.

Figure 4.13

The yeast α2 repressor protein is a homeodomain protein with weak DNA binding activity. However, it can form DNA binding heterodimers with either the a1 homeodomain protein or the MCM1 protein. As these heterodimers have different DNA binding specificities, they bind different genes which are then repressed by the α2 protein.

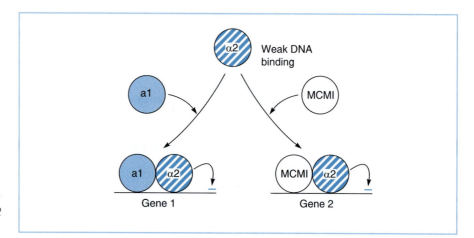

4.2.5 HOMEODOMAIN TRANSCRIPTION FACTORS IN OTHER ORGANISMS

The critical role played by the homeobox genes in the regulation of *Drosophila* development suggests that they may also play a similar role in other organisms. Thus, in the nematode *C. elegans*, homeoboxes have been identified in several genes whose mutation affects development such as the *mec-3* gene which controls the terminal differentiation of specific sensory cells (Way and Chalfie, 1988).

As in *Drosophila*, studies in the nematode have been facilitated by the availability of well characterized mutations affecting development, allowing the corresponding genes to be isolated and the homeobox identified. In higher organisms where such genetic evidence was unavailable, numerous investigators have used Southern blot hybridization with labelled probes derived from *Drosophila* homeoboxes in an attempt to identify homeobox-containing genes in these species. Thus, for example, Holland and Hogan (1986) used a probe from the Antennapedia homeobox to identify homeobox genes in a wide range of species including not only other invertebrates such as the molluscs but also chordates such as the sea urchin and vertebrates including the mouse (Fig. 4.14). Subsequent studies have resulted in the identification of a large number of different homeobox-containing genes from a wide variety of organisms including both mouse and human and many of these genes have been isolated and their DNA sequences obtained (for reviews see Kenyon, 1994; Krumlauf, 1994).

It is clear from these studies that homeobox-containing genes are not confined to invertebrates such as *Drosophila* or yeast but are found also in vertebrates including mammals such as mouse and human. Interestingly, this evolutionary conservation is not confined to the homeobox portion of these

Figure 4.14
Southern blot of mouse DNA hybridized with a probe from the *Drosophila* Antennapedia gene (A), a mouse Antennapedia-like gene (M) and mouse ribosomal DNA (R). Note the presence of DNA fragments which hybridize to both Antennapedia-like DNAs but not to ribosomal DNA and which represent Antennapedia-like sequences in the mouse genome.

genes. Thus homologues of individual homeobox genes of *Drosophila* such as engrailed and deformed have been identified in mouse and human, the fly and mammalian proteins showing extensive sequence homology which extends beyond the homeobox to include other regions of the proteins.

Moreover, the similarity between the *Drosophila* and mammalian systems extends also to the manner in which the homeobox-containing genes are organized in the genome. Thus, in both *Drosophila* and mammals these genes are organized into clusters containing several homeobox-containing genes with homologous genes in the different organisms occupying equivalent positions in the clusters. For example, in a detailed comparison of the genes in the *Drosophila* Bithorax and Antennapedia complexes with those of one mouse homeobox gene complex Hoxb (Hox2), Graham *et al.* (1989) showed that the first gene in the mouse complex, Hoxb-9 (2.5) was most homologous to the first gene in the *Drosophila* Bithorax complex, Abd-B and so on across the complex (Fig. 4.15). Hence both the homeobox genes and their arrangement are highly conserved in evolution, the common ancestor of mammals and insects having presumably possessed a similar cluster of homeobox-containing genes. Interestingly, the DNA sequences and arrangement in the genome of different homeobox genes has been used as a means of determining evolutionary relationships amongst multicellular organisms (for review see Martindale and Kourakis, 1999).

As well as the simple homeobox/homeodomain proteins we have discussed so far, other families of transcription factor exist which contain the homeodomain as part of a larger, more complex, DNA binding structure. Two such families are discussed in the next two sections.

Figure 4.15

Comparison of the Bithorax/Antennapedia complex on *Drosophila* chromosome 3 with the Hoxb complex on mouse chromosome 11. Individual genes are indicated by boxes. Note that each gene in the *Drosophila* complex is most homologous to the equivalent gene in the mouse complex as indicated by the vertical lines. The *Drosophila* Abd-A, Ubx and Antp genes are too closely related each other to be individually related to a particular mouse gene but are most closely related to the Hoxb-6, b-7 and b-8 genes which occupy the equivalent positions in the Hoxb cluster as indicated by the brackets. The two alternative nomenclatures for mouse Hox genes are indicated.

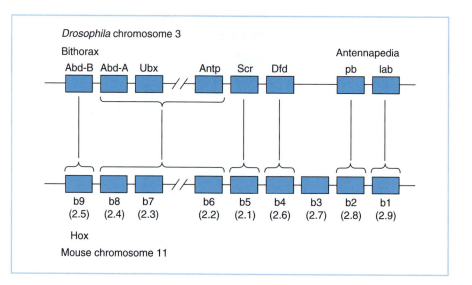

4.2.6 POU PROTEINS

As discussed above, the homeobox-containing genes were first identified in *Drosophila* and only subsequently in other organisms. The reverse is true, however, for another set of transcription factors which possess a homeobox as part of a much larger motif and which were first identified in mammalian cells. Thus, the transcription factors Oct-1 and Oct-2, which bind to the octamer motif ATGCAAAT, play an important role in regulating the expression of specific genes such as those encoding histone H2B, the SnRNA molecules and the immunoglobulins. Similarly, the transcription factor Pit-1, which binds to a sequence two bases different from the octamer sequence, plays a critical role in pituitary-specific gene expression (Chapter 1, section 1.3.3).

When the genes encoding these factors were cloned, they were found to share a 150–160 amino acid sequence which was also found in the protein encoded by the nematode gene *unc*-86 whose mutation affects sensory neuron development. This common POU (Pit-Oct-Unc) domain contains both a homeobox sequence and a second conserved domain, the POU-specific domain (Fig. 4.16, for reviews see Verrijzer and Van der Vliet, 1993; Ryan and Rosenfeld, 1997).

Interestingly, while the homeoboxes of the different POU proteins are closely related to one another (53 out of 60 homeobox residues are the same in Oct-1 and Oct-2 and 34 out of 60 in Oct-1 and Pit-1), they show less similarity to the homeoboxes of other mammalian genes lacking the POU-specific domain, sharing at best only 21 out of 60 homeobox residues. Hence they represent a distinct class of homeobox proteins containing both a POU-specific domain and a diverged homeodomain.

As with the *Drosophila* homeobox proteins, however, the isolated homeodomains of the Pit-1 and Oct-1 proteins are capable of mediating sequence specific DNA binding in the absence of the POU-specific domain. The affinity and specificity of binding by such an isolated homeodomain is much lower, however, than that exhibited by the intact POU domain indicating that the POU-specific domain plays a critical role in producing high affinity binding to specific DNA sequences. Hence the POU homeodomain and the POU-specific domain form two parts of a DNA binding element which are held together by a flexible linker sequence.

The crystal structure of the Oct-1 POU domain bound to DNA (Klemm *et al.*, 1994) has shown that the Oct-1 homeodomain binds in a similar manner to the classical homeobox proteins, with the recognition helix lying in the major groove and the N-terminal arm in the minor groove. Like the homeodomain, the POU-specific domain forms a helix-turn-helix motif, which allows it to bind to the adjacent bases within the DNA to those contacted by the

Figure 4.16

Amino acid sequences of the POU proteins. The homeodomain and the POU-specific domain are indicated. Solid boxes indicate regions of identity between the different POU proteins. The final line shows a consensus sequence obtained from the four proteins. Note the highly conserved sequences near each end of the POU-domain which have been used as a method of isolating novel POU proteins (see Chapter 2, section 2.3.2c and Fig. 2.14).

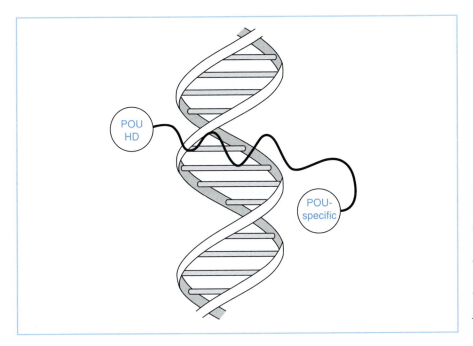

Figure 4.17
Binding of the POU-specific domain and POU-homeodomain to opposite sides of the DNA double helix. Note the flexible linker region joining the two DNA binding motifs.

homeodomain with binding of the two regions occurring on opposite sides of the DNA double helix (Fig. 4.17).

The POU domain appears to allow factors which contain it to bind to highly divergent DNA sequences. Thus, Oct-1 binds to a sequence in the SV40 enhancer which shares less than thirty per cent homology (four out of fourteen bases) or little more than a random match with another Oct-1-binding sequence in the herpes simplex virus (HSV) immediate-early (IE) gene promoters (Fig. 4.18). By analysing a series of other Oct-1 binding elements,

Figure 4.18
Relationship between the various diverse sequences bound by the Oct-1 transcription factor in the simian virus 40 enhancer, the immunoglobulin IgH chain gene enhancer (IgH), the U2 snRNA gene, clone 12 (a mutated version of a site in the SV40 enhancer which binds Oct-1) and the herpes simplex virus immediate-early genes (HSV IE).

however, Baumruker *et al.* (1988) were able to show that the two apparently unrelated Oct-1-binding sites could be linked by a smooth progression via a series of other binding sites which were related to one another (Fig. 4.18). This suggests therefore that Oct-1 can bind to very dissimilar sequences because there are few, if any, obligatory contacts with specific bases in potential binding sites. Rather, specific binding to a particular sequence can occur via many possible independent interactions with DNA, only some of which will occur with any particular binding site. Hence the binding to apparently unrelated sequences does not reflect two distinct binding specificities but indicates that the protein can make many different contacts with DNA, the sequences which can specifically bind the protein being those with which it can make a certain proportion of these possible contacts.

Interestingly, it has been shown that the secondary structure of Oct-1 bound to these sites differs so that its configuration when bound to the HSV IE sequence is different to that observed when it is bound to the other sequences (Walker *et al.*, 1994). Moreover, this configurational change allows the Oct-1 bound to the HSV promoter to be recognized by the HSV VP16 (Vmw 65) protein whereas this does not occur with Oct-1 bound to other sequences. As VP16 is a much stronger transactivator than Oct-1 alone, this therefore results in the strong activation of the HSV IE promoters by the Oct-1/VP16 complex whereas other promoters in which Oct-1 has bound to different sequences are insensitive to such transactivation by VP16. Hence this provides a novel example of gene regulation in which the nature of the sequence bound by a factor controls its recognition by another factor resulting in strong transactivation only from a subset of sequences bound by Oct-1 (Fig. 4.19).

As well as the different configuration Oct-1 adopts when binding to viral sequences, it has been shown that it can also adopt different configurations when binding to different cellular DNA targets and this also has consequences for its effect on gene transcription. Thus, when Oct-1 binds as a dimer to a DNA element known as the PORE sequence, it exposes a region of the POU domain which can recruit a cellular co-activator, OBF-1, resulting in strong activation of transcription. In contrast, when it binds to a distinct DNA sequence, known as the MORE sequence, this region of the POU domain is masked at the interface between the two Oct-1 molecules. Hence, in this case OBF-1 cannot be recruited and only weak transactivation results (Fig. 4.20) (Reményi *et al.*, 2001; Tomilin *et al.*, 2000; for review see Latchman, 2001).

A more extreme example of this effect of DNA binding sequence is seen in the case of the Pit-1 member of the POU family. When Pit-1 binds as a dimer to its binding site in the prolactin promoter, it activates transcription. However, its binding site in the growth hormone promoter contains two

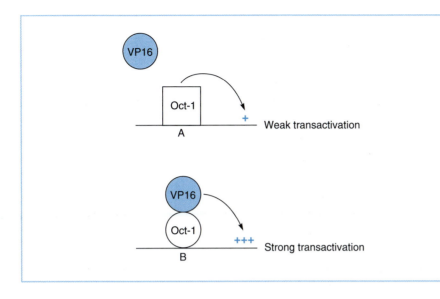

Figure 4.19
The octamer binding protein Oct-1 binds to most binding sites (A) in a configuration which is not recognized by VP16. This results in only the weak transactivation characteristic of Oct-1 alone. In contrast, when it binds to its binding sites in the HSV IE promoters (B), Oct-1 undergoes a conformational change allowing it to be recognized by the strong transactivator VP16 leading to strong transactivation.

extra T bases. This results in a different binding configuration of the Pit-1 dimer which allows it to recruit a co-repressor molecule and thereby inhibit rather than activate the growth hormone gene (Fig. 4.21) (Scully *et al.*, 2000, for review see Marx, 2000; Latchman, 2001).

Hence, the DNA binding sequence that is bound by a particular factor can have profound effects. Indeed, in the case of Pit-1 this is critical to its role in specifying the production of lactotrope cells in the pituitary gland, where expression of prolactin and not of growth hormone must occur. Clearly, in the cases described above, the effect of the binding site on the configuration of the DNA bound POU protein, affects its ability to recruit other molecules

Figure 4.20
Binding of the Oct-1 dimer to the PORE DNA target sequence exposes a region of Oct-1 (heavy line) which can recruit the cellular co-activator OBF-1 resulting in strong activation of transcription. In contrast, binding of the Oct-1 dimer to the MORE DNA sequence produces a configuration in which this region is hidden in the interface between the two Oct-1 molecules. Hence, OBF-1 cannot be recruited and only weak transactivation occurs.

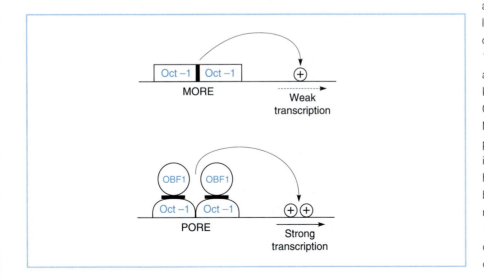

Figure 4.21

Binding of the Pit-1 dimer to its DNA binding site in the prolactin promoter allows it to activate transcription. In contrast, the extra two T bases in the binding site in the growth hormone promoter, result in a different configuration of the Pit-1 dimer, leading to recruitment of the N-CoR co-repressor

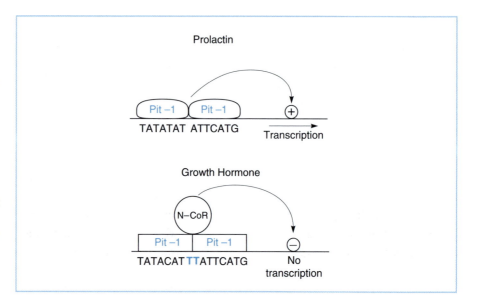

which induce activation (co-activators) or inhibition (co-repressors) (see Chapter 5, section 5.4.3 and Chapter 6, section 6.3.2, for further discussion of co-activators and co-repressors respectively).

As well as control of recruitment of such proteins at the level of a single factor, another level of control can operate by different POU proteins differing in their ability to recruit these factors. Thus, for example, the ability of Oct-1 and not Oct-2 to interact with the herpes simplex virus transactivator protein VP16 is controlled by a single difference in the homeodomain region of the POU domains in the two proteins. Thus the replacement of a single amino acid residue at position 22 in the homeodomain of Oct-2 with the equivalent amino acid of Oct-1 allows Oct-2 to interact with VP16 which is normally a property only of Oct-1 (Lai *et al.*, 1992) (Fig. 4.22).

Figure 4.22

Alteration of an alanine residue (A) in the homeodomain of Oct-2 to the glutamic acid residue (E) found at the equivalent position in the homeodomain of Oct-1 allows Oct-2 to interact with the herpes simplex virus transactivator Vmw65 which is normally a property of Oct-1 only.

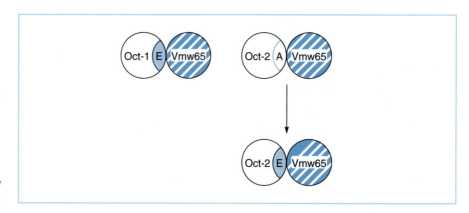

Interestingly, the key role of position 22 in the homeodomain is not confined to the interaction of Oct-1/Oct-2 with VP16. Thus, the closely related mammalian POU factors Brn-3a and Brn-3b differ in that Brn-3a activates the promoter of several genes expressed in neuronal cells whereas Brn-3b represses them. Alteration of the isoleucine residue found at position 22 in Brn-3b to the valine found in Brn-3a converts Brn-3b from a repressor into an activator, whereas the reciprocal mutation in Brn-3a converts it into a repressor (Dawson *et al.*, 1996). This effect suggests that the activating/repressing effects of Brn-3a/Brn-3b are mediated by their binding of cellular co-activator or co-repressor molecules whose binding to Brn-3a/Brn-3b is affected by the nature of the amino acid at position 22. More generally, this finding provides the first example of a single amino acid change which can reverse the functional activity of a transcription factor, from activator to repressor and vice versa.

As in the case of the homeobox-containing proteins, the POU proteins appear to play a critical role in the regulation of developmental gene expression and in the development of specific cell types. Thus the *unc*-86 mutation in the nematode results, for example, in the lack of touch receptor neurons or male-specific cephalic companion neurons indicating that this POU protein is required for the development of these specific neuronal cell types. Similarly, inactivation of the gene encoding Pit-1 leads to a failure of pituitary gland development resulting in dwarfism in both mice and humans (for review see Andersen and Rosenfeld, 1994). Interestingly, however, one type of dwarfism in mice (the Ames dwarf) is produced not by a mutation in Pit-1 but by a mutation in a gene encoding a homeobox-containing factor which was named Prophet of Pit-1 (Sornson *et al.*, 1996). This factor appears to control the activation of the Pit-1 gene in pituitary cells so that Pit-1 is not expressed when this factor is inactivated. This example illustrates how hierarchies of regulatory transcription factors are required in order to control the highly complex process of development.

Following the initial identification of the original four POU factors, a number of other members of this family have been described both in mammals and other organisms such as *Drosophila*, *Xenopus* and zebra fish. Like the original factors, these novel POU proteins also play a critical role in the regulation of developmental gene expression. Thus, for example, the *Drosophila* POU protein drifter (CFla) has been shown to be of vital importance in the development of the nervous system (Anderson *et al.*, 1995), while mutations in the gene encoding the Brn-4 factor appear to be the cause of the most common form of deafness in humans (de Kok *et al.*, 1995). Moreover, all the novel POU domain-containing genes isolated by He *et al.* (1989) from the rat, on the basis of their containing a POU domain (see Chapter 2, section

2.3.2c), are expressed in the embryonic and adult brain suggesting a similar role for these proteins in the regulation of neuronal-specific gene expression. Such a close connection of POU proteins and the central nervous system is also supported by studies using the original POU domain genes which revealed expression in the embryonic brain even in the case of Oct-2 which had previously been thought to be expressed only in B lymphocytes (He *et al.*, 1989).

It is clear therefore that, like the homeobox proteins, POU proteins occur in a wide variety of organisms and play an important role in the regulation of gene expression in development. Moreover, these proteins may be of particular importance in the development of the central nervous system.

4.2.7 PAX PROTEINS

As well as being found as part of the POU domain which gives the POU factors their name, a homeodomain is also found in some members of another family of transcription factors, the Pax factors (for reviews see Mansouri *et al.*, 1996; Chi and Epstein, 2002). These factors are defined on the basis that they contain a common DNA binding domain, known as the paired domain because it was originally identified in the *Drosophila* paired gene. In addition, however, some Pax proteins also contain a full size or truncated homeodomain while some, but not all, members of the family contain an eight amino acid element known as the octapeptide which is of unknown function. All combinations of the paired domain with or without a homeodomain and/or the octapeptide are found in the various mammalian Pax factors (Fig. 4.23).

Obviously in the Pax factors which lack the homeodomain, the paired domain is necessary and sufficient for DNA binding. Hence this case is dis-

Figure 4.23

Structure of the mammalian Pax factors which contain an N-terminal paired domain linked in some cases to an octapeptide (OP) of unknown function and/or a full length or truncated homeodomain.

tinct from that of the POU factors where the POU-specific and POU-homeo-domains are both necessary for high affinity DNA binding. Nonetheless, in factors such as Pax3, which have both a paired domain and a full length homeodomain, both domains participate in DNA binding. This produces very high affinity binding to a DNA binding site which contains the recognition sequence for both the DNA binding domains and the affinity of binding to such sites is greatly reduced when either the paired domain or the homeodomain is deleted. Interestingly, the paired domain itself is distantly related to the homeodomain in terms of its structure and mechanism of DNA binding.

Thus, like the homeodomain, the paired domain also binds to DNA via a helix-turn-helix motif. Structural analysis of this motif, however, reveals that it is more similar to that in the bacteriophage proteins (see section 4.2.3) than that in the eukaryotic homeodomain proteins with the residues at the N-terminus of the recognition helix being critical for DNA binding (Xu *et al.*, 1995). Indeed, one form of Waardenburg syndrome, which results from inactivation of Pax3 (see Chapter 9, section 9.1), is due to mutation in a glycine residue at the N-terminus of the Pax3 recognition helix resulting in a failure of the factor to bind to DNA. Hence the helix-turn-helix motif is a widely used DNA binding domain which exists in at least two different forms that differ in the manner in which the recognition helix contacts the DNA.

As with the POU proteins, Pax factors play a critical role in gene regulation during development particularly in the developing nervous system. Thus, for example, Pax6 has been shown to be of critical importance in specifying which cells will develop into different types of motor neurons during development (Ericson *et al.*, 1997) and also appears to play a critical role in eye development in a wide range of organisms (Gehring and Ikeo, 1999). In agreement with the critical role of these genes in development, knock out mice in which specific Pax genes have been inactivated show defects in the development of the nervous system while the naturally occurring mutant mouse strain splotch which exhibits spina bifida, exencephaly and neural crest and limb muscle defects is due to a mutation in the Pax3 gene. Interestingly, mutations in Pax3 in humans result in Waardenburg syndrome which is characterized by deafness and eye defects while mutations in Pax6 also result in severe eye defects such as aniridia (for review see Latchman, 1996).

Hence the Pax proteins play a particularly critical role in the development of the nervous system. In addition, however, they also play a role in other tissues with mice lacking functional Pax6 showing abnormalities in the development of the pancreas as well as of the nervous system (Sander *et al.*, 1997) while, as discussed in Chapter 7 (section 7.2.1), Pax3 is involved in activating the expression of the muscle determining factor, MyoD.

4.3 THE TWO CYSTEINE TWO HISTIDINE ZINC FINGER

4.3.1 TRANSCRIPTION FACTORS WITH THE TWO CYSTEINE TWO HISTIDINE FINGER

Transcription factor TFIIIA plays a critical role in regulating the transcription of the 5S ribosomal RNA genes by RNA polymerase III (see Chapter 3, section 3.4). When this transcription factor was purified, it was found to have a repeated structure and to be associated with between seven and eleven atoms of zinc per molecule of purified protein (Miller *et al.*, 1985). When the gene encoding TFIIIA was cloned, it was shown that this repeated structure consisted of the unit, Tyr/Phe-X-Cys-X-Cys-$X_{2,4}$-Cys-X_3-Phe-X_5-Leu-X_2-His-$X_{3,4}$-His-X_5 which is repeated nine times within the TFIIIA molecule. This repeated structure therefore contains two invariant cysteine and two invariant histidine residues which were predicted to bind a single zinc atom accounting for the multiple zinc atoms bound by the intact molecule.

This motif is referred to as a zinc finger on the basis of its proposed structure in which a loop of twelve amino acids containing the conserved leucine and phenylalanine residues as well as several basic amino acids projects from the surface of the molecule, being anchored at its base by the cysteine and histidine residues which tetrahedrally coordinate an atom of zinc (Fig. 4.24). The proposed interaction of zinc with the conserved cysteine and histidine residues in this structure was subsequently confirmed by X-ray adsorption spectroscopy of the purified TFIIIA protein.

Following its identification in the RNA polymerase III transcription factor TFIIIA, similar cys_2 his_2-containing zinc finger motifs were identified in a number of RNA polymerase II transcription factors such as Sp1, which contains three contiguous zinc fingers (Kadonaga *et al.*, 1987) and the *Drosophila*

Figure 4.24

Schematic representation of the zinc finger motif. The finger is anchored at its base by the conserved cysteine and histidine residues which tetrahedrally coordinate an atom of zinc.

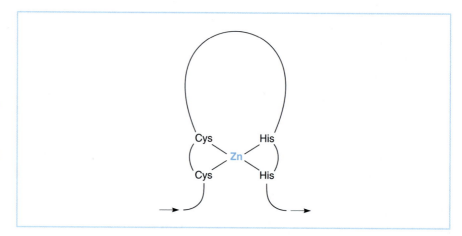

Kruppel protein, which contains four finger motifs (see section 4.2.1). A list of zinc finger-containing transcription factors is given in Table 4.1 (for reviews see Evans and Hollenberg, 1988; Klug and Schwabe, 1995; Turner and Crossley, 1999; Bieker, 2001).

Table 4.1

Transcriptional regulatory proteins containing Cys$_2$-His$_2$ zinc fingers

Organism	Gene	Number of fingers
Drosophila	Kruppel	4
	Hunchback	6
	Snail	4
	Glass	5
Yeast	ADR1	2
	SW15	3
Xenopus	TFIIIA	9
	Xfin	37
Rat	NGF-1A	3
Mouse	MK1	7
	MK2	9
	Egr 1	3
	Evi 1	10
Human	Sp1	3
	TDF	13

In all cases studied the zinc finger motifs have been shown to constitute the DNA binding domain of the protein, with DNA binding being dependent upon their activity. Thus, in the case of TFIIIA, DNA binding is dependent on the presence of zinc, allowing the finger structures to form while progressive deletion of more and more zinc finger repeats in the molecule results in a parallel loss of DNA binding activity. Similarly, in the case of Sp1, DNA binding is dependent on the presence of zinc and, most importantly, the sequence specific binding activity of the intact protein can be reproduced by a protein fragment containing only the zinc finger region (Kadonaga *et al.*, 1987).

A similar dependence of DNA binding on the zinc finger motif is also seen in the *Drosophila* Kruppel protein which is essential for correct thoracic and abdominal development. In this case a single mutation in one of the conserved cysteine residues in the finger, replacing it with a serine which cannot bind zinc, results in the production of a mutant fly indistinguishable from that produced by a complete deletion of the gene (Redemann *et al.*, 1988) indicating the vital importance of the zinc finger (Fig. 4.25).

Figure 4.25

Zinc finger in the *Drosophila* Kruppel protein indicating the cysteine to serine change which abolishes the ability to bind zinc and results in a mutant fly indistinguishable from that obtained when the entire gene is deleted.

As with the helix-turn-helix motif of the homeobox therefore, the zinc finger motif forms the DNA binding element of the transcription factors which contain it. Interestingly, however, a single zinc finger taken from the yeast ADR1 protein is unable to mediate sequence specific DNA binding in isolation, whereas a protein fragment containing both the two fingers present in the intact protein can do so. This suggests therefore that DNA binding by the zinc finger is dependent upon interactions with adjacent fingers and explains why zinc finger-containing transcription factors always contain multiple copies of the zinc finger motif (see Table 4.1).

4.3.2 DNA BINDING BY THE TWO CYSTEINE TWO HISTIDINE FINGER

In the zinc finger structure the zinc coordination via cysteine and histidine serves as a scaffold for the intervening region which makes direct contact with the DNA. Detailed structural analysis has shown that these intervening amino acids do not form a simple loop structure as proposed in the original model (for review see Rhodes and Klug, 1993; Klug and Schwabe, 1995). Rather, the finger region forms a motif consisting of two anti-parallel beta-sheets with an adjacent alpha-helix packed against one face of the beta-sheet (Fig. 4.26; see Plate 3; Lee *et al.*, 1989). Upon contact with DNA, the alpha-helix lies in the major groove of the DNA and makes sequence specific contacts with the bases of DNA while the beta-sheets lie further away from the helical axis of the DNA and contact the DNA backbone.

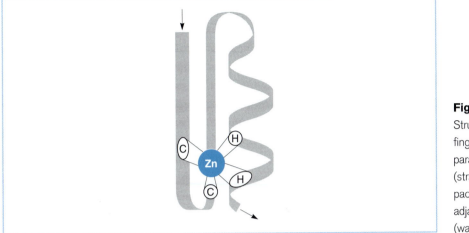

Figure 4.26
Structure of the zinc finger in which two anti-parallel beta sheets (straight lines) are packed against an adjacent alpha-helix (wavy line).

Most interestingly, this structure indicates that a critical role in sequence specific DNA binding will be played by amino acids at the amino terminus of the alpha-helix, most notably the amino acids immediately preceding the first histidine residue. In agreement with this idea, two amino acids in this region play a critical role in determining the DNA binding specificity of the *Drosophila* Krox-20 transcription factor (Nardelli *et al.*, 1991). Thus this factor contains three zinc fingers and interacts with the DNA sequence 5' GCGGGGGCG 3'. If each finger contacts three bases within this sequence, then the central finger must recognize the sequence GGG whereas the two outer fingers will each recognize the sequence GCG (Fig. 4.27).

When the amino acid sequence of each of the Krox-20 fingers was compared, it was found that the two outer fingers contain a glutamine residue at position 18 of the finger and an arginine at position 21, whereas the central finger differs in that it has histidine and threonine residues at these positions.

Figure 4.27
DNA binding specificity and amino acid sequence of the three cysteine-histidine zinc fingers in the *Drosophila* Krox 20 protein. Note that each finger binds to three specific bases in the recognition sequence and that finger 2, which differs from fingers 1 and 3 in the DNA sequence it recognizes, also differs in the amino acids at positions 18 and 21 in the finger (bold letters). Mutating these amino acids to their equivalents in fingers 1 and 3 changes the DNA binding specificity of finger 2 to that of fingers 1 and 3, indicating that these amino acids play a critical role in determining the DNA sequence that is recognized.

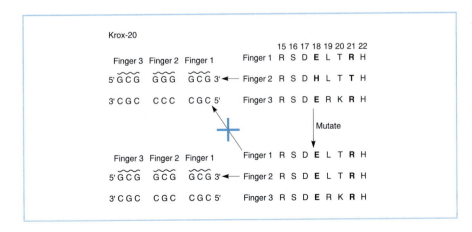

As expected, if these two amino acid differences are critical in determining the DNA sequence that is recognized, altering these two residues in the central finger to their equivalents in the outer two fingers resulted in a factor which failed to bind to the normal Krox-20 binding site but instead bound to the sequence 5' GCGGCGGCG 3 in which each finger binds the sequence GCG. This experiment therefore indicates the critical role of two amino acids at the amino terminus of the alpha helix in producing the DNA binding specificity of zinc fingers of this type and also shows that, at least in the case of Krox-20, each successive finger interacts with three bases of DNA within the recognition sequence.

The importance of these amino acids has also been confirmed in experiments in which the amino acids at different positions in the zinc finger were randomly altered and their interaction with a wide range of DNA sequences assessed (Choo and Klug, 1994; Rebar and Pabo, 1994). Clearly, such an important role for the amino acids at the amino terminus of an alpha helix, parallels the similar critical role for the equivalent amino acids in the recognition helix of the bacteriophage DNA recognition proteins and in the paired domain (see section 4.2).

Interestingly, using this type of information on the DNA binding properties of individual fingers, it has recently proved possible to create novel zinc finger transcription factors with a defined DNA binding specificity. In this way novel factors were created which could bind to and switch on the endogenous VEGF gene *in vivo*. As the VEGF protein is a growth factor able to induce enhanced blood vessel growth, this in turn resulted in the induction of such blood vessel growth due to the elevated level of VEGF (Fig. 4.28) (for review see Pasqualini *et al.*, 2002).

Clearly, as well as their implications for DNA binding studies, these findings have important potential therapeutic implications since they could allow specific genes to be switched on in human patients by delivery of a transcription factor with defined DNA binding specificity, inducing, for example, the growth of new blood vessels in patients suffering from a poor blood supply to specific regions. Interestingly, designer zinc fingers have also been produced and linked to an inhibitory domain (see Chapter 6, section 6.3.2) allowing them to repress transcription of the genes to which they bind. These have recently been used to block infection of cultured cells with specific human viruses, further reinforcing the therapeutic potential of this approach (Papworth *et al.*, 2003; Reynolds *et al.*, 2003).

Hence, like the helix-turn-helix motif, the cysteine-histidine zinc finger plays a critical role in mediating the DNA binding abilities of transcription factors which contain it, with sequence specific recognition of DNA being determined in both cases by amino acids within an alpha helix.

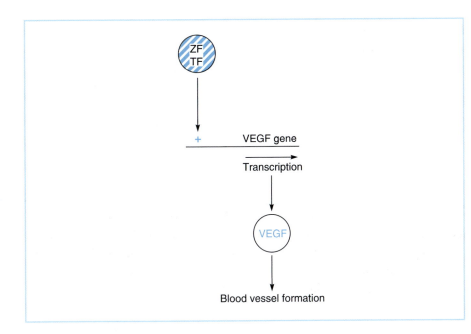

Figure 4.28

The synthesis of a zinc finger transcription factor (ZFTF) with a novel DNA binding specificity that allows it to bind to the VEGF gene results in VEGF gene transcription. The resulting VEGF protein then induces blood vessel formation.

4.4 THE MULTI-CYSTEINE ZINC FINGER

4.4.1 STEROID RECEPTORS

The steroid hormones are a group of substances derived from cholesterol which exert a very wide range of effects on biological processes such as growth, metabolism and sexual differentiation (for review see King and Mainwaring, 1974). Early studies using radioactively-labelled hormones showed that they act by interacting with specific receptor proteins. This binding of hormone to its receptor activates the receptor and allows it to bind to a limited number of specific sites in chromatin. In turn this DNA binding activates transcription of genes carrying the receptor binding site. Hence, these receptor proteins are transcription factors becoming activated in response to a specific signal and in turn activating specific genes (for reviews see Weatherman *et al.*, 1999; Khorasanizadeh and Rastinejad, 2001; Olefsky, 2001; McKenna and O'Malley, 2002). These receptor proteins were therefore among the earliest transcription factors to be identified, well before the techniques described in Chapter 2 were in routine use, simply on the basis of their ability to bind radioactively-labelled steroid ligand.

Genes that are induced by a particular steroid hormone contain a specific binding site for the receptor–hormone complex. The responses to different steroid hormones, such as glucocorticoids and oestrogen, are mediated by distinct palindromic sequences which are related to one another. In turn, such sequences are related to one of the sequences which mediates induction by other substances which are related to steroids such as thyroid hormone and retinoic acid. Similarly, repeated elements with different spacings between the repeats also mediate responses to these different substances (Table 4.2; see Gronemeyer and Moras, 1995, for review).

Table 4.2

Relationship of various hormone response elements

(a) Palindromic repeats	
Glucocorticoid	RGRACANNNTGTYCY
Oestrogen	RGGTCANNNTGACCY
Thyroid	RGGTCA - - - TGACCY
(b) Direct repeats	
9-cis retinoic acid	$AGGTCAN_1AGGTCA$
All -transretinoic acid	$AGGTCAN_2AGGTCA$
	$AGGTCAN_5AGGTCA$
Vitamin D_3	$AGGTCAN_3AGGTCA$
Thyroid hormone	$AGGTCAN_4AGGTCA$

N indicates that any base can be present at that position, R indicates a purine, i.e. A or G, Y indicates a pyrimidine, i.e. C or T, W indicates A or T. A dash indicates that no base is present, the gap having been introduced to align the sequence with the other sequences.

The basis of this binding site relationship was revealed when the genes encoding the receptor proteins were cloned. Thus, they were found to constitute a family of genes encoding closely related proteins of similar structure with particular regions being involved in DNA binding, hormone binding and transcriptional activation (Fig. 4.29). This has led to the idea that these receptors are encoded by an evolutionarily-related gene family which is known as the steroid-thyroid hormone receptor or nuclear receptor gene super family (for reviews see Weatherman et al., 1999; Khorasanizadeh and Rastinejad, 2001; Olefsky, 2001; McKenna and O'Malley, 2002). The structure of the thyroid hormone receptor bound to its ligand, thyroid hormone, is illustrated in Plate 4 (Wagner et al., 1995).

As shown in Figure 4.29, the most conserved region between the different receptors is the DNA binding domain explaining the ability of the receptors to bind to similar DNA sequences. Interestingly, both DNAseI protection and

Figure 4.29
Domain structure of individual members of the steroid-thyroid hormone receptor super family. The proteins are aligned on the DNA binding domain, which shows the most conservation between different receptors. The percentage homologies in each domain of the receptors to that of the glucocorticoid receptor are indicated.

methylation studies support the idea that the receptor binds to DNA as a dimer, each receptor molecule binding to one half of the recognition sequence.

4.4.2 DNA BINDING BY THE MULTI-CYSTEINE ZINC FINGER

Analysis of the nuclear receptor DNA binding domains identified a similar zinc binding motif to that discussed in section 4.3. As with the cysteine-histidine fingers, this motif has been shown by X-ray adsorption spectroscopy to bind zinc in a tetrahedral configuration. However, in this case, coordination is achieved by four cysteine residues rather than the two cysteine two histidine structure discussed above. Similar multi-cysteine motifs have also been identified in several other DNA binding transcription factors such as the yeast proteins GAL4, PPRI and LAC9 as well as in the adenovirus transcription factor E1A (Table 4.3; for review see Evans and Hollenberg, 1988; Klug and Schwabe, 1995) indicating that this type of motif is not confined to the nuclear receptors.

In the case of the nuclear receptors, the DNA binding domain has the consensus sequence $Cys-X_2-Cys-X_{13}-Cys-X_2-Cys-X_{15,17}-Cys-X_5-Cys-X_9-Cys-X_2-Cys-X_4-Cys$. This motif is therefore capable of forming a pair of fingers each with four cysteines coordinating a single zinc atom (Fig. 4.30) and, as with the cysteine-histidine finger proteins, DNA binding of the receptors is dependent on the presence of zinc.

Table 4.3

Transcriptional regulatory proteins with multiple cysteine fingers

Finger type	Factor	Species
Cys_4–Cys_5	Steroid, thyroid receptors	Mammals
Cys_4	E1A	Adenovirus
Cys_6	Gal4, PPRI, LAC9	Yeast

However, the multi-cysteine finger cannot be converted into a functional cysteine-histidine finger by substituting two of its cysteine residues with histidines indicating that the two types of finger are functionally distinct (Green and Chambon, 1987). Moreover, unlike the cysteine-histidine zinc finger which is present in multiple copies within the proteins which contain it, the unit of two multi-cysteine fingers present in the steroid receptors is found only once in each receptor. Interestingly, structural studies of the two multi-cysteine fingers in the glucocorticoid and oestrogen receptors (for review see Schwabe and Rhodes, 1991; Klug and Schwabe, 1995) have indicated that the two fingers form one single structural motif consisting of two alpha helices perpendicular to one another with the cysteine-zinc linkage holding the base of a loop at the N terminus of each helix (Fig. 4.31; see Plate 5; Hard *et al.*, 1990). This is quite distinct from the modular structure of the two cysteine two histidine finger where each finger constitutes an independent structural element whose configuration is unaffected by the presence or absence of adjacent fingers.

Figure 4.30

Schematic representation of the four cysteine zinc finger. Regions labelled A and B are of critical importance in determining respectively the DNA sequence which is bound by the finger and the optimal spacing between the two halves of the palindromic sequence which is recognized.

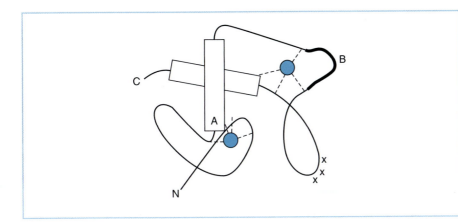

Figure 4.31
Schematic model of a pair of zinc fingers in a single molecule of the oestrogen receptor. Note the helical regions (indicated as cylinders) with the critical residues for determining the DNA sequence which is bound located at the terminus of the recognition helix (indicated as A), the zinc atoms (blue), conserved basic residues (+++) and the region that interacts with another receptor molecule and determines the optimal spacing between the two halves of the palindromic sequence that is recognized (indicated as B). Note that A and B indicate the same regions as in Figure 4.30.

Thus, although these two DNA binding motifs are similar in their coordination of zinc, they differ in the lack of histidines and of the conserved phenylalanine and leucine residues in the multi-cysteine finger, as well as structurally. It is clear therefore that they represent distinct functional elements and are unlikely to be evolutionarily related (for review see Schwabe and Rhodes, 1991; Rhodes and Klug, 1993; Klug and Schwabe, 1995).

Whatever the precise relationship between these motifs, it is clear that the multi-cysteine finger mediates the DNA binding of the nuclear receptors. Thus mutations which eliminate or alter critical amino acids in this motif interfere with DNA binding by the receptor (Fig. 4.32).

The role of the cysteine fingers in mediating DNA binding by the nuclear receptors can also be demonstrated by taking advantage of the observation that the different steroid receptors bind to distinct but related palindromic sequences in the DNA of hormone responsive genes (see Khorasanizadeh and Rastinejad, 2001 for review and Table 4.2 for a comparison of these binding sites). Thus, if the cysteine-rich region of the oestrogen receptor is replaced by that of the glucocorticoid receptor, the resulting chimaeric receptor has the DNA binding specificity of the glucocorticoid receptor but continues to bind oestrogen since all the other regions of the molecule are derived from the oestrogen receptor (Green and Chambon, 1987; Fig. 4.33). Hence the DNA binding specificity of the hybrid receptor is determined by its cysteine-rich region, resulting in the hybrid receptor inducing the expression of glucocorticoid responsive genes (which carry its DNA binding site) in response to oestrogen (to which it binds).

These so-called 'finger swop' experiments therefore provide further evidence in favour of the critical role for the multi-cysteine fingers in DNA binding, exchanging the fingers of two receptors exchanging the DNA bind-

Figure 4.32

Effect of various deletions or mutations on the DNA binding of the glucocorticoid receptor. Note that DNA binding is only prevented by deletions that include part of the DNA binding domain (shaded) or by mutations within it (arrows), but not by deletions in other regions such as the steroid-binding domain. Numbers indicate amino acid residues.

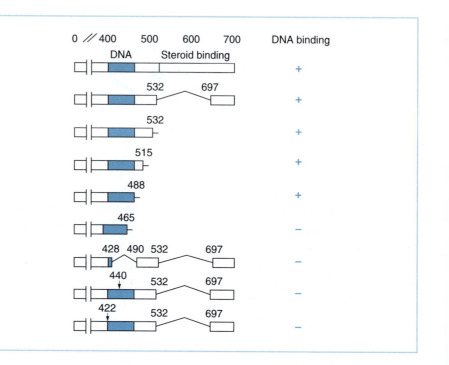

Figure 4.33

Effect of exchanging the DNA binding domain (shaded) of the oestrogen receptor with that of the glucocorticoid receptor on the binding of hormone and gene induction by the hybrid receptor.

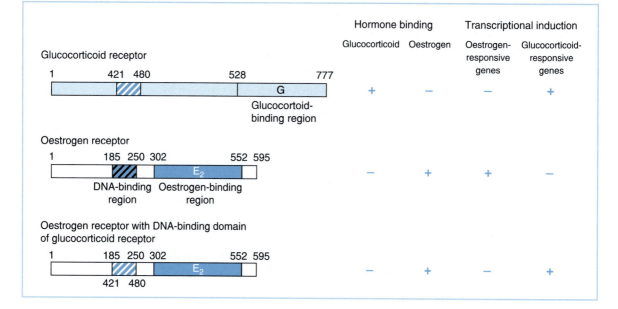

ing specificity. In addition, however, because of the existence of short distinct DNA binding regions of this type in receptors which bind to distinct but related DNA sequences, they provide a unique opportunity to dissect the elements in a DNA binding structure which mediate binding to specific sequences.

Thus by exchanging one or more amino acids between two different receptors it is possible to investigate the effects of these changes on DNA binding specificity and hence elucidate the role of individual amino acid differences in producing the different patterns of sequence specific binding. For example, the alteration of the two amino acids between the third and fourth cysteines of the N terminal finger in the glucocorticoid receptor for their equivalents in the oestrogen receptor changes the DNA binding specificity of the chimaeric receptor to that of the oestrogen receptor (Umesono and Evans, 1989; Fig. 4.34). Hence the exchange of two amino acids in a critical region of a protein of 777 amino acids (indicated as A in Fig. 4.30) can completely change the DNA binding specificity of the glucocorticoid receptor resulting in it binding to and activating genes that are normally oestrogen responsive. The specificity of this hybrid receptor for such oestrogen responsive genes can be further enhanced by exchanging another amino acid located between the two fingers (Fig. 4.34) indicating that this region also plays a role in controlling the specificity of DNA binding.

As noted above (section 4.4.1), the steroid receptors bind to palindromic recognition sequences within DNA, with the receptor binding to DNA as a homodimer in which each receptor molecule interacts with one half of the palindrome. In addition to differences in the actual sequence recognized, steroid/thyroid hormone receptors can also differ in the optimal spacing between the two separate halves of the palindromic DNA sequence that is recognized (see Table 4.2a). Thus the oestrogen receptor and the thyroid

Figure 4.34

Effect of amino acid substitutions in the zinc finger region of the glucocorticoid receptor on the ability to bind to and activate genes that are normally responsive to different steroid hormones.

| | Gene activation | | |
	Glucocorticoid-responsive genes	Oestrogen-responsive genes	Thyroid hormone-responsive genes
GR VLTCGSC KVFF // HNYL CAGRNDCIID	++++	–	–
EG	+	++++	–
EG A	–	++++	–
EG A KYEGK	–	++++	++++

hormone receptor both recognize the identical palindromic sequence in the DNA but differ in that in the thyroid receptor binding sites the two halves of the palindrome are adjacent whereas in the oestrogen receptor binding sites they are separated by three extra bases. The further alteration of the chimaeric receptor illustrated in Figure 4.34 by changing five amino acids in the second finger to their thyroid hormone receptor equivalents is sufficient to allow the receptor to recognize thyroid hormone receptor binding sites (Umesono and Evans, 1989; Fig. 4.34). These amino acids in the second finger (indicated as B in Fig. 4.30) appear to play a critical role therefore in determining the optimal spacing of the palindromic sequence that is recognized.

As discussed above, structural studies of the two zinc fingers in the oestrogen and glucocorticoid receptors suggest that they form a single structural motif with two perpendicular alpha helices (see Fig. 4.31). In this structure, the critical amino acids for determining the spacing in the palindromic sequence recognized are located on the surface of the molecule allowing them to interact with equivalent residues on another receptor monomer during dimerization (indicated as B in Fig. 4.35; see Plate 6; Schwabe *et al.*, 1993). Hence differences in the interaction of these regions in the different receptors determine the spacing of the two monomers within the receptor dimer and thus the optimal spacing in the palindromic DNA sequence that is recognized.

Interestingly, within this structure, the critical residues for determining the precise DNA sequence that is recognized are located at the N terminus of the first alpha helix (indicated as A in Fig. 4.31 and Fig. 4.35), further supporting the critical role of such helices in DNA binding. Moreover, in the proposed

Figure 4.35

Interaction of two oestrogen receptor molecules to form a DNA binding dimer. Compare with Figure 4.31 and note the interaction of the B regions on each molecule. The resulting dimer has a spacing of 34 Angstroms between the two DNA-binding regions allowing binding in successive major grooves of the DNA molecule.

structure of the oestrogen receptor dimer, the DNA binding helices in each monomer will be separated by 34 Angstroms allowing each of these recognition helices to make sequence specific contacts in adjacent major grooves of the DNA molecule.

Differences in the DNA binding domain also regulate the binding of members of the nuclear receptor family to directly repeated sequences with different spacings between the two halves of the repeat (see Table 4.2b). Thus, when the direct repeats are separated by only one base, they can bind a homodimer of the retinoid X-receptor (RXR) and hence confer a response to 9-cis retinoic acid which binds to this receptor (Fig. 4.36). In contrast the RXR homodimer cannot bind to the direct repeats when they are separated by between two and five base pairs. Rather, on these elements RXR forms a heterodimer with other members of the nuclear receptor family (Fig. 4.36).

Figure 4.36
Binding of different nuclear receptor heterodimers to directly repeated elements with different spacings between the repeats determines the response mediated by each element.

Moreover, the nature of the heterodimers that form on a particular response element controls the response it mediates with the nature of the non-RXR component determining the response. Thus a spacing of two or five base pairs binds a heterodimer of RXR and the retinoic acid receptor (RAR) and therefore mediates responses to all transretinoic acid that binds to RAR. In contrast, a spacing of four base pairs binds a heterodimer of RXR and the thyroid hormone receptor (TR) and therefore can mediate responses to thyroid hormone.

As on the palindromic repeats, it is the DNA binding domain of the receptors that controls which heterodimers can form on particular spacings of the direct repeat. Interestingly, the crystal structure of the RXR-TR heterodimer bound to a direct repeat with a four base spacing indicates that the dimerization interface involves amino acids in the first finger of the thyroid hormone receptor and the second finger of RXR rather than only residues in the second finger as occurs for homodimerization of receptors on palindromic repeats (Rastinejad *et al.*, 1995) (Fig. 4.37).

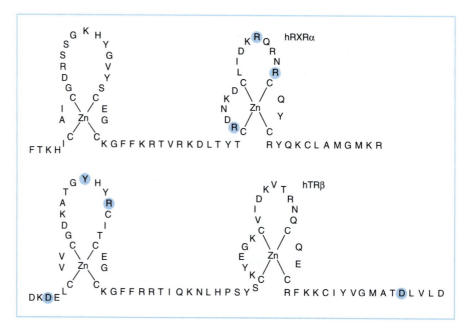

Figure 4.37

Zinc fingers in the retinoid X-receptor α and the thyroid hormone receptor β. The residues in each receptor that are involved in heterodimer formation with the other receptor are indicated.

The definition of the DNA binding domain of the nuclear receptors as a short sequence containing two multi-cysteine fingers has therefore allowed the elucidation of the features in this motif which mediate the different sequence specificities of the different receptors and their relationship to the structure of the motif. In particular, a helical region of the first finger plays a critical role in determining the precise DNA sequence that is recognized by binding in the major groove of the DNA. Similarly, other regions in either the first or second fingers control the spacing of adjacent palindromic or directly repeated sequences which is optimal for the binding of receptor homo- or heterodimers by interacting with another receptor monomer and hence affecting the structure of the receptor dimer that forms.

4.5 THE BASIC DNA BINDING DOMAIN

4.5.1 THE LEUCINE ZIPPER AND THE BASIC DNA BINDING DOMAIN

As discussed in the preceding sections of this chapter, the study of motifs common to several different transcription factors has led to the identification of the role of these motifs in DNA binding. A similar approach led to the identification of the leucine zipper motif (for reviews see Lamb and McKnight, 1991; Hurst, 1996; Kerppola and Curran, 1995). Thus this structure has been detected in several different transcription factors such as the CAAT box binding protein C/EBP, the yeast factor GCN4 and the oncogene products Myc, Fos and Jun (see Chapter 9, sections 9.3.1 and 9.3.3). It consists of a leucine-rich region in which successive leucine residues occur every seventh amino acid (Fig. 4.38).

```
C/EBP    L T S D N D R L R K R V E Q L S R E L D T L R G I F R Q L
Jun B    L E D K V K T L K A E N A G L S S A A G L L R E Q V A Q L
Jun      L E E K V K T L K A Q N S E L A S T A N M L R E Q V A Q L
GCN 4    L E D K V E E L L S K N Y H L E H E V A R L K K L V G E R
Fos      L Q A E T D Q L E D E K S A L Q T E I A N L L K E K E K L
Fra 1    L Q A E T D K L E D E K S G L Q R E I I E L Q K Q K E R L
c-Myc    V Q A E E Q K L I S E E D L L R K R R E Q L K H K L E Q L
n-Myc    L Q A E E H Q L L L L E K E K L Q A R Q Q Q L L K K I E H A
l-Myc    L V G A E K K M A T E K R Q L R C R Q Q Q L Q K R I A Y L
```

Figure 4.38

Alignment of the leucine-rich region in several cellular transcription factors. Note the conserved leucine residues (L) which occur every seven amino acids.

In all these cases, the leucine-rich region can be drawn as an alpha-helical structure in which adjacent leucine residues occur every two turns on the same side of the helix. Moreover, these leucine residues appear to play a critical role in the functioning of the protein. Thus, with one exception (a single methionine in the Myc protein), the central leucine residues of the motif are conserved in all the factors that contain it (Fig. 4.38). It was therefore proposed (Landshultz *et al.*, 1988) that the long side chains of the leucine residues extending from one polypeptide would interdigitate with those of the analogous helix of a second polypeptide, forming a motif known as the leucine zipper which would result in the dimerization of the factor (Fig. 4.39). This effect could also be achieved by a methionine residue which, like leucine, has a long side chain with no lateral methyl groups but not by other hydrophobic amino acids such as valine or isoleucine which have methyl groups extending laterally from the beta carbon atom.

In agreement with this idea, substitutions of individual leucine residues in C/EBP or other leucine zipper-containing proteins such as Myc, Fos and Jun

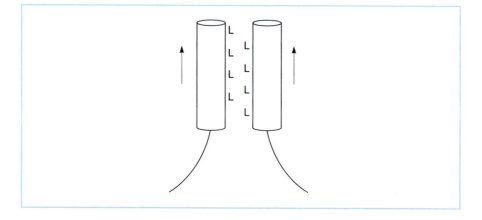

Figure 4.39

Model of the leucine zipper and its role in the dimerization of two molecules of a transcription factor.

with isoleucine or valine, abolish the ability of the intact protein to form a dimer, indicating the critical role of this region in dimerization. A comparison of the effects of various mutations of this type on the ability of the mutant protein to dimerize, suggested that the two leucine-rich regions associate in a parallel manner with both helices oriented in the same direction (as illustrated in Fig. 4.39) rather than in an anti-parallel configuration as originally suggested (Landshultz *et al.*, 1989). This idea was confirmed by structural studies of the leucine zipper regions in GCN4 and in the Fos/Jun dimer bound to DNA (Glover and Harrison, 1995). These studies indicated that each zipper motif forms a right-handed alpha-helix with dimerization occurring via the association of two parallel helices that coil around each other to form a coiled coil motif similar to that found in fibrous proteins such as the keratins and myosins (Fig. 4.40).

In addition to its role in dimerization, the leucine zipper is also essential for DNA binding by the intact molecule. Thus mutations in the zipper which

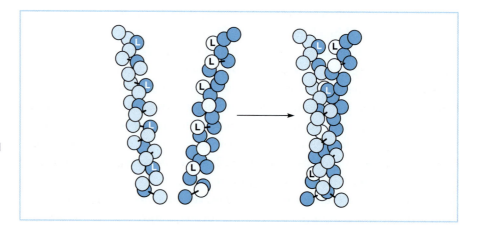

Figure 4.40

Coiled coil structure of the leucine zipper formed by two helical coils wrapping around each other. L indicates a leucine residue.

prevent dimerization also prevent DNA binding from occurring (Landshultz *et al.*, 1989). Unlike the zinc finger or helix-turn-helix motifs, however, the zipper is not itself the DNA binding domain of the molecule and does not directly contact the DNA. Rather it facilitates DNA binding by an adjacent region of the molecule which in C/EBP, Fos and Jun is rich in basic amino acids and can therefore interact directly with the acidic DNA. The leucine zipper is believed therefore to serve an indirect structural role in DNA binding, facilitating dimerization which in turn results in the correct positioning of the two basic DNA binding domains in the dimeric molecule for DNA binding to occur (Fig. 4.41).

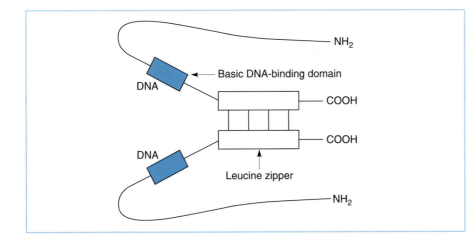

Figure 4.41
Model for the structure of the leucine zipper and the adjacent DNA binding domain following dimerization of the transcription factor C/EBP.

In agreement with this idea mutations in the basic domain abolish the ability to bind to DNA without affecting the ability of the protein to dimerize as expected for mutations that directly affect the DNA binding domain (Landshultz *et al.*, 1989). Similarly, exchange of the basic region of GCN4 for that of C/EBP results in a hybrid protein with the DNA binding specificity of C/EBP while exchange of the leucine zipper region has no effect on the DNA binding specificity of the hybrid molecule (Fig. 4.42).

Hence the DNA binding specificity of leucine zipper-containing transcription factors is determined by the sequence of their basic domain with the leucine zipper allowing dimerization to occur and hence facilitating DNA binding by the basic domain. As expected from this idea, the basic DNA binding domain can interact with DNA in a sequence specific manner in the absence of the leucine zipper if it is first dimerized via an intermolecular disulphide bond (Fig. 4.43). Interestingly, the basic DNA binding domain can bind to DNA as a monomer in the case of the Skn-1 factor which lacks a leucine zipper (Blackwell *et al.*, 1994). In this factor, however, the basic

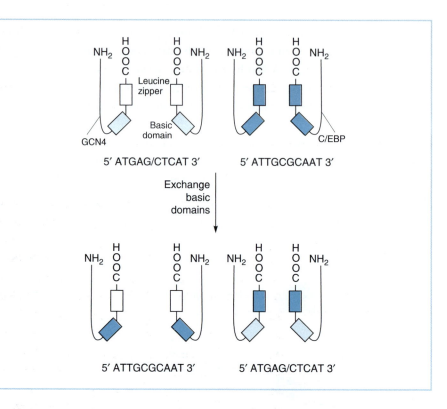

Figure 4.42

Effect of exchanging the basic domains of GCN4 and C/EBP on the DNA binding specificity. Note that the DNA binding specificity is determined by the origin of the basic domain and not that of the leucine zipper.

domain is part of a composite DNA binding domain which also contains a region homologous to the N-terminal arm of the homeobox (see section 4.2).

In factors having a simple basic DNA binding domain, following dimerization via the leucine zipper, the intact transcription factor will form a rotationally symmetric dimer that contacts the DNA via the bifurcating basic regions (see Fig. 4.41) which form alpha-helical structures. These two helices then track along the DNA in opposite directions corresponding to the dyad symmetric structure of the DNA recognition site and form a clamp or scissors grip around the DNA, similar to the grip of a wrestler on his opponent, resulting in

Figure 4.43

DNA binding of molecules containing basic DNA binding domains can occur following dimerization mediated by leucine zippers or by a disulphide bridge (S-S) but cannot be achieved by unlinked monomeric molecules.

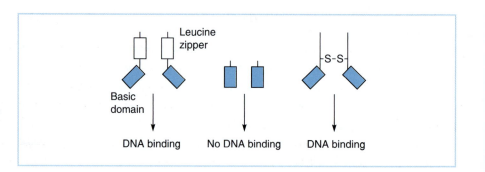

very tight binding of the protein to DNA (Glover and Harrison, 1995). Most interestingly, structural studies have suggested that the basic region does not assume a fully alpha-helical structure until it contacts the DNA when it undergoes a configurational change to a fully alpha-helical form. Hence the association of the transcription factor with the appropriate DNA sequence results in a conformational change in the factor leading to a tight association with that sequence (for discussion see Sauer, 1990).

4.5.2 THE HELIX-LOOP-HELIX MOTIF AND THE BASIC DNA BINDING DOMAIN

Although originally identified in the leucine zipper-containing proteins the basic DNA binding domain has also been identified by a homology comparison in a number of other transcription factors which do not contain a leucine zipper (Prendergast and Ziff, 1989). These factors include the MyoD transcription factor which plays a key role in activating specific genes in skeletal muscle (see Chapter 7, section 7.2.1) and the E12 and E47 factors which play a key role in the development of immunoglobulin-producing B lymphocytes.

In these cases, the basic DNA binding domain is juxtaposed to a region which can form a helix-loop-helix motif (for review see Littlewood and Evan, 1995). This helix-loop-helix motif is distinct from the helix-turn-helix motif in the homeobox (section 4.2) in that it can form two amphipathic helices, containing all the charged amino acids on one side of the helix, which are separated by a non-helical loop (Murre *et al.*, 1989a). This helix-loop-helix motif plays a similar role to the leucine zipper, allowing dimerization of the transcription factor molecule and thereby facilitating DNA binding by the basic motif (Murre *et al.*, 1989b; for discussion see Jones, 1990).

In agreement with this, deletion or mutations in the basic domain of the MyoD protein do not abolish dimerization but do prevent DNA binding, paralleling the effect of similar mutations in C/EBP (Fig. 4.44). Similarly, mutations or deletions in the helix-loop-helix region abolish both dimerization and DNA binding paralleling the effects of similar mutations in leucine zipper-containing proteins. Moreover, the DNA binding ability of MyoD from which the basic DNA binding domain has been deleted can be restored by substituting the basic domain of the E12 protein (Davis *et al.*, 1990). However, such substitution does not allow the hybrid protein to activate muscle-specific gene expression suggesting that, in addition to mediating DNA binding, the basic region of MyoD also contains elements involved in the activation of muscle-specific genes (Davis *et al.*, 1990, Fig. 4.44).

Interestingly, it has been shown that the conversion of three amino acids within the E12 basic region to their MyoD equivalents allows the E12 basic

Figure 4.44

Effect of deleting the basic domain or the adjacent helix-loop-helix motif on dimerization, DNA binding and activation of muscle-specific gene expression by the MyoD transcription factor. Note that deletion of any part of the helix-loop-helix motif abolishes dimerization and consequent DNA binding and gene activation, while deletion of the basic domain directly abolishes DNA binding and consequent gene activation. Substitution of the basic domain of the constitutive factor E12 for that of MyoD restores DNA binding but not the ability to activate muscle-specific gene expression.

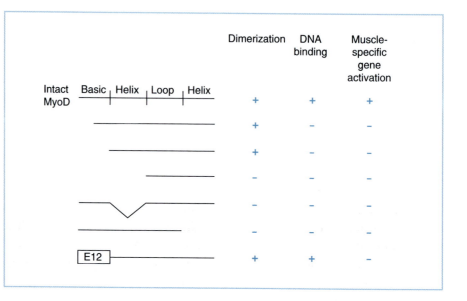

region to activate muscle specific gene expression following DNA binding (Fig. 4.45) (Davis and Weintraub, 1992). The crystal structure of MyoD bound to DNA (Ma *et al.*, 1994) suggests that these amino acids may play a critical role in allowing the MyoD basic region to assume a particular structural configuration in which it can interact with other activating transcription factors. In agreement with this idea, the substitution of these same three amino acids in E12 for their MyoD equivalents allows the mutant E12 protein to bind to another muscle-specific transcription factor MEF2A which is normally a property of MyoD alone (Fig. 4.45; Kanshal *et al.*, 1994). Hence, like the POU domain (see section 4.2.6), the basic domain appears to function both as a DNA binding domain and as a site for protein–protein interactions critical for transcriptional activation.

4.5.3 DIMERIZATION OF BASIC DNA BINDING DOMAIN-CONTAINING FACTORS

Both the leucine zipper and the helix-loop-helix motif therefore act by causing dimerization, allowing DNA binding by the adjacent basic motif. Interestingly, the Myc oncoproteins contain both a helix-loop-helix motif and a leucine zipper region adjacent to the basic DNA binding region (Landshultz *et al.*, 1988; Murre *et al.*, 1989a). Moreover, the leucine zipper can also be found as a dimerization motif in proteins which use DNA binding motifs other than the basic region. For example, in the *Arabidopsis* Athb-1 and 2 proteins, the leucine zipper facilitates dimerization with DNA binding being produced by the

				DNA binding	Muscle-specific gene activation	Interaction with MEF2A
		114 115	124			
MyoD	———	A T ———	K ———	+	+	+
E12	------	N N -------	D ------	+	−	−
E12(M)	------	A T -------	K ------	+	+	+

Figure 4.45
Alterations of three amino acids (positions 114, 115 and 124) in the E12 basic domain to their MyoD equivalents confers on the resulting protein (E12(M)) the ability to interact with the muscle specific transcription factor MEF2A and activate muscle-specific genes following DNA binding which are normally properties of MyoD alone.

adjacent homeobox (Sessa *et al.*, 1993). Thus individual DNA binding and dimerization motifs can be combined in different combinations to produce molecules capable of dimerizing and binding to DNA.

The essential role of dimerization (mediated by the leucine zipper or the helix-loop-helix motifs) in allowing DNA binding by basic DNA binding domain proteins provides an additional aspect to the regulation of these factors (for discussion see Jones, 1990; Lamb and McKnight 1991). Thus, in addition to the formation of homodimers, it is possible to hypothesize that heterodimers will also form between two different leucine zipper or two different helix-loop-helix-containing factors allowing the production of dimeric factors with novel DNA binding specificities or affinities for different sites.

One example of this process is seen in the oncogene products Fos and Jun. Thus, as discussed in Chapter 9 (section 9.3.1) the Fos protein cannot bind to AP-1 sites in DNA when present alone but can form a heterodimer with the Jun protein that is capable of binding to such sites with thirtyfold greater affinity than a Jun homodimer (Fig. 4.46). The formation of Jun homodimers and Jun/Fos heterodimers is dependent upon the leucine zipper regions of the proteins. Moreover, the failure of Fos to form homodimers is similarly dependent on its leucine zipper region. Thus, if the leucine zipper domain of Fos is replaced by that of Jun, the resulting protein can dimerize and the chimaeric protein can bind to DNA through the basic DNA binding region of Fos which is therefore a fully functional DNA binding domain. Hence the ability of leucine zipper proteins to bind to DNA is determined both by the nature of the leucine zipper which facilitates homodimerization and/or heterodimerization as well as by the basic DNA binding motif which allows DNA binding following dimerization (for discussion see Kerppola and Curran, 1995).

In addition to its positive role in allowing DNA binding by factors which cannot do so as homodimers, heterodimerization between two related factors can also have an inhibitory role. Thus, the DNA binding ability of functional

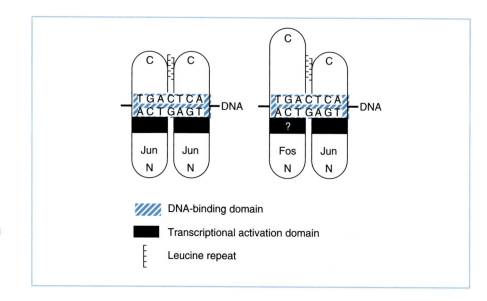

Figure 4.46

Model for DNA binding by the Jun homodimer and the Fos-Jun heterodimer.

helix-loop-helix proteins which contain a basic DNA binding domain can be inhibited by association with the Id protein. This protein contains a helix-loop-helix motif allowing it to associate with other members of this family but lacks the basic DNA binding domain. The heterodimer of Id and a functional protein therefore lacks the dimeric basic regions necessary for DNA binding and the activity of the functional transcription factor is thereby inhibited by Id (see Chapter 6, section 6.2.2 for further discussion of transcriptional repression by Id).

Hence the role of the leucine zipper and helix-loop-helix motifs in dimerization can be put to use in gene regulation in both positive and negative ways either allowing DNA binding by factors which could not do so in isolation or inhibiting the binding of fully functional factors.

4.6 OTHER DNA-BINDING MOTIFS

Although the majority of DNA binding domains that have been identified in known transcription factors fall into the families we have discussed in the preceding sections, not all do so. Thus, for example, the DNA binding domains of transcription factors such as AP2, and the CAAT box binding factor CTF/NFI are distinct from the known motifs and from each other. As more and more factors are cloned, it is likely that other factors with DNA binding motifs similar to those of these proteins will be identified and that they will become founder members of new families of DNA binding motifs. Indeed, this process is already under way, for example, the UBF ribosomal RNA transcription factor (see Chapter 3, section 3.3) contains a DNA binding domain that has also been identified in several other factors including high mobility group (HMG) proteins and which is therefore known as the HMG box (Grosschedel *et al.*, 1994), while the DNA binding domain in the p53 protein discussed in Chapter 9 (section 9.4.2) has been shown to be related to that of the NFκB family (Muller *et al.*, 1995; for review see Baltimore and Beg, 1995).

Interestingly, however, as the structure of more and more DNA binding domains is understood, relationships have emerged between different domains which were originally thought to be entirely distinct. For example, structural analysis of the Ets DNA binding domain which is found in the Ets-1 proto-oncogene protein (see Chapter 9, section 9.3.1) and the mouse PU-1 factor has revealed it to be identical to the winged helix-turn-helix motif originally identified in the *Drosophila* fork head factor and in the mammalian liver transcription factor HNF-3 (Donaldson *et al.*, 1996).

Moreover, as its name suggests, this domain contains a helix-turn-helix motif which is similar to that found in the homeobox proteins discussed in section 4.2. However, the winged helix-turn-helix motif also contains an additional β-sheet structure with loops that appear as wings protruding from the DNA bound factor, giving this motif its name (for review see Brennan, 1993). In the majority of winged helix-containing proteins, the helix-turn-helix motif is responsible for DNA binding. However, in the hRFX1 member of the family, it is the β-sheet structure which binds to DNA rather than the helix-turn-helix motif indicating that members of this family can use one of two distinct structures to bind to DNA (Gajiwala *et al.*, 2000).

As discussed in section 4.2 both the POU-specific domain of the POU factors and the paired box of the Pax proteins also bind to DNA via helix-turn-helix motifs indicating that this is one of the most commonly used motifs mediating the DNA binding of factors whose DNA binding domains appear distinct at first sight.

4.7 CONCLUSIONS

In this chapter we have discussed a number of different DNA binding motifs common to several different transcription factors which can mediate DNA binding. These motifs are listed in Table 4.4.

Interestingly, it is also possible for the same DNA sequence to be bound by more than one factor. Although in many cases, the factors binding to a particular DNA sequence share a common DNA binding domain, this is not always the case. Thus, while the transcription factors CTF/NFI and C/EBP both bind

Table 4.4

DNA binding motifs

Motif	Structure	Factors containing domain	Comments
Homeobox	Helix-turn-helix	Numerous *Drosophila* homeotic genes, related genes in other organisms	Structurally related to similar motif in bacteriophage proteins
POU	Helix-turn-helix and adjacent helical region	Mammalian Oct-1, Oct-2, Pit-1, nematode *unc86*	Related to homeodomain
Paired	Helix-turn-helix	Mammalian Pax factors, *Drosophila* paired factor	Often found in factors which also contain a homeobox
Cysteine-histidine zinc finger	Multiple fingers, each coordinating a zinc atom	TFIIIA, Kruppel, Sp1, etc.	May form β-sheet and adjacent α-helical structure
Cysteine-cysteine zinc finger	Single pair of fingers each coordinating a zinc atom	Steroid-thyroid hormone receptor family	Related motifs in EIA, GAL4, etc.
Basic domain	α-Helical	C/EBP *c-fos, c-jun, c-myc*, MyoD, etc.	Associated with leucine zipper and/or helix-loop-helix dimerization motifs
Winged HTH	Helix-turn-helix	Fork head, HNF 3A *c-ets, c-erg, Drosophila* E74, PU.1	Binds purine rich sequences

to the CAAT box sequence they do so via completely different DNA binding domains with C/EBP having a basic DNA binding domain (section 4.5) while CTF/NFI has a DNA binding domain distinct from that of any other factor (section 4.6).

It is unlikely therefore that the existence of several distinct DNA binding domains reflects the need of the factors that contain them to bind to distinct types of DNA sequences. Rather it seems perfectly possible that one DNA binding motif could be present in all factors with variations of it in different factors producing the observed binding to different DNA sequences. This is particularly so in view of the fact that in diverse DNA binding motifs such as the helix-turn-helix, the basic DNA binding domain and the two types of zinc fingers, the amino acids which determine sequence specific binding to DNA are all located within similar alpha-helical structures. This idea evidently begs the question of why different DNA binding motifs exist.

It is possible that this situation has arisen simply by different motifs which could produce DNA binding having arisen in particular factors during evolution and having been retained since they efficiently fulfilled their function. Alternatively, it may be that the existence of different motifs reflects other differences in the factors containing them other than the specific DNA sequence that is recognized. For example, the highly repeated zinc finger motif may be of particular use where, as in the case of transcription factor TFIIIA, the factor must contact a large regulatory region in the DNA. Similarly, a motif such as the basic domain which can only bind to DNA following dimerization will be of particular use where the activity of the factor must be regulated whether positively or negatively via dimerization with another factor.

Whatever the case, it is clear that DNA binding by transcription factors is dependent upon specific domains of defined structure within the molecule. Following such DNA binding, the bound factor must influence the rate of transcription either positively or negatively. The manner in which this occurs and the regions of the factors which achieve this effect are discussed in the next two chapters.

REFERENCES

Andersen, B. and Rosenfeld, M.G. (1994) Pit-1 determines cell types during development of the interior pituitary gland. Journal of Biological Chemistry 269, 335–338.

Anderson, M.G., Perkins, G.L., Chittick, P. *et al.* (1995) Drifter, a *Drosophila* POU-domain transcription factor, is required for correct differentiation and migration of tracheal cells and midline glia. Genes and Development 9, 123–127.

Andrews, B. and Donoviel, M.S. (1995) A heterodimeric transcriptional repressor becomes clear. Science 270, 251–253.

Baltimore, D. and Beg, A.E. (1995) DNA-binding proteins: a butterfly flutters by. Nature 373, 287–288.

Baumruker, T., Sturm, R. and Herr, W. (1988) OBP 100 binds remarkably degenerate octamer motifs through specific interaction with flanking sequences. Genes and Development 2, 1400–1413.

Bieker, J. J. (2001) Krüppel-like factors: three fingers in many pies. Journal of Biological Chemistry 276, 34355–34358.

Blackwell, T.K., Bowerman, B., Priess, J.R. and Weintraub, H. (1994) Formation of a monomeric DNA binding domain by Skn-1 bZip and homeodomain elements. Science 266, 621–628.

Brennan, R. G. (1993) The winged-helix DNA-binding motif: another helix-turn-helix take off. Cell 74, 773–776.

Chi, N. and Epstein, J. A. (2002) Getting your Pax straight: Pax proteins in development and disease. Trends in Genetics 18, 41–47.

Choo, Y. and Klug, A. (1994) Toward a code for the interaction of zinc fingers with DNA: Selection of randomized fingers displayed on phage. Proceedings of the National Academy of Sciences USA 91, 11163–11167.

Davis, R.L. and Weintraub, H. (1992) Acquisition of myogenic specificity by replacement of three amino acid residues from MyoD into E12. Science 256, 1027–1030.

Davis, R.L., Cheng, P-F., Lassar, A.B. and Weintraub, H. (1990) The MyoD DNA binding domain contains a recognition code for muscle-specific gene activation. Cell 60, 733–746.

Dawson, S.J., Morris, P.J. and Latchman, D.S. (1996) A single amino acid change converts a repressor into an activator. Journal of Biological Chemistry 271, 11631–11633.

de Kok, Y.J.M., Van der Maarel, S.M., Bitner-Glindzicz, M. et al. (1995) Association between X-linked mixed deafness and mutations in the POU domain gene POU3F4. Science 267, 685–688.

Dolan, J. K. and Fields, S. (1991) Cell type-specific transcription in yeast. Biochimica et Biophysica Acta 1088, 155–169.

Donaldson, L.W., Peterson, J.M., Graves, B.J. and McIntosh, P. (1996) Solution structure of the ETS domain from murine Ets-1: a winged helix-turn-helix DNA binding motif. EMBO Journal 15, 125–134.

Ericson, J., Rashbass, P., Schedl, A. et al. (1997) Pax6 controls progenitor cell identity and neuronal fate in response to graded Shh signalling. Cell 90, 169–180.

Evans, R.M. and Hollenberg, S.M. (1988) Zinc fingers: gilt by association. Cell 52, 1–3.

Gajiwala, K. S., Chen, H., Cornille, F. et al. (2000) Structure of the winged-helix protein hRFX1 reveals a new mode of DNA binding. Nature 403, 916–921.

Garvie, C. W. and Wolberger, C. (2001) Recognition of specific DNA sequences. Molecular Cell 8, 937–946.

Gehring, W. J. and Ikeo, K. (1999) Pax 6 mastering eye morphogenesis and eye evolution. Trends in Genetics 15, 371–377.

Gehring, W.J., Affolter, M. and Burglin, T. (1994a) Homeodomain proteins. Annual Review of Biochemistry 63, 487–526.

Gehring, W.J., Qian, Y.Q., Billeter, M. et al. (1994b) Homeodomain-DNA recognition. Cell 78, 211–223.

Glover, J.N.M. and Harrison, S.C. (1995) Crystal structure of the heterodimeric bZip transcription factor c-Fos-c-Jun bound to DNA. Nature 373, 257–261.

Graham, A., Papalopulu, N. and Krumlauf, R. (1989) The murine and *Drosophila* homeobox gene complexes have common features of organization and expression. Cell 57, 367–378.

Gronemeyer, H. and Moras, D. (1995) How to finger DNA. Nature 375, 190–191.

Green, S. and Chambon, P. (1987) Oestradiol induction of a glucocorticoid-response gene by a chimaeric receptor. Nature 325, 75–78.

Grosschedel, R., Giese, K. and Pagel, J. (1994) HMG proteins: architectural elements in the assembly of nucleoprotein structures. Trends in Genetics 10, 94–100.

Hadorn, E. (1968) Transdetermination in cells. Scientific American 219 (Nov), 110–120.

Hard, T., Kellenbach, E., Boelens, R. *et al.* (1990) Solution structure of the glucocortoid receptor DNA-binding domain. Science 249, 157–160.

Harrison, S.C. (1991) A structural taxonomy of DNA binding domains. Nature 353, 715–719.

Hayashi, S. and Scott, M.P. (1990) What determines the specificity of action of *Drosophila* homeodomain proteins. Cell 63, 883–894.

He, X., Treacy, M.N., Simmons, D.M. *et al.* (1989) Expression of a large family of POU-domain regulatory genes in mammalian brain development. Nature 340, 35–42.

Hoey, T. and Levine, M. (1988) Divergent homeobox proteins recognize similar DNA sequences in *Drosophila*. Nature 332, 858–861.

Holland, P.W.H. and Hogan, B.L.M. (1986) Phylogenetic distribution of Antennapedia-like homeoboxes. Nature 321, 251–253.

Hurst, H.C. (1996) bZIP proteins. Protein profile 3, 1–72.

Ingham, P.W. (1988) The molecular genetics of embryonic pattern formation in *Drosophila*. Nature 335, 25–34.

Jones, N. (1990) Transcriptional regulation by dimerization: two sides to an incestuous relationship. Cell 61, 9–11.

Kadonaga, J.T., Carner, K.R., Masiarz, F.R. and Tjian, R. (1987) Isolation of cDNA encoding the transcription factor Sp1 and functional analysis of the DNA binding domain. Cell 51, 1079–1090.

Kanshal, S., Schneider, J.W., Nudal-Ginard, B. and Mahdavi, V. (1994) Activation of the myogenic lineage by MEF2A, a factor that induces and cooperates with MyoD. Science 266, 1236–1240.

Kenyon, C. (1994) If birds can fly, why can't we? Homeotic genes and evolution. Cell 78, 175–180.

Kerppola, T. and Curran, T. (1995) Zen and the art of Fos and Jun. Nature 373, 199–200.

King, R.J.B. and Mainwaring, W.I.P. (1974) Steroid Cell Interactions. Butterworths.

Khorasanizadeh, S. and Rastinejad, F. (2001) Nuclear-receptor interactions on DNA-response elements. Trends in Biochemical Sciences 26, 384–390.

Klemm, J.D., Rould, M.A., Aurora, R. *et al.* (1994) Crystal structure of the Oct-1 POU domain bound to an octamer site: DNA recognition with tethered DNA-binding molecules. Cell 77, 21–23.

Klug, A. and Schwabe, J.R. (1995) Zinc fingers. FASEB Journal 9, 597–604.

Kornberg, T.B. (1993) Understanding the homeodomain. Journal of Biological Chemistry 268, 26813–26816.

Krumlauf, R. (1994) Hox genes in vertebrate development. Cell 78, 191–201.

Lai, J.S., Cleary, M.A. and Herr, W. (1992) A single amino acid exchange transfers VP16-induced positive control from the Oct-1 to the Oct-2 homeo domain. Genes and Development 6, 2058–2065.

Lamb, P. and McKnight, S.L. (1991) Diversity and specificity in transcriptional regulation: the benefits of heterotypic dimerization. Trends in Biochemical Sciences 16, 417–422.

Landschulz, W.H., Johnson, P.F. and McKnight, S.L. (1988) The leucine zipper: a hypothetical structure common to a new class of DNA binding proteins. Science 240, 1759–1764.

Landschulz, W.H., Johnson, P.F. and McKnight, S.L. (1989) The DNA binding domain of the rat liver nuclear protein C/EBP is bipartite. Science 243, 1681–1688.

Latchman, D.S. (1996) Transcription factor mutations and human disease. New England Journal of Medicine 334, 28–33.

Latchman, D. S. (2001) Transcription factors: bound to activate or repress. Trends in Biochemical Sciences 26, 211–213.

Lawrence, P.A. and Morata, G. (1994) Homeobox genes: their function in *Drosophila* segmentation and pattern formation. Cell 78, 181–189.

Lee, M.S., Gippert, G.P., Soman, K.V. *et al.* (1989) Three-dimensional solution structure of a single zinc finger DNA binding domain. Science 245, 635–637.

Li, T., Stark, M.R., Johnson, A.D. and Wolberger, C. (1995) Crystal structure of the MATa1/MAT alpha 2 homeodomain heterodimer bound to DNA. Science 270, 262–269.

Littlewood, T. and Evan, G. (1995) Helix-loop-helix. Protein profile 2, 621–702.

Ma, P.C.M., Rould, M.A., Weintraub, H and Pabo, C.O. (1994) Crystal structure of MyoD bHLH domain – DNA complex: Perspectives on DNA recognition and implications for transcriptional activation. Cell 77, 451–459.

Mann, R.S. and Chan, S-K. (1996) Extra specificity from extradenticle: the partnership between Hox and PBX/EXD homeodomain proteins. Trends in Genetics 12, 258–262.

Mansouri, A., Hallone, T.M. and Gruss, P. (1996) Pax genes and their roles in cell differentiation and development. Current Opinion in Cell Biology 8, 851–857.

Martindale, M. Q. and Kourakis, M. J. (1999) Size doesn't matter. Nature 399, 730–731.

Marx, J. (2000) New clues to how genes are controlled. Science 290, 1066–1067.

McKenna, N. J. and O'Malley, B. W. (2002) Combinational control of gene expression by nuclear receptors and coregulators. Cell 108, 465–474.

Miller, J., McLachlan, A.D. and Klug, A. (1985) Repetitive zinc-binding domains in the protein transcription factor III A from *Xenopus* oocytes. EMBO Journal 4, 1609–1614.

Muller, C.W., Rey, F.A., Sodeoka, M. *et al.* (1995) Structure of the NF-κB p50 homodimer bound to DNA. Nature 373, 311–317.

Murre, C., McCaw, P.S. and Baltimore, D. (1989a) A new DNA binding and dimerization motif in immunoglobulin enhancer binding, daughterless, MyoD and myc proteins. Cell 56, 777–783.

Murre, C., McCaw, P.S., Vaessin, H. *et al.* (1989b) Interactions between heterologous helix-loop-helix proteins generate complexes that bind specifically to a common DNA sequence. Cell 58, 537–544.

Nardelli, J., Gibson, T.J., Vesque, C. and Charnay, P. (1991) Base sequence discrimination by zinc-finger DNA binding domains. Nature 349, 175–178.

Olefsky, J. M. (2001) Nuclear receptor minireview series. Journal of Biological Chemistry 276, 36863–36864.

Pabo, C. and Sauer, R.T. (1992) Transcription factors: structural families and principles of DNA recognition. Annual Review of Biochemistry 61, 1053–1095.

Papworth, M., Moore, M., Isalan, M. *et al.* (2003) Inhibition of herpes simplex virus 1 gene expression by designer zinc-finger transcription factors. Proceedings of the National Academy of Sciences USA 100, 1621–1626.

Pasqualini, R., Barbas, C. F. and Arap, W. (2002) Vessel manoeuvres: Zinc fingers promote angiogenesis. Nature Medicine 8, 1353–1354.

Passner, J. M., Ryoo, H. D., Shen, L. *et al.* (1999) Structure of a DNA-bound ultrabithorax-extradenticle homeodomain complex. Nature 397, 714–719.

Prendergast, G.C. and Ziff, E.B. (1989) DNA-binding motif. Nature 341, 392.

Rastinejad, F., Perlmann, T., Evans, R.M. and Sigler, P.B. (1995) Structural determinants of nuclear receptor assembly on DNA direct repeats. Nature 375, 203–211.

Rebar, E.J. and Pabo, C.O. (1994) Zinc finger phage: affinity selection of fingers with new DNA-binding specificities. Science 263, 671–673.

Redemann, N., Gaul, U. and Jackle, H. (1988) Disruption of a putative Cys-zinc interaction eliminates the biological activity of the Kruppel finger protein. Nature 332, 90–92.

Reményi, A., Tomilin, A., Pohl, E. *et al.* (2001) Differential dimer activities of the transcription factor Oct-1 by DNA-induced interface swapping. Molecular Cell 8, 569–580.

Reynolds, L., Ullman, C., Moore, M. *et al.* (2003) Repression of the HIV-1 5' LTR promoter and inhibition of HIV-1 replication by using engineered zinc-finger transcription factors. Proceedings of the National Academy of Sciences USA 100, 1615–1620.

Rhodes, D. and Klug, A. (1993) Zinc finger structure. Scientific American 268, 32–39.

Ryan, A.K. and Rosenfeld, M.G. (1997) POU domain family values: flexibility, partnerships and developmental codes. Genes and Development 11, 1207–1225.

Sakai, D.D., Helms, S., Carlstedt-Duke, J. *et al.* (1988) Hormone-mediated repression: a negative glucocorticoid response element from the bovine prolactin gene. Genes and Development 2, 1144–1154.

Sander, M., Neubuser, A., Kalamara, J. *et al.* (1997) Genetic analysis reveals that PAX6 is required for normal transcription of pancreatic hormone genes and islet development. Genes and Development 11, 1662–1673.

Sauer, R.T. (1990) Scissors and helical forks. Nature 347, 514–515.

Schwabe, J.W.R. and Rhodes, D. (1991) Beyond zinc fingers: steroid hormone receptors have a novel structural motif for DNA recognition. Trends in Biochemical Sciences 16, 291–296.

Schwabe, J.W.R., Chapman, L. Finch, T. and Rhodes, D. (1993) The crystal structure of the estrogen receptor DNA binding domain bound to DNA – how receptors discriminate between their response elements. Cell 75, 567–578.

Scott, M. P. (1999) Hox proteins reach out round DNA. Nature 397, 649–651.

Scully, K. M., Jacobson, E. M., Jepsen, K. *et al.* (2000) Allosteric effects of Pit-1 DNA sites on long-term repression in cell type specification. Science 290, 1127–1131.

Sessa, G., Morelli, G and Ruberti, I. (1993) The Athb-1 and -2 HD-ZIP domains homodimerise forming complexes of different DNA binding specificities. EMBO Journal 12, 3507–3517.

Sornson, M.W., Wu, W., Dasen, J.S. *et al.* (1996) Pituitary lineage determination by the prophet of Pit-1 homeodomain factor defective in Ames dwarfism. Nature 384, 327–333.

Tomilin, A., Reményi, A., Lins, K. *et al.* (2000) Synergism with the coactivator OBF-1 (OCA-B, BOB-1) is mediated by a specific POU dimer configuration. Cell 103, 853–864.

Travers, A. (1993) DNA-protein interactions. Chapman and Hall.

Treisman, J., Harris, E., Wilson, D. and Desplan, C. (1992) The homeodomain: a new face for the helix-turn-helix. Bio essays 14, 145–150.

Turner, J. and Crossley, M. (1999) Mammalian Krüppel-like transcription factors: more than just a pretty finger. Trends in Biochemical Sciences 24, 236–240.

Umesono, K. and Evans, R. H. (1989) Determinants of target gene specificity for steroid/thyroid hormone receptors. Cell 57, 1139–1146.

Verrijzer, C.R. and Van der Vliet, P.C. (1993) POU domain transcription factors. Biochimica et Biophysica Acta 1173, 1–21.

Wagner, R. L., Apriletti, J. W., McGrath, M. E. et al. (1995) A structural role for hormone in the thyroid hormone receptor. Nature 378, 690–697.

Walker, S., Hayes, S. and O'Hare, P. (1994) Site-specific conformational alteration of the Oct-1 POU domain -DNA complex as the basis for differential recognition by Vmw65 (VP16). Cell 79, 841–852.

Way, J.C. and Chalfie, M. (1988) mec-3, a homeobox-containing gene that specifies differentiation of the touch receptor neurons in C. elegans. Cell 54, 5–16.

Weatherman, R. V., Fletterick, R. J. and Scanlan, T. S. (1999) Nuclear-receptor ligands and ligand-binding domains. Annual Reviews of Biochemistry 68, 559–581.

Xu, W. Rould, M.A., Jun, S. et al. (1995) Crystal structure of a paired domain-DNA complex at 2.5A resolution reveals structural basis for Pax developmental mutations. Cell 80, 639–650.

CHAPTER 5

ACTIVATION OF GENE EXPRESSION BY TRANSCRIPTION FACTORS

5.1 ACTIVATION DOMAINS

Extensive studies on a variety of transcription factors have shown that they have a modular structure in which distinct regions of the protein mediate particular functions such as DNA binding (see Chapter 4) or interaction with specific effector molecules such as steroid hormones. It is likely therefore that a specific region of each individual transcription factor will be involved in its ability to activate transcription following DNA binding. As described in Chapter 2 (section 2.4.1) such activation domains have been identified by so-called 'domain swap' experiments in which various regions of one factor are linked to the DNA binding domain of another factor and the ability to activate transcription assessed.

In general, these experiments have confirmed the modular nature of transcription factors with distinct domains mediating DNA binding and transcriptional activation. Thus, in the case of the yeast factor GCN4 two distinct regions, each of sixty amino acids, have been identified which mediate respectively DNA binding and transcriptional activation (Fig. 5.1a, Hope and Struhl, 1986). Similarly, domain swap experiments have identified two regions of the glucocorticoid receptor, one at the N terminus of the molecule and the other near the C terminus which can independently mediate gene activation when linked to the DNA binding domain of another transcription factor (Hollenberg and Evans, 1988) and both of these are distinct from the DNA binding domain of the molecule. Interestingly, the C terminal activation domain is located close to the hormone binding domain of the receptor (Fig.5.1b) and can mediate the activation of transcription only following hormone addition. It therefore plays an important role in the steroid-dependent activation of transcription following hormone addition (see Chapter 8, section 8.2.2).

Studies on a variety of transcription factors have therefore strongly indicated their modular nature with distinct regions of the molecule mediating DNA binding and transcriptional activation. An extreme example of this

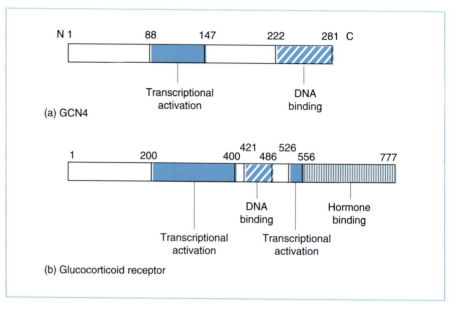

Figure 5.1

Domain structure of the yeast GCN4 transcription factor (panel a) and the mammalian glucocorticoid receptor (panel b). Note the distinct domains which are active in DNA binding or transcriptional activation.

modularity is provided by the interaction of the cellular transcription factor Oct-1 (see Chapter 4, section 4.2.6) and the herpes simplex virus transactivating protein VP16 (for review see Goding and O'Hare, 1989). Thus, although VP16 contains a very strong activating region which can strongly induce transcription when artificially fused to the DNA binding domain of the yeast GAL4 transcription factor, it contains no DNA binding domain and cannot therefore bind to DNA itself. Transcriptional activation by VP16 following viral infection is therefore dependent upon its ability to form a protein–protein complex with the cellular Oct-1 protein. This complex then binds to the octamer-related TAATGARAT (R = purine) motif in the viral immediate-early genes via the DNA binding domain of Oct-1 and transcription is activated by the activation domain of VP16. Hence in this case, the DNA binding and transcriptional activation domains are actually located on different proteins in the DNA binding complex (Fig. 5.2). A similar example in which the constitutively expressed Oct-1 recruits a non-DNA binding cellular co-activator molecule, OCA-B, to the promoter resulting in its activation was also discussed in Chapter 4 (section 4.2.6) indicating that this effect is not confined to viral transactivating molecules.

5.2 NATURE OF ACTIVATION DOMAINS

Following the identification of activation domains in different transcription factors, it rapidly became clear that they fell into several distinct families with

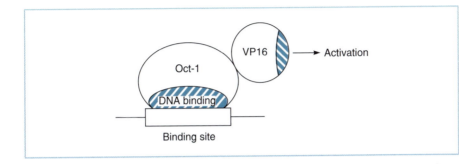

Figure 5.2

Activation of gene transcription by interaction of the cellular factor Oct-1, which contains a DNA-binding domain, and the herpes simplex virus VP16 protein, which contains an activation domain but cannot bind to DNA.

common features which will be discussed in turn (for a typical example of each of the major classes of activation domain see Fig. 5.3) (for review see Mitchell and Tjian, 1989; Triezenberg, 1995).

GAL4 acidic domain

D S A A A H H D N S T I P L D F M P R D
A L H G F D W S E E D D M S D G L P F L
K T D P N N N G F

Sp1 glutamine-rich domain B

Q G Q T P Q R V S G L Q G S D A L N I Q
Q N Q T S G G S L Q A G Q Q K E G E Q N
Q Q T Q Q Q Q I L I Q P Q L V Q G G Q A
L Q A L Q A A P L S G Q T F T T Q A I S
Q E T L Q N L Q L Q A V P N S G P I I I
R T P T V G P N G Q V S W Q T L Q L Q N
L Q V Q N P Q A Q T I T L A P M Q G V S
L G Q

CTF/NFI Proline-rich domain

P P H L N P Q D P L K D L V S L A C D P
A S Q Q P G R L N G S G Q L K M P S H C
L S A Q M L A P P P P G L P R L A L P P
A T K P A T T S E G G A T S P S Y S P P
D T S P

Figure 5.3

Structure of typical members of each of the three classes of activation domains. Acidic, glutamine or proline residues are highlighted in the appropriate case.

5.2.1 ACIDIC DOMAINS

Comparison of several different activation domains, including those of the yeast factors GCN4 and GAL4 as well as the activation domain at the N terminus of the glucocorticoid receptor and that of VP16 which were discussed above (section 5.1), indicated that, although they do not show any strong amino acid sequence homology to each other, they all have a large proportion of acidic amino acids producing a strong net negative charge (see Fig. 5.3a). Thus the 82 amino acid activating region of the glucocorticoid receptor contains 17 acidic residues (Hollenberg and Evans, 1988) while the same number of negatively charged amino acids is found within the sixty amino acid activating region of GCN4 (Hope and Struhl, 1986). These findings indicated therefore that these activation regions consist of so-called 'acid blobs' or 'negative noodles' with a high proportion of negatively charged amino acids which are involved in the activation of transcription (for review see Hahn, 1993a).

In agreement with this idea, mutations in the activation domain of GAL4, which increase its net negative charge, increase its ability to activate transcription. Similarly, if recombination is used to create a GAL4 protein with several more negative charges, the effect on gene activation is additive, a mutant with four more negative charges than the parental wild-type, activating transcription ninefold more efficiently than the wild-type. Thus the acidic nature of these domains is likely to be important in their function. It has been suggested that, in the case of VP16, the negative charge of its acidic domain allows it to establish long range electrostatic interactions with the $TAF_{II}31$ component of TFIID (see section 5.4.2) with which it interacts to stimulate transcription (Uesugi *et al.*, 1997; Fig. 5.4a).

Figure 5.4

(a) The negatively charged acidic residues in the VP16 activation domain allow its initial long distance interaction with $TAF_{II}31$. (b) Interaction with $TAF_{II}31$ induces a conformational charge in the domain to α-helical structure in which the hydrophobic residues asparagine (D) at position 472, phenylalanine (F) at position 479 and leucine (L) at position 483 are brought close to one another and bind to $TAF_{II}31$.

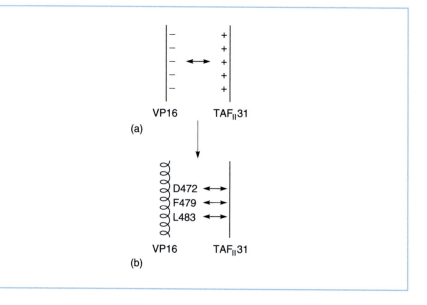

Although the acidic nature of the activation domain is clearly important for its function, it is not the only feature required since it is possible to decrease the activity of the GAL4 activation domain without reducing the number of negatively charged residues. Indeed, recent evidence indicates that conserved hydrophobic residues in the acidic activation domains play a key role in their ability to stimulate transcription. Thus, when the VP16 activation domain interacts with the $TAF_{II}31$ component of TFIID it undergoes a conformational change from a random coil to an α-helix which brings together three hydrophobic residues within the acidic domain which then interact directly with $TAF_{II}31$ (Uesugi et al., 1997; Fig. 5.4b). Hence the acidic domain would interact with $TAF_{II}31$ via a two-step process in which the initial long range attraction produced by the acidic residues allows a subsequent structural change, facilitating a close interaction of the hydrophobic residues within the acidic domain with $TAF_{II}31$. Hence both the acidic and hydrophobic residues are of importance for the activity of this domain.

Although activation domains of the acidic type form the majority of the activation domains so far identified in eukaryotic transcription factors, from yeast to mammals, other types of activation domains have been identified in a number of different transcription factors in higher eukaryotes and these will be discussed in turn.

5.2.2 GLUTAMINE-RICH DOMAINS

Analysis of the constitutive transcription factor Sp1, which binds to the Sp1 binding site found in many gene promoters (see Chapter 1, section 1.3.2), revealed that the two most potent activation domains contained approximately twenty-five per cent glutamine residues and very few negatively charged residues (Courey and Tjian, 1988; see Fig. 5.3b). These glutamine-rich motifs are essential for the activation of transcription mediated by these domains since their deletion abolishes the ability to activate transcription. Most interestingly, however, transcriptional activation can be restored by substituting the glutamine-rich regions of Sp1 with a glutamine-rich region from the *Drosophila* homeobox transcription factor Antennapedia which has no obvious sequence homology to the Sp1 sequence. Hence, as with the acidic activation domains, the activating ability of a glutamine-rich domain is not defined by its primary sequence but rather by its overall nature in being glutamine rich. In agreement with this a continuous run of glutamine residues with no other amino acids has been shown to act as a transcriptional activation domain (Gerber et al., 1994).

Similar glutamine-rich regions have been defined in transcription factors other than Sp1 and Antennapedia, including the N terminal activation domains of the octamer binding proteins Oct-1 and Oct-2, the *Drosophila* homeobox proteins ultra-bithorax and zeste and the yeast HAP1 and HAP2 transcription factors, indicating that this motif is quite widespread, being found in different transcription factors in different species (for review see Mitchell and Tjian, 1989).

5.2.3 PROLINE-RICH DOMAINS

Studies on the constitutive factor CTF/NF1 which binds to the CCAAT box motif (see Chapter 1, section 1.3.2) defined a third type of activation domain distinct from those previously discussed. Thus the activation domain located at the C terminus of CTF/NF1 is not rich in acidic or glutamine residues but, instead, contains numerous proline residues forming approximately one-quarter of the amino acids in this region (Mermod *et al.*, 1989; see Fig. 5.3c). As with the other classes of activation domains, this region is capable of activating transcription when linked to the DNA binding domains of other transcription factors. Moreover, as with the glutamine-rich domain, a continuous run of proline residues can mediate activation, indicating that the function of this type of domain depends primarily on its richness in proline (Gerber *et al.*, 1994). Similar proline-rich domains have been identified in several other transcription factors, such as the oncogene product Jun, AP2 and the C terminal activation domain of Oct-2 (for review see Mitchell and Tjian, 1989). Thus, as with the glutamine-rich domains, proline-rich domains are not confined to a single factor while a single factor such as Oct-2 can contain two activation domains of different types.

In summary therefore it is clear that as with DNA binding, several distinct protein motifs can activate transcription (see Fig. 5.3).

5.2.4 FUNCTIONAL RELATIONSHIP OF THE DIFFERENT ACTIVATION DOMAINS

The existence of at least three distinct classes of activation domain raises the question of whether these three domains are functionally equivalent or whether they differ in their ability to activate transcription. This question was investigated by Seipel *et al.* (1992) who linked each of the activation domains to the DNA binding domain of the GAL4 factor and tested the ability of these chimaeric proteins to activate transcription in mammalian cells when the GAL4 DNA binding site was placed at different positions relative to the start site of transcription (Fig. 5.5). In these experiments all three domains

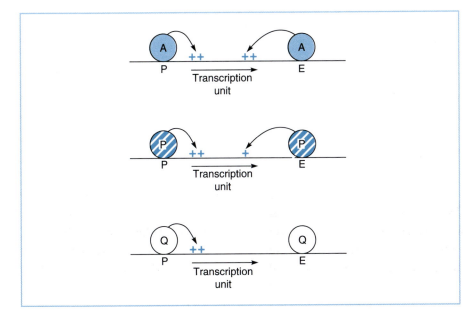

Figure 5.5

An acidic activation domain (A) can stimulate transcription when bound to DNA in the promoter (P) close to the transcriptional start site or when bound at a distant enhancer (E). In contrast, a proline-rich domain (P) stimulates only weakly from an enhancer position and a glutamine-rich domain (Q) does not stimulate at all from this position.

were able to activate transcription when the DNA binding site was placed close to the start site of transcription in the promoter region. In contrast, the glutamine-rich domain was unable to activate transcription when the binding site was located downstream of the transcription unit mimicking a position within an enhancer element (see Chapter 1, section 1.3.4). The acidic domain was strongly active from this enhancer position while the proline-rich domain could also activate transcription from this position but only weakly.

These findings indicate therefore that clear differences exist in the abilities of the different activation domains to activate transcription when bound to the DNA at different positions relative to the promoter. Such differences are likely to be important in determining the functional activity of different factors. In addition, such differences in the activity of different activation domains are likely to reflect differences in the mechanisms by which these factors act. In agreement with this idea, acidic or proline-rich activation domains derived from mammalian factors can also activate transcription when introduced into yeast cells, whereas glutamine-rich domains cannot do so (Kinzler *et al.*, 1994).

In the next sections we will consider the mechanisms by which activation domains act, focusing particularly on the acidic domains where most information is available. Similarities and differences in the mode of action of the other activation domains will be discussed where this information is available.

5.3 INTERACTION OF ACTIVATION DOMAINS WITH THE BASAL TRANSCRIPTIONAL COMPLEX

5.3.1 ACTIVATORS AND THE BASAL TRANSCRIPTIONAL COMPLEX

The widespread interchangeability of acidic activation domains from yeast, *Drosophila* and mammalian transcription factors discussed above, strongly suggests that a single common mechanism may mediate transcriptional activation by acidic activation domains in a wide range of organisms. This idea is supported by the finding noted above that mammalian transcription factors carrying such domains, such as the glucocorticoid receptor, can activate a gene carrying their appropriate DNA binding site in yeast cells while the yeast GAL4 factor can do so in cells of *Drosophila*, tobacco plants and mammals (reviewed by Guarente, 1988; Ptashne, 1988).

These considerations suggest that the target factor or factors with which these activators interact is likely to be highly conserved in evolution. A number of experiments have indicated that in many cases this target factor is likely to be required for the transcription of a number of different genes and not solely for that of the activated gene. Thus the over-expression of the yeast GAL4 protein which contains a strong activation domain results in the down regulation of genes which lack GAL4-binding sites such as the CYC1 gene as well as activating genes which do contain GAL4-binding sites. This phenomenon, which has been noted for a number of transcription factors with strong activation domains, is known as squelching (for review see Ptashne, 1988). Although the degree of squelching by any given factor is proportional to the strength of its activation domain, squelching differs from activation in that it does not require DNA binding, and can be achieved with truncated factors containing only the activation domain and lacking the DNA binding domain. This phenomenon can therefore be explained on the basis that a transcriptional activator, when present in high concentration, can interact with its target factor in solution as well as on the DNA. If this target factor is present at limiting concentrations it will therefore be sequestered away from other genes that require it for transcription, resulting in their inhibition (Fig. 5.6).

The existence of squelching indicates therefore that in many cases the target factor for activation domains is likely to be a component that is required for the transcription of a wide range of genes and which is conserved from yeast to mammals allowing yeast activators to work in mammalian cells and vice versa. Obviously, such a common component could be part of the basal transcriptional complex required for transcription of a wide range of genes in different organisms. Clearly, an activating factor could act by stimulating the binding of such a component so that the basal complex assembled

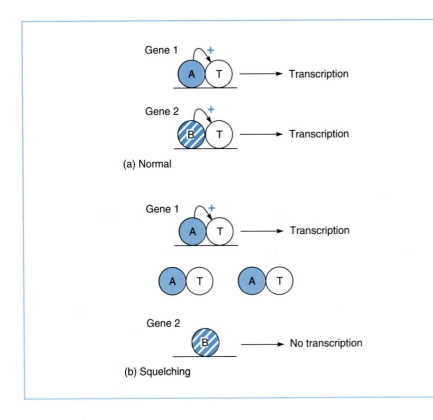

Figure 5.6
The process of squelching. In the normal case, illustrated in panel (a), two distinct activator molecules A and B involved in the activation of genes one and two respectively, both act by interacting with the general transcription factor T and both genes are transcribed. In squelching, illustrated in panel (b), factor A is present at high concentration and hence interacts with T both on gene one and in solution. Hence factor T is not available for transcription of gene two and therefore only gene one is transcribed while transcription of gene two is squelched.

more efficiently. Alternatively, it could act by interacting with a factor which had already bound so that the activity or stability of the assembled complex was stimulated. It appears that both these mechanisms are used and they will be discussed in turn.

5.3.2 STIMULATION OF FACTOR BINDING

As described in Chapter 3 (section 3.5.1) the basal transcriptional complex can assemble in a stepwise manner with the binding of TFIID being followed by the binding of TFIIB and then the binding of RNA polymerase in association with TFIIF. Clearly an activator could increase the rate of complex assembly by enhancing any one of these assembly steps. Indeed, there is evidence that activators target several of these steps in the assembly process (Fig. 5.7). Thus, for example, it appears that acidic activators interact directly with TFIID (see Chapter 3, section 3.5.1) to stimulate the binding of TFIID to the promoter (Fig. 5.7a). Interestingly, this enhanced recruitment of TFIID, induced by transcriptional activators, which was initially observed in the test tube, has been confirmed in intact cells using the ChIP assay described in Chapter 2 (section 2.4.3) (Kuras and Struhl, 1999; Li *et al.*, 1999).

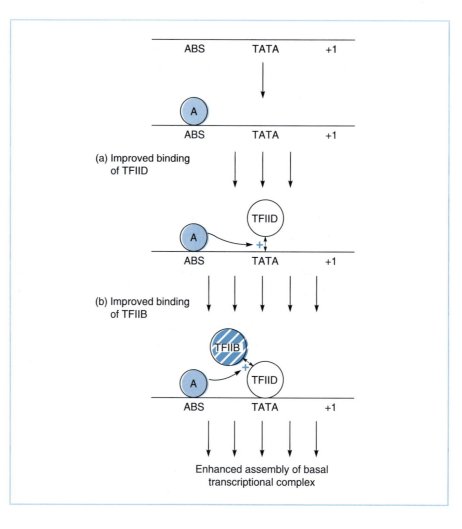

Figure 5.7

The binding of an activating molecule (A) to its binding site (ABS) can enhance both the binding of TFIID to the TATA box (a) and the recruitment of the TFIIB factor (b) so enhancing the rate of basal complex assembly and of transcription.

Although increased binding of TFIID to the promoter will directly enhance the assembly of the complex by allowing TFIIB to bind, there is evidence that activators can also act directly to improve the recruitment of TFIIB independent of their effect on TFIID (Fig. 5.7b). Thus it has been shown that both an acidic activator and glutamine or proline-rich activators can greatly stimulate the binding of TFIIB to the promoter (Choy and Green, 1993). Hence activators can enhance the assembly of the basal transcriptional complex by independently enhancing the binding of both TFIID and TFIIB. This ability of activators to act at these two independent steps results in a strong synergistic activation of transcription in the presence of different activators targeting either TFIID or TFIIB (Gonzalez-Couto *et al.*, 1997).

As with TFIID, it has been shown that TFIIB interacts directly with activating molecules. Thus TFIIB can be purified on a column containing a bound acidic activator and interactions of TFIIB with non-acidic activators have also been reported. Moreover, mutations in the activator that abolish this interaction with TFIIB prevent it from activating transcription (for review see Hahn, 1993b). Thus the effect of activators on TFIIB is mediated via a direct protein–protein interaction which is essential for their ability to stimulate transcription.

In addition to the stepwise pathway of complex assembly, it has also been proposed that the basal transcriptional complex can assemble in a much simpler manner with binding of TFIID being followed by binding of the RNA polymerase holoenzyme which contains the polymerase itself, TFIIB, TFIIF and TFIIH as well as a number of other proteins (see Chapter 3, section 3.5.2). There is evidence that activators can also act in this pathway not only by enhancing the recruitment of TFIID as described above but also by directly enhancing the binding of the RNA polymerase holoenzyme itself (Fig. 5.8a). Thus, for example, if a DNA binding domain is linked to the yeast protein Gal11, which is a component of the RNA polymerase holoenzyme, the holoenzyme is recruited to the DNA via this DNA binding domain and transcription is activated (Fig. 5.8b) (Barberis *et al.*, 1995). Hence the need for activators can be bypassed by recruiting the RNA polymerase holoenzyme to DNA via an artificial DNA binding domain.

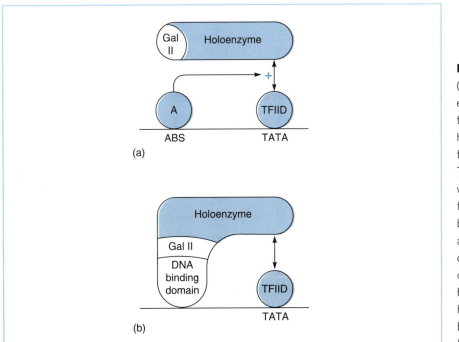

Figure 5.8
(a) Activators can act by enhancing the binding of the RNA polymerase holoenzyme complex following binding of TFIID. (b) In agreement with this idea, the need for an activator can be bypassed by artificially attaching a DNA binding domain to the Gal 11 component of the holoenzyme so enhancing holoenzyme recruitment by allowing it to bind to DNA directly.

Indeed, on the basis of experiments of this type, Ptashne and Gann (1997) have argued that the sole role of activators is to enhance the assembly of the basal complex by interacting with one or other of its specific components so facilitating their recruitment to the DNA. However, while such enhanced recruitment of specific components of the complex clearly plays a major part in the action of transcriptional activators and operates in both pathways of complex assembly, it is likely that other effects are also involved in the action of transcriptional activators. These effects are discussed in the next section.

5.3.3 STIMULATION OF FACTOR ACTIVITY

In addition to their effects on complex assembly, it is clear that activators can also stimulate transcription at a subsequent step following assembly of the complex, resulting in its enhanced stability or increased activity (Choy and Green, 1993) (Fig. 5.9).

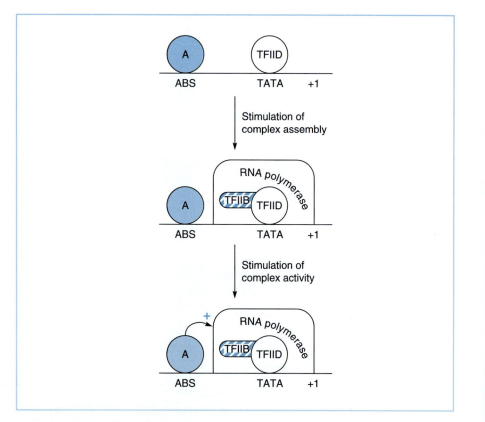

Figure 5.9

An activator can stimulate transcription both by promoting the assembly of the basal transcription complex and by stimulating its activity following assembly.

An obvious mechanism for activation would be for activating domains to interact directly with the RNA polymerase itself to increase its activity (Fig. 5.10a). As discussed in Chapter 3 (section 3.1), the largest subunit of RNA polymerase II contains at its C terminus multiple copies of a sequence whose consensus is Tyr-Ser-Pro-Thr-Ser-Pro-Ser, which is highly conserved in evolution and is essential for its function. This motif is very rich in hydroxyl groups and lacks negatively charged acidic residues.

It has therefore been suggested that this motif could interact directly either with a negatively charged acidic domain or via hydrogen bonding with amide groups in a glutamine-rich activation domain. This would provide a mechanism for direct interaction between activating domains and RNA polymerase itself while the evolutionary conservation of the target region within the polymerase would explain why yeast activators work in mammalian cells and vice versa. In agreement with this idea it has been shown that yeast mutants containing a reduced number of copies of the heptapeptide repeat in RNA polymerase II are defective in their response to activators such as GAL4.

Although these results are consistent with a direct interaction between transcriptional activators and the RNA polymerase, they are equally consistent with an indirect interaction in which the activator contacts another component of the transcriptional machinery which then interacts with the repeated motif in the polymerase. Indeed, despite the attractiveness of a model involving direct interaction between activating factors and the polymerase itself,

Figure 5.10

Two possible mechanisms by which an activating factor (A) could stimulate the activity of the basal transcriptional complex. This could occur via direct interaction with the RNA polymerase itself (a) or by interaction with another transcription factor such as TFIID which in turn interacts with the polymerase (b).

it is unlikely to be correct and it appears that activators interact with the polymerase indirectly via other factors (Fig. 5.10b).

TFIID is one potential candidate for the component with which activating factors interact since this factor is both required for the transcription of a wide variety of genes both with and without TATA boxes (see Chapter 3, section 3.6) and is highly conserved in evolution, with the yeast factor being able to promote transcription in mammalian cell extracts and vice versa. Evidence for an effect of activating factors on TFIID has been obtained in the case of the yeast acidic activating factor GAL4 (Horikoshi *et al.*, 1988). Thus, in the absence of GAL4, TFIID was shown to be bound only at the TATA box of a promoter containing both a TATA box and GAL4 binding sites. In contrast, in the presence of GAL4 bound to its upstream binding sites in the promoter, the conformation of TFIID was altered such that it now covered both the TATA box and the start site for transcription (Fig. 5.11). Moreover, no change in TFIID conformation was observed in the presence of a truncated GAL4 molecule which can bind to DNA but lacks the acidic activation domain. Hence, an acidic activator can produce a change in TFIID conformation resulting in its binding to the start site for transcription and this effect correlates with the ability of GAL4 to activate transcription rather than being a consequence of its binding to DNA. It is clear therefore that activating molecules can alter the configuration of TFIID bound to the promoter by interacting with it.

As well as interacting with TFIID to change its configuration, activators can also interact with TFIIB changing its conformation and enhancing its ability to recruit the complex of RNA polymerase II and TFIIF (Roberts and Green,

Figure 5.11
Effect of GAL4 binding on the binding of TFIID.

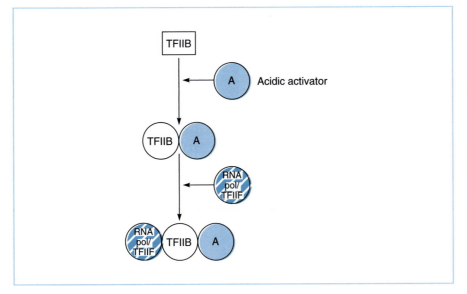

Figure 5.12
The binding of an acidic activator (A) to TFIIB produces a conformational change which enhances the ability of TFIIB to interact with the RNA polymerase/TFIIF complex, thereby enhancing its recruitment to the promoter.

1994; Fig. 5.12). Hence activators appear to target both TFIID and TFIIB in two ways. First, as described in the previous section, they enhance their binding to the promoter and secondly, they alter their conformation so as to enhance their activity (Fig. 5.13).

Together with TFIIB, TFIID constitutes a major target for transcriptional activators. Interestingly, however, other components of the basal complex such as TFIIA (Ozer *et al.*, 1994), TFIIF (Joliot *et al.*, 1995) and TFIIH (Xiao *et al.*, 1994) have also been shown to interact with transcriptional activators. Hence a number of different factors within the basal transcriptional complex serve as targets for direct interactions with transcriptional activators. It is clear, however, that in many cases, activators interact with the basal complex only indirectly via other factors and such interactions are discussed in the next section.

5.4 INTERACTION OF ACTIVATION DOMAINS WITH OTHER REGULATORY PROTEINS

5.4.1 THE MEDIATOR COMPLEX

As noted in section 5.3.1, the existence of the squelching phenomenon indicates that activators act by contacting a factor which is involved in the transcription of a wide range of genes. Although this could be a component of the basal transcriptional complex (see section 5.3) studies in yeast resulted in the purification of a multi-protein complex (distinct from the basal transcriptional

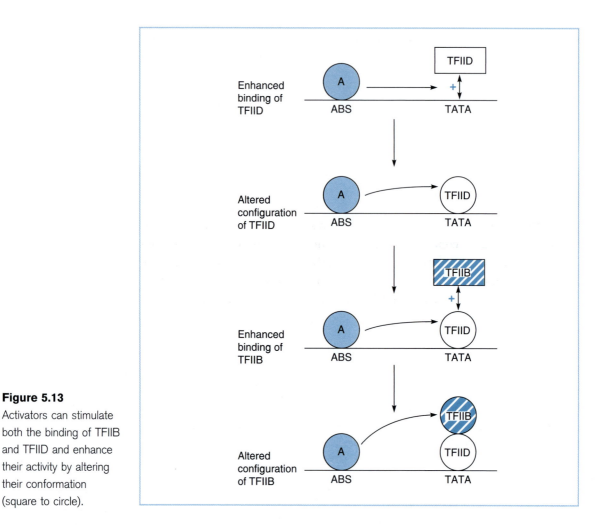

Figure 5.13

Activators can stimulate both the binding of TFIIB and TFIID and enhance their activity by altering their conformation (square to circle).

complex) which could prevent squelching when added in excess. This so-called 'mediator' complex therefore represents a target for transcriptional activators which is present in limiting amounts so that activators compete for it. Hence, its addition in excess relieves this competition and prevents squelching.

The mediator complex consists of over twenty proteins and, following its original identification in yeast, has now been found in a wide range of multi-cellular organisms including humans. It therefore appears to be a conserved component of the transcriptional machinery involved in activation of a wide range of genes (for reviews see Malik and Roeder, 2000; Myers and Kornberg, 2000; Boube *et al.*, 2002).

As well as interacting with activators, the mediator also interacts with RNA polymerase II itself. Indeed, the mediator is part of the RNA polymerase holoenzyme discussed in Chapter 3 (section 3.5.2) which therefore consists not only of RNA polymerase II, basal factors such as TFIIB, TFIIE, TFIIF and TFIIH and a chromatin remodelling activity, but also contains the mediator complex. Hence, the mediator serves as a bridge by which activating signals are transmitted from DNA-bound transcriptional activators to RNA polymerase II (Fig. 5.14). Indeed, structural studies suggest that the mediator partially envelops the polymerase, allowing it to receive signals from transcriptional activators and transmit them to the polymerase (Asturias *et al.*, 1999) (Fig. 5.15).

Interestingly, the mediator has been shown to contact the C-terminal domain of RNA polymerase II. Hence, the involvement of this motif in activation of the polymerase, which was discussed in section 5.3.3, can be accounted for by the mediator contacting this motif and transmitting the signal from transcriptional activators. Indeed, it appears that one of the roles of the mediator is to stimulate TFIIH to phosphorylate the C-terminal domain of RNA polymerase II which, as discussed in Chapter 3 (sections 3.1 and 3.5.1), is necessary for it to begin transcribing the gene.

5.4.2 TAFS

As described in Chapter 3 (section 3.6), TFIID consists of the TBP protein which binds to the TATA box and a number of other proteins known as TAFs

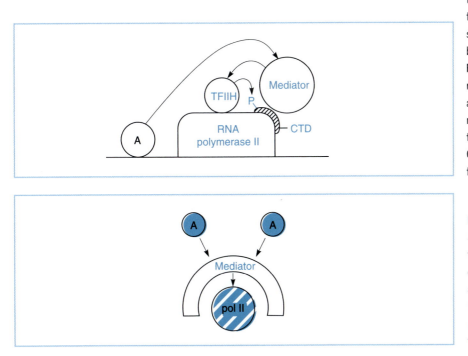

Figure 5.14

The mediator binds to the C-terminal domain (CTD) of RNA polymerase II and thereby acts as a bridge transmitting the activating signal between DNA-binding activators and RNA polymerase. One mechanism for such activation involves the mediator inducing TFIIH to phosphorylate the CTD, thereby stimulating transcription.

Figure 5.15

Structural studies suggest that the mediator partially envelops the polymerase allowing it to serve as a bridge between the polymerase and transcriptional activators.

(TBP-associated factors). In some cases where activators interact with TFIID, such interactions can be reproduced with purified TBP. Moreover, mutations in specific acidic activators which interfere with their ability to interact with TBP also abolish their ability to activate transcription, indicating an important functional role for these interactions.

Although there is thus evidence that the ability to interact with TBP appears to be essential for transcriptional activation in some cases (Fig. 5.16a), there is also evidence that in some circumstances such activation requires interaction of the activator with the TAFs rather than with TBP. Thus, in many cases, stimulation of transcription *in vivo* by activator molecules does not occur with purified TBP but is dependent upon the presence of the TFIID complex and hence of the TAFs. This suggests a model in which the interaction of activators with TBP occurs indirectly via TAFs with the TAFs being co-activator molecules linking the activators with the basal transcriptional complex (Fig. 5.16b) (for reviews see Hahn, 1998; Green, 2000).

Interestingly, there is evidence that different classes of activation domain may interact with different TAFs (Chen *et al.*, 1994). Thus, while acidic activation domains have been shown to interact directly with $TAF_{II}31$ (also known as $TAF_{II}40$), the glutamine-rich domain of Sp1 interacts with $TAF_{II}110$, while multiple activators including proline-rich activators target $TAF_{II}55$. Hence different types of activation domains may have different targets within the TFIID complex (Fig. 5.17).

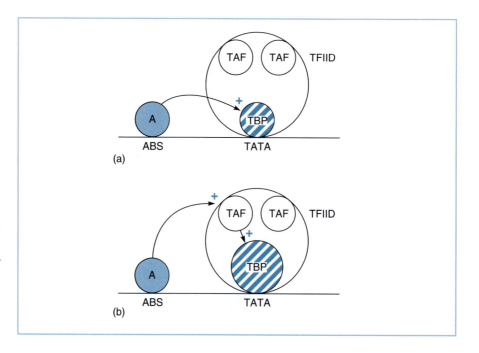

Figure 5.16

Interaction of an activator molecule with TBP can occur either directly (a) or indirectly (b) via an intermediate TBP-associated adaptor molecule (TAF).

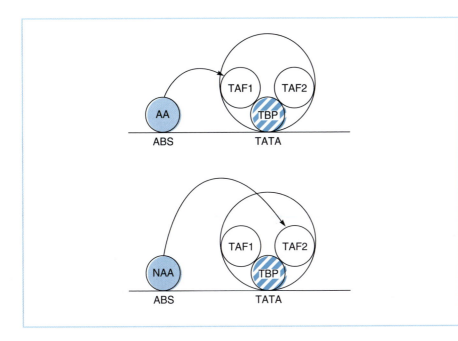

Figure 5.17
Acidic (AA) and non-acidic (NAA) activator molecules may interact with different TBP associated factors (TAFS) within the TFIID complex.

In agreement with this idea, the acidic activation domain of VP16 is not capable of squelching gene activation by the non-acidic activation domain of the oestrogen receptor, whereas the oestrogen receptor activation domain is capable of squelching gene activation mediated both by its own activation domain and by the acidic domain of VP16 indicating that they contact different molecules. Moreover, these findings suggest that a series of TAFs within TFIID may mediate activation, with the acidic activation domain of VP16 contacting a factor which is located earlier in the series than that contacted by the non-acidic activation domain of the oestrogen receptor (Fig. 5.18). Hence the factor contacted by the activation domain of the oestrogen receptor would also be essential for activation by VP16 (factor 4 in Fig. 5.18), whereas the factor contacted by the acidic activation domain of VP16 (factor 1 in Fig. 5.18) would not be required for activation by the oestrogen receptor.

The functional differences that exist between different factors in their ability to activate transcription from different positions and in different species (see section 5.2.4) are therefore paralleled by differences in their ability to interact with different TAFs. This ability of different activation domains to interact with different TAFs can produce a strong synergistic activation of transcription which is far stronger than the sum of that observed with either activation domain alone. Thus, the ability of different activators to bind to different TAFs in the TFIID complex would result in greatly enhanced recruitment of TFIID compared to the effect of either activator alone (Fig. 5.19) (for review see Buratowski, 1995).

Figure 5.18

Interaction of different activator molecules with different adaptor molecules (1–4) which each activate each other and ultimately activate TBP. Note that the ability of the non-acidic activation domain of the oestrogen receptor to squelch activation by the acidic activation domain of VP16 but not vice versa can be explained if the oestrogen receptor interacts with an adaptor molecule (4) closer to TBP in the series than that with which VP16 interacts (1).

These findings thus suggest that the TAFs are of importance for transcriptional activation and mediate some of the interactions between activators and TFIID which were described in section 5.3. However, it is clear that their importance varies between different species and on different promoters. Thus, while TAFs appear to be of central importance in transcriptional activation in higher eukaryotes such as humans and *Drosophila*, they are not essential for transcriptional activation at most promoters in yeast (Kuras *et al.*, 2000; Li *et al.*, 2000). Similarly, even in higher eukaryotes, specific TAFs appear to be of key importance at particular types of promoters. Thus mutation of $TAF_{II}250$ inhibits the expression of specific genes and results in cell cycle arrest in mammalian cells without affecting the transcription of other genes (Wang and Tjian, 1994).

This idea that particular TAFs may play a critical role in mediating the response to activators at specific genes, has been extended by findings suggesting that TAFs also function in promoter selectivity. Thus it appears that

Figure 5.19

The ability of different activators (A1 and A2) to interact with different TAFs will result in a strong synergistic enhancement of TFIID recruitment and hence of transcriptional activation.

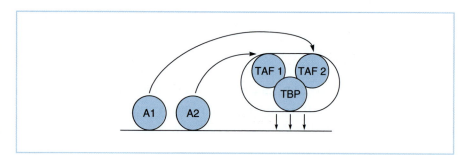

TFIID complexes containing particular TAFs assemble preferentially at particular promoters. This effect may be mediated by particular TAFs binding preferentially to particular core promoters (see Chapter 1, section 1.3.1) containing different sequences between the TATA box and the start site of transcription (Fig. 5.20). Thus, as noted above, most yeast genes do not require TAFs for the activation of transcription. However, a few genes involved in cell cycle progression, such as the cyclin genes, have been shown to be dependent upon TAF$_{II}$145 for their transcription. This dependence upon TAF$_{II}$145 is not due to the nature of the activator sequences in the promoter but is dependent upon the nature of the core promoter (Shen and Green, 1997) (Fig. 5.21). Although the yeast promoters used in this study contain a TATA box, the ability of TAFs to interact with specific core promoter sequences may be of particular importance on promoters lacking a TATA box and containing an initiator element where, as discussed in Chapter 3 (section 3.6) TBP is brought to the promoter by factors binding to the initiator element rather than by TBP binding to the TATA box.

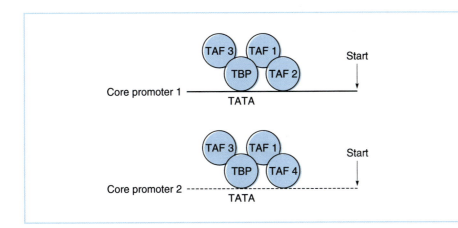

Figure 5.20
TFIID complexes containing different TAFs bind preferentially to different core promoters containing different sequences between the TATA box and the transcriptional start site.

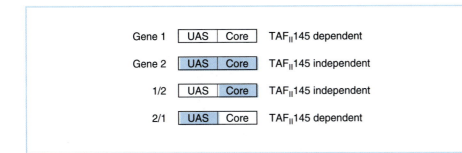

Figure 5.21
The dependence of particular yeast promoters on TAF$_{II}$145 for transcription is determined by the nature of the core promoter not by the upstream activator binding sites (UAS).

Thus, particular TFIID complexes containing specific combinations of TAFs may bind selectively to specific promoters rather than only responding to transcriptional activators following binding. This idea has been supported by the finding of a cell type specific form of TAF$_{II}$130, known as TAF$_{II}$105 which is expressed only in B lymphocytes (Dikstein *et al.*, 1996). Hence different forms of TFIID containing different TAFs may exist in different tissues and may thus play a role in the cell type specific regulation of gene expression (for review see Verrijzer, 2001) (Fig. 5.22). This is reinforced by the finding (discussed in Chapter 3, section 3.2.6) of TBP-like factors which are expressed in specific cell types.

Obviously, the different TFIID complexes formed in this manner may also differ in their responses to different transcriptional activators. Thus, for example, TAF$_{II}$30 which mediates transcriptional activation by the oestrogen receptor is found in only some TFIID complexes. In others it is replaced by TAF$_{II}$18 which does not mediate activation by the receptor (for review see Chang and Jaehning, 1997). Therefore, the ability of an activator to stimulate transcription may depend not only on its pattern of synthesis or activation (see Chapters 7 and 8) but also on its ability to interact with different TAFs or with TBP and TBP-like factors.

Hence the TAF factors play a key role in transcription, by acting as co-activators mediating the response to specific activators and by regulating the

Figure 5.22
The ability of an activator (A) to stimulate transcription may be controlled by the expression pattern of the TAFs with which it interacts. Hence, an activator which interacts with a tissue-specific TAF will produce tissue-specific gene transcription, even if the activator itself is ubiquitously expressed.

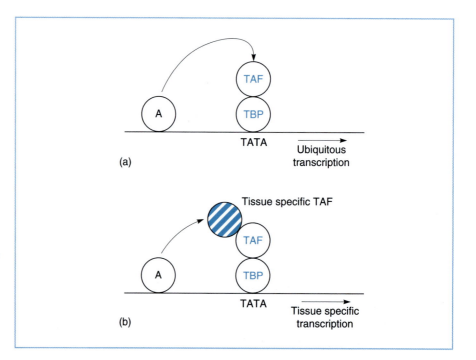

binding of TFIID to specific promoters containing particular sequences adjacent to the TATA box (Fig. 5.23). This ability of the TAFs to act as an intermediate between the basal transcriptional complex and transcriptional activators evidently parallels the role of the mediator complex which acts as an intermediate between activators and the RNA polymerase itself within the RNA polymerase holoenzyme complex (see section 5.4.1).

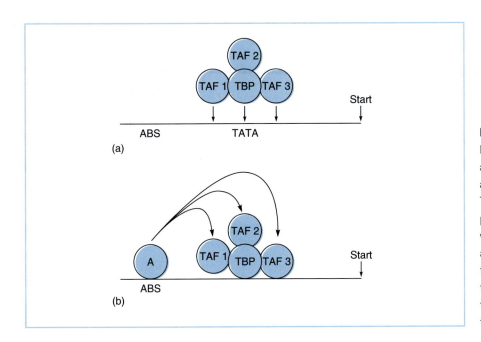

Figure 5.23

Mechanisms of TAF action. (a) The TAFs may act to enhance binding of TFIID to specific promoters by interacting with DNA sequences adjacent to the TATA box to which TBP binds; (b) the TAFs can mediate the response of TFIID to transcriptional activators.

5.4.3 CBP AND OTHER CO-ACTIVATORS

In addition to factors such as the TAFs and the mediator, which were originally defined via their association with the basal transcriptional complex, other co-activators exist which were originally defined on the basis of their essential role in transcriptional activation mediated by a specific transcriptional activator. Thus, cyclic AMP inducible genes contain a short sequence in their regulatory regions which can confer responsiveness to cyclic AMP when it is transferred to another gene that is not normally cyclic AMP inducible. This sequence, which is known as the cyclic AMP response element (CRE), consists of the eight base pair palindromic sequence TGACGTCA.

The first transcription factor that was shown to bind to this site was a 43 kilo-dalton protein which was named CREB (cyclic AMP response element binding protein). This factor has a basic DNA binding domain with adjacent leucine zipper dimerization motif (Fig. 5.24) (see Chapter 4, section 4.5 for

Figure 5.24

Structure of the CREB transcription factor indicating the glutamine-rich activation domains (Q_1 and Q_2), the phosphorylation box (P) containing the serine 133 residue, and the basic DNA binding domain (BD) with associated leucine zipper (LZ).

further discussion of this motif) and binds to the palindromic CRE as a dimer with each CREB monomer binding to one half of the palindrome (for review of CREB see Shaywitz and Greenberg, 1999; de Cesare and Sassone-Corsi, 2000).

The CREB factor plays a key role in the activation of gene expression via the CRE following cyclic AMP treatment. The CREB factor is present in cells in an inactive form prior to exposure to the activating stimulus. Moreover, CREB is actually bound to the CRE prior to exposure to cyclic AMP but this DNA bound CREB does not activate transcription. Elevated levels of cyclic AMP result in the activation of the protein kinase A enzyme which, in turn, phosphorylates CREB on the serine amino acid at position 133 in the molecule. This serine residue is located in a region of CREB known as the phosphorylation box (P-box), which is flanked on either side by regions rich in glutamine amino acids which act as transcriptional activation domains (see section 5.2) (Fig. 5.24). The phosphorylation of CREB on serine 133 results in a change in the structure of the molecule which now allows it to activate transcription (Fig. 5.25).

To identify the mechanism of this effect, Chrivia *et al.* (1993) screened a cDNA expression library with CREB protein phosphorylated on serine 133 to identify proteins which interact with phosphorylated CREB. This resulted in the isolation of cDNA clones encoding CBP (CREB binding protein). CBP is a

Figure 5.25

Activation of the CREB factor by cyclic AMP-induced phosphorylation. The ability of DNA-bound CREB to activate transcription is produced by the cyclic AMP dependent activation of protein kinase A which phosphorylates the CREB protein resulting in its activation.

265 kilo-dalton protein which associates only with phosphorylated CREB and not with the unphosphorylated form (for review see Shikama *et al.*, 1997; Giordano and Avantaggiati, 1999; Goodman and Smolik, 2000). This pattern of association immediately suggests that CBP plays a critical role in the ability of CREB to activate transcription only after phosphorylation. In agreement with this, injection of cells with antibodies to CBP prevents gene activation in response to cyclic AMP, indicating that CBP is essential for this effect. Hence, CBP is a co-activator molecule whose binding to phosphorylated CREB is essential for transcriptional activation to occur (Fig. 5.26).

Although the CBP factor was originally defined as a co-activator essential for cyclic AMP stimulated transcription mediated via the CREB factor, it was subsequently shown that CBP and its close relative p300 are essential co-activators for a vast range of other factors such as the nuclear receptors (Chapter 4, section 4.4), MyoD (Chapter 7, section 7.2.1), AP1 (Chapter 9, section 9.3.1, p53 (Chapter 9, section 9.4.2) and a number of others (for review see Shikama *et al.*, 1997; Giordano and Avantaggiati, 1999; Goodman and Smolik, 2001) (Fig. 5.27).

Figure 5.26
The phosphorylation of CREB on serine 133 allows it to bind the CBP co-activator which then stimulates transcription.

Figure 5.27
Some transcription factors which interact with the CBP/p300 co-activators and the signalling pathways which activate them.

This ability of CBP and p300 to interact with a vast array of transcription factors places them at the centre of a whole range of signalling pathways in the cell and they thus play a critical role in gene activation via these pathways. The relatively low abundance of CBP/p300 in the cell means that different signalling pathways compete for them and results in mutual antagonism between different competing pathways, such as the inflammation mediated by the AP1 pathway and the anti-inflammatory effects of glucocorticoids (see Chapter 6, section 6.5) or the growth promoting effects of the AP1 pathway compared to the growth arresting effects of the p53 pathway (see Chapter 9, section 9.4.2). Interestingly, the activation domain of CREB undergoes a structural transition from a coiled structure to form two α-helices when it interacts with CBP (Radhakrishnan *et al.*, 1997). This evidently parallels the change in the activation domain of VP16 when it interacts with $TAF_{II}31$ (see section 5.2.1) suggesting that the formation of a specific helical structure may be a general feature which occurs when many activation domains interact with their targets.

Although the p300/CBP proteins are the best defined co-activators, other co-activators have also been defined on the basis of their association with particular activators. Thus, for example, the nuclear receptors discussed in Chapter 4 (section 4.4) interact not only with CBP but also with a range of other co-activators such as TIF1, TIF2, SRC-1 and Sug1 (for review see Rosenfeld and Glass, 2001; McKenna and O'Malley, 2002). Moreover, several of these co-activators associate with the receptors only after they have been activated by binding their ligand, indicating that they are likely to play a key role in the ability of the receptors to activate transcription only following ligand binding (see Chapter 8, section 8.2.2 for a discussion of the mechanisms producing ligand-dependent activation of the nuclear receptors).

The key role of CBP/p300 and other co-activators obviously leads to the question of how they act. Two possible mechanisms by which CBP/p300 achieve their effects have been described. Thus, CBP/p300 have been shown to interact via a protein–protein interaction with several components of the basal transcriptional complex such as TFIIB (see Chapter 3, section 3.2.4) and have been identified as part of the RNA polymerase II holoenzyme complex (which also contains RNA polymerase II, components of the basal transcriptional complex and other regulatory proteins) (Nakajima *et al.*, 1997). Hence, like the TAFs, CBP/p300 may serve as a bridge between CREB and the basal transcriptional complex either interacting with components of the complex to enhance their activity or serving to recruit the RNA polymerase holoenzyme to the DNA by the CBP component binding to CREB (Fig. 5.28).

As well as this mechanism, however, it is also possible that CBP acts via a mechanism involving alterations in chromatin structure. Thus, several co-activators such as CBP/p300 and SRC-1 have been shown to have histone

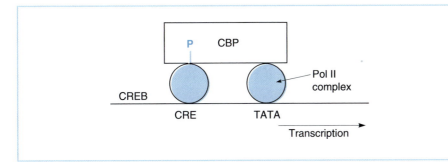

Figure 5.28
CBP can bind to both CREB and the basal transcriptional complex. It may therefore act as a bridge between CREB and the complex allowing transcriptional activation to occur.

acetyltransferase activity (Ogryzko *et al.*, 1996). As discussed in Chapter 1 (section 1.2.3), acetylated histones are associated with the more open chromatin structure that is required for transcription. Hence, the binding of CBP to CREB which recruits it to DNA may then result in the acetylation of histones leading to a chromatin structure compatible with transcription (Fig. 5.29).

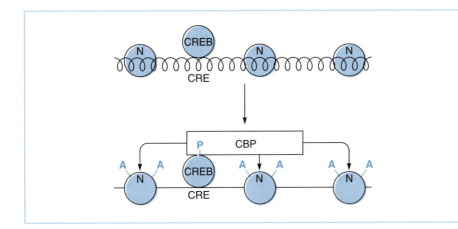

Figure 5.29
CBP has histone acetyltransferase activity. Therefore following binding to phosphorylated CREB it can acetylate (A) histones within the nucleosome (N) resulting in a more open chromatin structure (wavy versus solid line) compatible with transcription.

Similar histone acetyltransferase activity is also observed for the $TAF_{II}250$ component of TFIID (see Chapter 3, section 3.2.5) indicating that some TAFs may also act via the alteration of chromatin structure (for review see Struhl and Moqtaderi, 1998; Brown *et al.*, 2000). Such activation of transcription via changes in chromatin structure is discussed further in section 5.5.

5.4.4 A MULTITUDE OF TARGETS FOR TRANSCRIPTIONAL ACTIVATORS

There thus exists an array of target factors which are contacted by transcriptional activators and these include components of the basal complex such as TFIIB and TBP as well as the mediator, various TAFs and other co-activators, with some factors being contacted by activators of all classes and others by activators of only one class. Even when the finding that some of these targets, such as individual TAFs, can interact with only one class of activation domain is taken into account, there still remains a bewildering number of targets within the basal complex. Thus, for example, in the most extreme case described so far, the acidic activation domain of VP16 has been reported to interact with TFIIB, TFIIH, TBP, TAF$_{II}$40, TAF$_{II}$31 and the RNA polymerase holoenzyme (for review see Chang and Jaehning, 1997). Moreover, although these interactions of VP16 were originally defined in the test tube, several of them have recently been confirmed in the intact cell (Hall and Struhl, 2002) using the ChIP assays described in Chapter 2 (section 2.4.3). It should be noted, however, that the various possible targets are not mutually exclusive. Indeed, the ability of different molecules of the same factor or different activating factors to interact with different components within the basal transcriptional complex is likely to be essential for the strong enhancement of transcription which is the fundamental aim of activating molecules (Fig. 5.30) (for review see Carey, 1998).

Figure 5.30
The ability of multiple activating molecules (A) to contact different factors allows strong activation of transcription.

5.5 OTHER TARGETS FOR TRANSCRIPTIONAL ACTIVATORS

Although the basal transcriptional complex which initiates transcription is the best characterized target for transcriptional activators and co-activators (as discussed in the preceding sections) at least two other stages of the transcriptional process can be targeted by such activators and these will be discussed in turn.

5.5.1 MODULATION OF CHROMATIN STRUCTURE

As discussed in Chapter 1 (section 1.2), the DNA molecule is associated with histones and other proteins to form particles known as nucleosomes which are the basic unit of chromatin structure. Prior to the onset of transcription, the chromatin structure becomes altered thus allowing the subsequent binding of the factors which actually stimulate transcription. This alteration in chromatin structure can itself be produced by the binding of a specific transcription factor. This results in a change in the nucleosome pattern of DNA/ histone association thereby allowing other activating factors access to their specific DNA binding sites (Fig. 5.31).

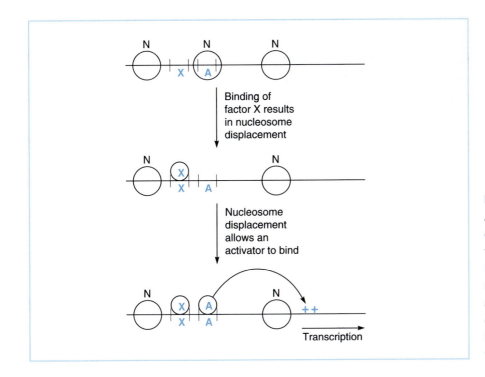

Figure 5.31

A transcription factor (X) can stimulate transcription by binding to DNA and displacing a nucleosome (N) so allowing a constitutively expressed activator (A) to bind and activate transcription.

Thus, as discussed in Chapter 1 (section 1.3.3), genes whose transcription is induced by elevated temperature share a common DNA sequence which, when transferred to another gene, can render the second gene heat inducible. This sequence is known as the heat shock element (HSE). The manner in which a *Drosophila* HSE, when introduced into mammalian cells, functioned at the mammalian rather than the *Drosophila* heat shock temperature suggested that this sequence acted by binding a protein rather than by acting directly as a thermosensor (see Chapter 1, Fig. 1.8).

Direct evidence that this was the case, was provided by studying the proteins bound to the promoters of the hsp genes before and after heat shock. Thus, prior to heat shock, the TFIID complex (see Chapter 3, sections 3.5 and 3.6) is bound to the TATA box and another transcription factor known as GAGA is bound upstream (Fig. 5.32a); (Wu, 1985; Tsukiyama *et al.*, 1994).

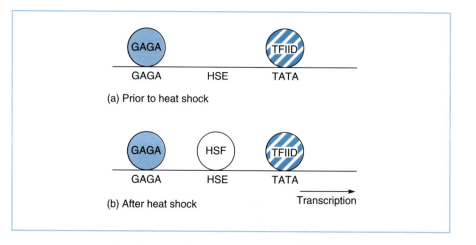

Figure 5.32

Proteins binding to the
promoter of the hsp70
gene before (a) and after
(b) heat shock.

Following heat-shock, however, an additional factor is observed which is
bound to the HSE (Fig. 5.32b) and it is this heat shock factor (HSF) which
produces activation of the genes in response to the stimulus of elevated
temperature.

However, prior to heat shock, the heat shock genes are poised for transcription. Thus, while the bulk of cellular DNA is associated with histone
proteins to form a tightly packed chromatin structure, the binding of the
GAGA factor to the heat shock gene promoters has resulted in the displacement of the histone-containing nucleosomes from the promoter region (for
review see Schumacher and Magnusson, 1997; Wilkins and Lis, 1997; Simon
and Tamkun, 2002). This opens up the chromatin and renders the promoter
region exquisitely sensitive to digestion with the enzyme DNAseI.

Although such a DNAseI hypersensitive site marks a gene as poised for
transcription (for review see Latchman, 2002), it is not in itself sufficient for
transcription. The binding of the GAGA factor thus opens up the gene and
renders it poised for transcription in response to a suitable stimulus. This
role for the GAGA factor in chromatin remodelling is not confined to the
heat shock genes. Thus, mutations in the gene encoding GAGA result in the
Drosophila mutant trithorax in which a number of homeobox genes (which
control the formation of the correct body plan – see Chapter 4, section 4.2)
are not converted from an inactive to an active chromatin state and are
hence not transcribed (for review see Shumacher and Magnusson, 1997;
Simon and Tamkun, 2002). This mutation thus produces a fly with an abnormal body pattern and thus has a similar effect to the brahma mutation in the
SWI 2 component of the SWI/SNF chromatin remodelling complex which
was discussed in Chapter 1 (section 1.2.2). Indeed, the GAGA factor has
been shown to be associated with a multi-protein complex known as

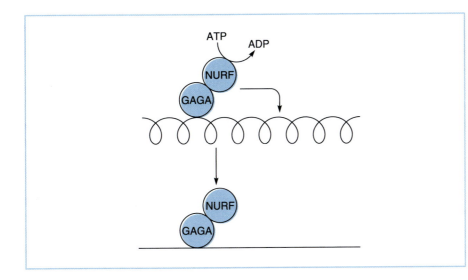

Figure 5.33

The GAGA factor can bind to DNA which is in a tightly packed chromatin structure (wavy line) and recruit the nucleosome remodelling factor (NURF). NURF then hydrolyses ATP and uses the energy to remodel the chromatin to a more open structure (solid line) to which other activating proteins can bind.

nucleosome remodelling factor (NURF) which, like SWI/SNF, can hydro-lyse ATP and alter chromatin structure (Fig. 5.33) (for review see Tsukiyama and Wu, 1997).

Hence, following binding of GAGA, the gene is in a state poised for the binding of an activating transcription factor which, in turn, will result in transcription of the gene. In the case of the heat shock genes, this is achieved following heat shock by the binding of the HSE to the HSF (Fig. 5.34). This factor then interacts with TFIID and other components of the basal transcription complex resulting in the activation of transcription. The manner in which

Figure 5.34

Activation of HSF by heat is followed by its binding to a pre-existing nucleosome-free region in the heat shock gene promoters which is marked by a DNAseI hypersensitive site and was produced by the prior binding of the GAGA factor. Binding of HSF then results in the activation of heat shock gene transcription.

HSF is activated in response to heat and can therefore mediate heat-inducible transcription is discussed in Chapter 8 (section 8.3.1).

A similar modulation of chromatin structure to that produced by the GAGA factor is also seen in the case of members of the steroid receptor family (see Chapter 4, section 4.4). In this case, the receptors are activated by treatment with the appropriate steroid and then bind to the DNA (see Chapter 8, section 8.2.2 for a discussion of the mechanism of this effect) allowing them to mediate steroid-inducible transcription.

In a number of cases, steroid hormone treatment has been shown to cause the induction of a DNAseI hypersensitive site located at the DNA sequence to which the receptor binds. Hence, the binding of the receptor may activate transcription by displacing or altering the structure of a nucleosome within the promoter of the gene creating the hypersensitive site. In turn this would facilitate the binding of other transcription factors necessary for gene activation whose binding sites would be exposed by the change in the position or structure of the nucleosome. These factors would be present in the cell in an active form prior to steroid treatment but could not bind to the gene because their binding sites were masked by a nucleosome (Fig. 5.35) (for review see Beato and Eisfeld, 1997). In agreement with this idea, the binding sites for TFIID and CTF/NFI in the glucocorticoid-responsive mouse mammary tumour virus promoter are occupied only following hormone treatment,

Figure 5.35

Binding of the steroid receptor-steroid hormone complex to the promoter of a steroid-inducible gene results in a change in chromatin structure creating a hypersensitive site and allowing pre-existing constitutively expressed transcription factors to bind and activate transcription. Note that the binding of these constitutive factors may occur because the receptor totally displaces a nucleosome from the DNA as shown in the diagram or because it alters the structure of the nucleosome so as to expose specific binding sites in the DNA.

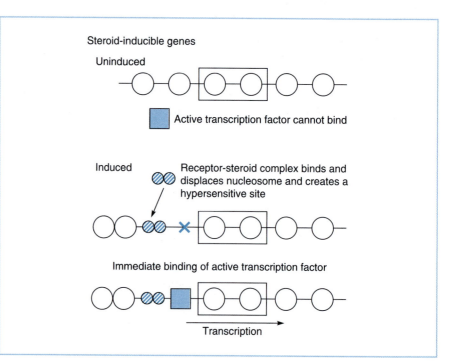

although these factors are present in an active DNA binding form at a similar level in treated and untreated cells.

Interestingly, the nuclear receptors may alter chromatin structure by both the mechanisms described in Chapter 1 (section 1.2). Thus, as discussed earlier (section 5.4.3) following ligand binding, the receptors bind co-activator molecules such as CBP, PCAF, SRC-1 and ACTR which are known to have histone acetyltransferase activity. Hence, the receptor-induced change in chromatin structure may be brought about by histone acetylation as discussed in Chapter 1, section 1.2.3. In addition, however, it appears that the glucocorticoid receptor can stimulate the activity of the SWI/SNF complex (Inoue *et al.*, 2002) allowing it to fulfill its role of hydrolysing ATP and unwinding chromatin (see Chapter 1, section 1.2.2) (Fig. 5.36).

This mechanism, in which the receptor acts by altering chromatin structure allowing constitutive factors access to their binding sites, is clearly in contrast to the binding of HSF to a promoter which already lacks a nucleosome and contains bound GAGA factor and TFIID. In this latter case, activation of transcription must occur not via alteration in chromatin structure but via interaction with the components of the constitutive transcriptional apparatus. It should be noted, however, that these two mechanisms are not exclusive. Thus, as discussed above (section 5.4.3) the CBP co-activator can also interact with components of the basal transcriptional complex to increase transcription. This finding indicates therefore that the steroid receptors and their associated co-activators such as CBP, promote transcription both by altering chromatin structure to allow constitutive factors to bind and also by interacting directly with other transcription factors such as components of the basal transcriptional complex (Fig. 5.37).

Activation by steroid hormones would therefore be a two stage-process involving first alteration of chromatin structure and secondly stimulation of

Figure 5.36

By binding histone acetyltransferases such as CBP and chromatin remodelling factors such as SWI/SNF, the steroid receptors can alter chromatin structure from a tightly packed (wavy line) to a more open (solid line) configuration.

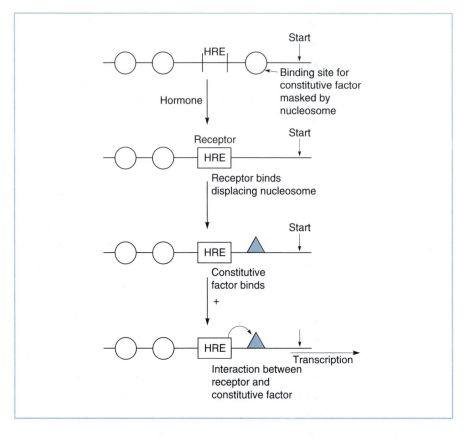

Figure 5.37

Activation of steroid-inducible genes by steroid receptors. As well as altering chromatin structure and allowing constitutive factors to bind, the hormone receptor and its associated co-factors are also able to interact directly with constitutive factors such as the basal transcriptional complex and directly activate transcription.

the basal transcriptional complex (Jenster *et al.*, 1997). In agreement with this idea, chromatin disruption following binding of the thyroid hormone receptor to DNA is necessary but not sufficient for transcriptional activation to occur (Wong *et al.*, 1997). Similarly, recruitment of specific multi-protein complexes by nuclear receptors can result in chromatin opening without transcription induction or both chromatin opening and transcriptional induction (King and Kingston, 2001).

Hence, in the case of the steroid receptors, a single factor and its associated co-factors can alter the chromatin structure and then activate transcription. In contrast, in heat shock gene activation, these functions are performed by separate factors with the GAGA factor displacing a nucleosome allowing HSE to bind and activate transcription following a subsequent heat shock.

Both cases illustrate, however, how factors which alter chromatin structure can prepare the way for the binding of the factors which actually activate transcription with such binding occurring either immediately following the change in chromatin structure as in the glucocorticoid receptor/NFI case or following a subsequent stimulus as in the GAGA/HSF case (Fig. 5.38).

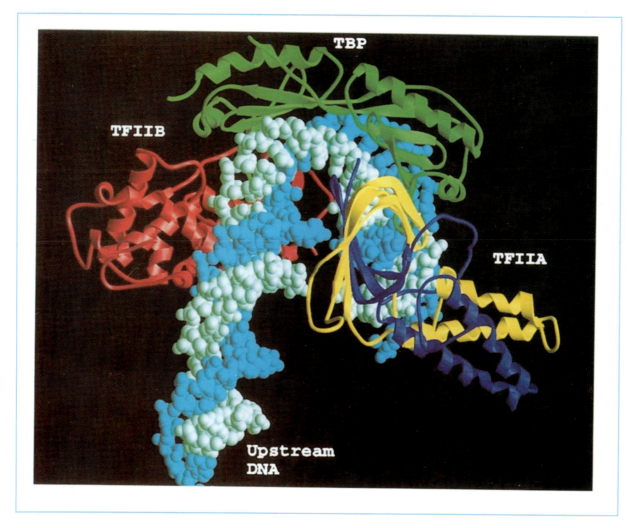

Plate 1 Schematic diagram illustrating the structure of the TFIIB/TBP/TFIIA complex bound to DNA. Note the bending of the DNA induced by TBP binding and the positions of TFIIB and TFIIA relative to TBP.

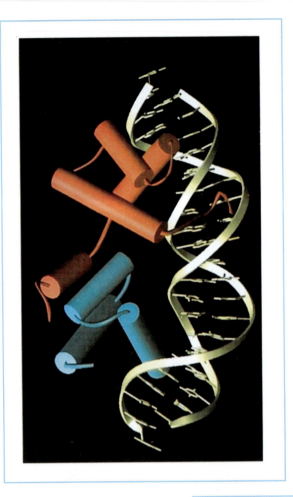

Plate 2 Binding of the a1 (blue)/$\alpha2$ (red) homeodomain heterodimer to DNA. α-helices are shown as cylinders. Note the three helical structure of the homeodomains of a1 and $\alpha2$ and the C-terminal region of $\alpha2$ which forms an additional α-helix in the presence of a1 and packs against the $\alpha2$ homeodomain forming the dimerization interface.

Plate 3 Structure of the $Cys_2 His_2$ zinc finger from Xfin. The Cys residues are shown in yellow and the His residues in dark blue.

Plate 4 Two views of the structure of the thyroid hormone receptor ligand binding domain (blue) with bound thyroid hormone ligand (red). Note that the ligand is completely buried in the interior of the protein.

Plate 5 Structure of the two Cys$_4$ zinc fingers in a single molecule of the glucocorticoid receptor. The first finger is shown in red and the second finger in green with the zinc atoms shown white.

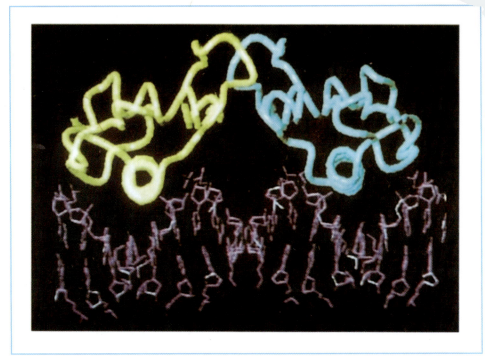

Plate 6 Structure of the oestrogen receptor dimer consisting of two receptor molecules bound to DNA. The two molecules of the receptor are shown green and blue respectively and the DNA is shown in purple.

(7a)   (7b)

Plate 7 Structure of (a) the RXRα receptor in the absence of ligand and (b) the closely related RARα receptor following binding of ligand (light blue atoms joined by white bonds). Note the structural change induced by the binding of ligand involving the movement of the H12 helix towards the ligand binding core so creating a sealed pocket in which the ligand is trapped.

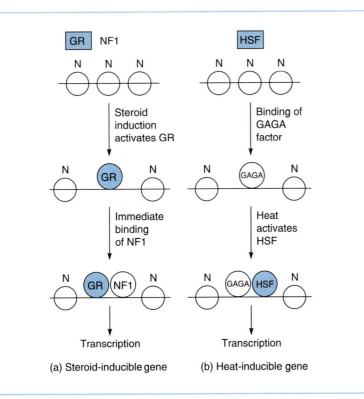

Figure 5.38

Two stage induction of genes by steroid treatment or heat shock involving alteration of chromatin structure and the subsequent binding of an activator molecule. In the steroid case (a), the steroid activates the glucocorticoid receptor (GR) resulting in a change in chromatin structure and immediate binding of the constitutively expressed NF1 factor. In contrast in the heat-inducible case, the chromatin structure has already been altered prior to heat shock by binding of the GAGA factor, but transcription only occurs when heat shock activates the heat shock factor (HSF) to a DNA binding form.

Interestingly, as well as interacting with the basal transcriptional apparatus, activation domains have also been shown to be involved in the ability of specific factors to alter chromatin structure. Thus, the activation of the yeast PHO5 gene promoter following phosphate starvation is mediated by the binding of the PHO4 factor to the PHO5 promoter resulting in nucleosome displacement. Surprisingly, a truncated PHO4 molecule lacking the acidic activation domain, but retaining the DNA binding domain, is incapable of nucleosome displacement while this ability can be restored by linking the truncated PHO4 molecule to the acidic activation domain of VP16 (Fig. 5.39). Hence the ability of PHO4 to disrupt chromatin structure is dependent upon the acidic activation domain which also interacts with the basal transcription complex to stimulate transcription (for reviews see Lohr, 1997; Svaren and Horz, 1997). This dual function is also seen in the case of the glucocorticoid receptor which, as well as altering chromatin structure thereby facilitating NF1 binding and consequent transcriptional activation, can also itself directly stimulate transcription via interaction with other transcription factors.

These findings indicate therefore that, the remodelling of chromatin structure can be achieved both by specific factors, such as the GAGA factor and by

Figure 5.39

Both the disruption of chromatin and the activation of the basal transcription complex by the yeast PHO4 factor are dependent on its acidic activation domain. They are therefore lost when this domain is deleted but can be restored by addition of the acidic activation domain of VP16.

the activation domains of other factors which can modulate chromatin structure as well as activate transcription directly by interacting with the basal transcriptional apparatus. In both these types of cases, the ability to alter chromatin structure is likely to depend upon the ability of these factors to recruit other factors which then actually alter chromatin structure either via recruitment of ATP-dependent chromatin remodelling complexes such as the SWI/SNF complex (Fig. 5.40a) or via recruiting factors with histone acetyl-transferase activity (Fig. 5.40b). It is clear therefore that the alteration of chromatin structure by specific factors is of considerable importance in the control of transcription.

5.5.2 STIMULATION OF TRANSCRIPTIONAL ELONGATION

In most genes, once transcription has been initiated, the RNA polymerase continues to transcribe the DNA until it has produced a complete RNA transcript. In some genes, however, some transcripts terminate prematurely and do not produce an RNA capable of encoding the appropriate protein. In such cases, activators could stimulate transcription at the level of transcriptional elongation by releasing the block to elongation and allowing full length transcripts to be produced (for review see Uptain *et al.*, 1997; Conaway *et al.*, 2000).

One case of this type involves the c-*myc* oncogene (see Chapter 9, section 9.3.3 for discussion of this oncogene). Thus, when the c-*myc* gene is transcribed in the pro-myeloid cell line HL-60 most transcripts terminate near the end of the first exon and do not produce a functional mRNA encoding the complete Myc protein. When the HL-60 cells are differentiated to form granulocytes, however, the majority of transcripts pass through this block and full length mRNA is produced (Fig. 5.41). Hence in this case an increased level of functional c-*myc* mRNA able to produce the Myc protein is obtained

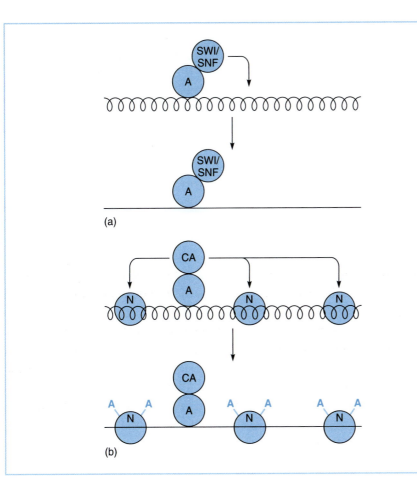

Figure 5.40

A transcriptional activator (A) can alter chromatin structure from a tightly packed (wavy line) to a more open (solid line) structure either (a) by recruiting the SWI/SNF chromatin remodelling complex or (b) by recruiting a co-activator (CA) with histone acetyltransferase activity. N = nucleosome.

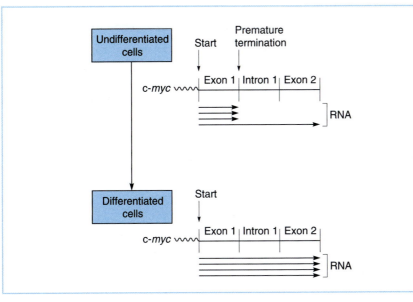

Figure 5.41

In undifferentiated HL-60 cells most transcripts from the c-myc gene terminate prematurely at the end of the first exon. In differentiated cells, however, this does not occur resulting in an increase in full length functional transcripts without increased initiation.

without an increase in transcriptional initiation (for review see Spencer and Groudine, 1990; Greenblatt *et al.*, 1993).

Regulation at the level of transcriptional elongation has also been demonstrated in the human immunodeficiency virus (HIV). Thus, in this case only short prematurely truncated transcripts are produced from the HIV promoter in the absence of the viral Tat protein. When Tat is present, however, it stimulates both the rate of initiation of transcription and also the proportion of full length transcripts which are produced so overcoming the block to transcriptional elongation (Fig. 5.42) (for review see Greenblatt *et al.*, 1993; Jones, 1997). A similar role for HSF in stimulating transcriptional elongation in the hsp70 gene has also been proposed (for review see Lis and Wu, 1993). Hence activating factors can act to stimulate the proportion of full length RNA transcripts which are produced.

It has been shown that Tat acts to stimulate elongation by recruiting the CDK9 kinase enzyme which phosphorylates the C-terminal domain (CTD) of RNA polymerase II (Kim *et al.*, 2002). As discussed in Chapter 3 (section 3.5.1), phosphorylation of this C-terminal domain is critical for transcriptional elongation by the polymerase after it has bound to the basal transcriptional complex. Hence, stimulation of phosphorylation in this way will enhance the production of full length transcripts (Fig. 5.43).

Such an effect on transcriptional elongation via the CTD of RNA polymerase II is not confined to the viral Tat protein but is observed also for cellular proteins. Thus, as discussed in section 5.4.1, the mediator complex stimulates the ability of TFIIH to stimulate the CTD, indicating that the mediator can transmit signals mediating enhanced transcriptional elongation as well as initiation.

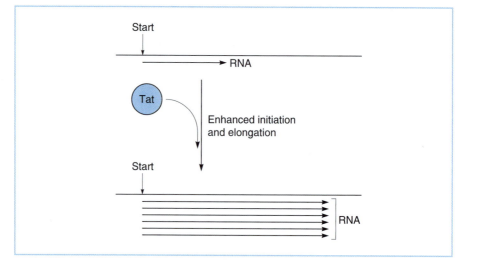

Figure 5.42

The human immunodeficiency virus Tat protein stimulates transcriptional initiation so that more RNA is made and also enhances the proportion of full length RNA species capable of encoding viral proteins.

Figure 5.43

The HIV Tat protein can recruit a cellular kinase which phosphorylates the C-terminal domain (CTD) of RNA polymerase allowing transcription to proceed.

Interestingly, as in the case of transcriptional initiation and the disruption of chromatin, there is evidence that the stimulation of transcriptional elongation involves activation domains. Thus when binding sites for the yeast transcriptional activator GAL4 were placed upstream of the c-*myc* promoter, both the rate of initiation and the proportion of full length transcripts were greatly stimulated by the binding of hybrid transcription factors containing the DNA binding domain of GAL4 linked to an acidic or non-acidic activation domain (Yankulov *et al.*, 1994). This effect was not observed in the presence of the GAL4 DNA binding domain alone or when the GAL4 binding sites were deleted (Fig. 5.44). Hence the ability of activating factors to act at the level of transcriptional elongation is dependent on the same activation domains which act to stimulate transcription at other stages. It is therefore possible for transcriptional activators to act by enhancing the proportion of full length

Figure 5.44

The binding of a chimaeric transcription factor containing the GAL4 DNA-binding domain linked to an activation domain (shaded) stimulates both transcriptional initiation and the proportion of full length transcripts produced by a c-myc gene carrying a DNA binding site for GAL4 (a). This effect is not observed when only the DNA binding domain of GAL4 is present (b), or when the GAL4 binding sites are deleted so that the activator cannot bind (c).

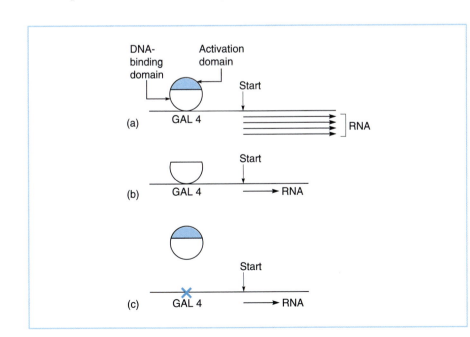

transcripts which are produced as well as by stimulating the number of transcripts which are initiated (for review see Bentley, 1995).

5.6 CONCLUSIONS

It is clear therefore that activation of gene expression by transcription factors can occur at three distinct stages to stimulate transcription. Thus activating factors can disrupt the chromatin structure to allow other activating factors to bind, stimulate the rate of transcriptional initiation so that more RNA transcripts are initiated and, in some genes, can increase the proportion of full length transcripts that are produced (Fig. 5.45).

Figure 5.45

An activation domain (hatched) within an activating transcription factor can act (a) by disrupting the nucleosome (N) arrangement of the DNA so that an activating factor can bind; (b) by stimulating the rate of transcriptional initiation; and (c) by increasing the proportion of initiated transcripts which go on to produce a full length RNA.

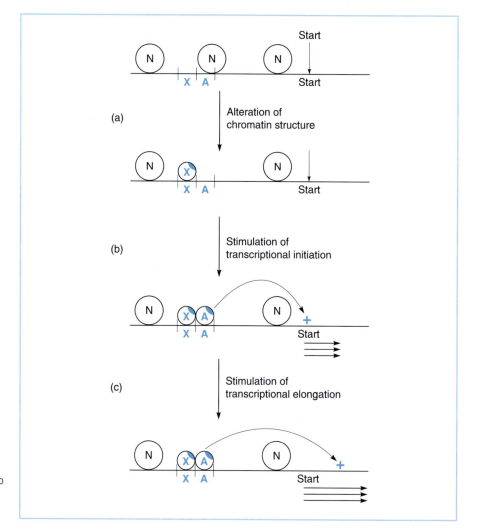

As described in section 5.5.1, these processes can be combined together in different ways, for example, in genes modulated by GAGA/HSF and the glucocorticoid receptor/NFI. An interesting example of this is provided by the activation of the interferon-β (IFN-β) gene by viral infection. Thus, as noted in Chapter 1 (section 1.3.6), the enhancer of the IFN-β gene binds a multi-protein enhanceosome complex. The binding of this complex (which includes the DNA binding protein HMGI(Y) and transcriptional activators such as NFκB) to the enhancer is the first stage in the activation of the IFN-β gene (for reviews see Cosma, 2002; Fry and Peterson, 2002) (Fig. 5.46a).

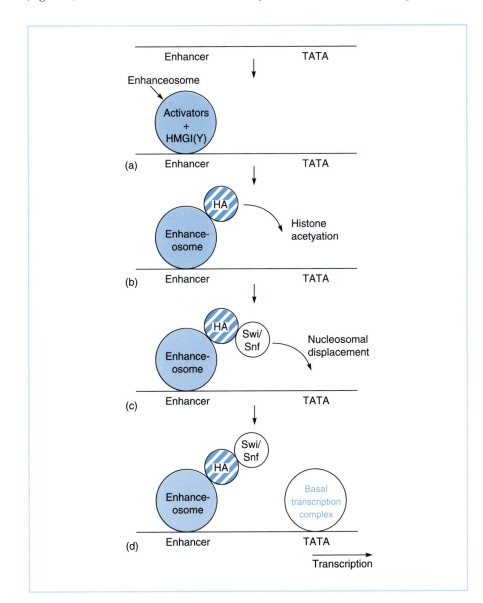

Figure 5.46

Multi-step activation of the interferon β promoter by viral infection. Binding of the enhanceosome multi-protein complex to the enhancer (panel a) is followed by successive recruitment of a histone acetylase (HA) (panel b), the SWI/SNF chromatin remodelling complex (panel c) and finally the recruitment of the basal transcriptional complex leading to transcriptional activation (panel d).

This enhanceosome complex then stimulates the recruitment of a histone acetyltransferase complex which acetylates nucleosomes (Fig. 5.46b), allowing the subsequent recruitment of the SWI/SNF complex (Fig. 5.46c). Further chromatin remodelling by this complex then displaces a nucleosome exposing the core promoter, allowing the basal transcriptional complex to bind and transcription begins (Fig. 5.46d).

Interestingly, simple displacement of the core promoter nucleosome by artificial means is not sufficient to induce transcriptional activation, indicating that this multi-step process is required for correct transcriptional regulation (Lomvardas and Thanos, 2002). Similarly, distinct histone acetylation events mediate subsequent stages of activation on this promoter. Thus, acetylation of the lysine at position 8 on histone H4 allows the recruitment of the SWI/SNF complex while subsequent acetylation of the lysines at position 9 and 14 on histone H3 allows the subsequent recruitment of TFIID (Agalioti et al., 2002) (Fig. 5.47). This case therefore illustrates how distinct parts of the 'histone code' (see Chapter 1, section 1.2.3) can mediate the recruitment of different activating factors, as well as how the two processes of histone modification and ATP-dependent chromatin remodelling discussed in Chapter 1 (section 1.2) interact with one another to produce transcriptional activation.

Hence, the ordered recruitment of chromatin remodelling factors, histone modifiers, transcriptional activators and the basal transcriptional complex is likely to be crucial for the correct regulation of a variety of different genes, although the order in which these factors are recruited is likely to vary for different genes (for review see Cosma, 2002; Fry and Peterson, 2002).

Figure 5.47

On the β-interferon gene promoter, acetylation (Ac) of histone H4 on the lysine (K) at position 8 facilitates recruitment of the chromatin remodelling complex SWI/SNF. Subsequent acetylation of histone H3 on the lysines at positions 9 and 14 facilitates recruitment of the basal transcription factor TFIID.

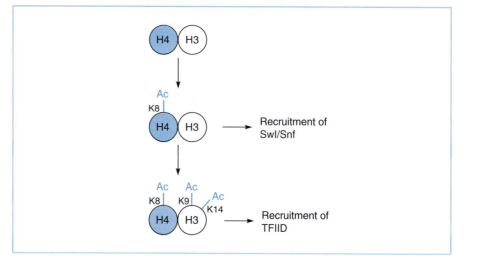

A variety of means are therefore used to activate gene transcription involving modulation of transcriptional initiation, transcriptional elongation and chromatin structure with a number of factors including components of the basal transcriptional complex, co-activators, the mediator complex and chromatin remodelling factors being targeted by transcriptional activators. Taken together these effects allow transcriptional activators to fulfil their function and strongly stimulate transcription.

REFERENCES

Agalioti, T., Chen, G. and Thanos, D. (2002) Deciphering the transcriptional histone acetylation code for a human gene. Cell 111, 381–392.

Asturias, F. J., Jiang, Y. W., Myers, L. C. *et al.* (1999) Conserved structures of mediator and RNA polymerase II holoenzyme. Science 283, 985–987.

Barberis, A., Pearlberg, J., Simkovich, N. *et al.* (1995) Contact with a component of the polymerase II holoenzyme suffices for gene activation. Cell 81, 359–368.

Beato, M. and Eisfeld, K. (1997) Transcription factor access to chromatin. Nucleic Acids Research, 25, 3559–3563.

Bentley, D. (1995) Regulation of transcriptional elongation by RNA polymerase II. Current Opinion in Genetics and Development 5, 210–216.

Boube, M., Joulia, L., Cribbs, D. L. and Bourbon, H-M. (2002) Evidence for a mediator of RNA polymerase II transcriptional regulation conserved from yeast to man. Cell 110, 143–151.

Brown, C. E., Lechner, T., Howe, L. *et al.* (2000) The many HATs of transcription coactivators. Trends in Biochemical Sciences 25, 15–19.

Buratowski, S. (1995) Mechanisms of gene activation. Science 270, 1773–1774.

Carey, M. (1998) The enhanceosome and transcriptional synergy. Cell 92, 5–8.

Chang, M. and Jaehning, J.A. (1997) A multiplicity of mediators: alternative forms of transcription complexes communicate with transcriptional regulators. Nucleic Acids Research.

Chen, J-L., Attardi, L.D., Verrijzer, C.P. *et al.* (1994) Assembly of recombinant TFIID reveals different co-activator requirements for different transcriptional activators. Cell 79, 93–105.

Choy, B. and Green, M.R. (1993) Eukaryotic activators function during multiple steps of pre-initiation complex assembly. Nature 366, 531–536.

Chrivia, J. C., Kwok, R. P. S., Lamb, N. *et al.* (1993) Phosphorylated CREB binds specifically to the nuclear protein CBP. Nature 365, 855–859.

Conaway, J. W., Shilatifard, A., Dvir, A. and Conaway, R. C. (2000) Control of elongation by RNA polymerase II. Trends in Biochemical Sciences 25, 375–380.

Cosma, M. P. (2002) Ordered recruitment: gene-specific mechanism of transcription activation. Molecular Cell 10, 227–236.

Courey, A.J. and Tjian, R. (1988) Analysis of Sp1 in vivo reveals multiple transcriptional domains including a novel glutamine-rich activation motif. Cell 55, 887–898.

de Cesare, D. and Sassone-Corsi, P. (2000) Transcriptional regulation by cyclic AMP-responsive factors. Progress in Nucleic Acids Research and Molecular Biology 64, 343–369.

Dikstein, R., Zhou, S. and Tjian, R. (1996) Human $TAF_{II}105$, is a cell type-specific TFIID subunit related to $TAF_{II}130$. Cell 87, 137–146.

Fry, C. J. and Peterson, C. L. (2002) Unlocking the gates to gene expression. Science 295, 1847–1848.

Gerber, H-P., Seipel, K., Georgiev, O. et al. (1994) Transcriptional activation modulated by homopolymeric glutamine and proline residues. Science 263, 808–811.

Giordano, A. and Avantaggiati, M. L. (1999) p300 and CBP: Partners for life and death. Journal of Cellular Physiology 181, 218–230.

Goding, C.R. and O'Hare, P. (1989) Herpes simplex virus Vmw65-octamer binding protein interaction: a paradigm for combinatorial control of transcription. Virology 173, 363–367.

Gonzalez-Couto, E.K., Klages, N. and Strubin, M. (1997) Synergistic and promoter-selective activation of transcription by recruitment of transcription factors TFIID and TFIIB. Proceedings of the National Academy of Sciences USA 94, 8036–8041.

Goodman, R. H. and Smolik, S. (2000) CBP/p300 in cell growth, transformation and development. Genes and Development 14, 1553–1577.

Green, M. R. (2000) TBP-associated factors ($TAF_{II}s$): multiple, selective transcriptional mediators in common complexes. Trends in Biochemical Sciences 25, 59–63.

Greenblatt, J., Nodwell, J.R. and Mason, S.W. (1993) Transcriptional anti-termination. Nature 364, 401–406.

Guarente, L. (1988) USAs and enhancers: common mechanism of transcriptional activation in yeast and mammals. Cell 52, 303–305.

Hahn, S. (1998) The role of TAFs in RNA polymerase II transcription. Cell 95, 579–582.

Hahn, S. (1993a) Structure (?) and function of acidic transcription activators. Cell 72, 481–483.

Hahn, S. (1993b) Efficiency in activation. Nature 363, 672–673.

Hall, D.B. and Struhl, K. (2002) The VP16 activation domain interacts with multiple transcriptional components as determined by protein–protein cross-linking in vivo. Journal of Biological Chemistry 277, 46043–46050.

Hollenberg, S.M. and Evans, R.M. (1988) Multiple and cooperative trans-activation domains of the human glucocorticoid receptor. Cell 55, 899–906.

Hope, I.A. and Struhl, K. (1986) Functional dissection of a eukaryotic transcriptional activator, GCN4 of yeast. Cell 46, 885–894.

Horikoshi, M., Carey, M.F., Kakidani, H. and Roeder, R.G. (1988) Mechanism of action of a yeast activator: direct effect of GAL4 derivatives on mammalian TFIID promoter interactions. Cell 54, 665–669.

Inoue, H., Furukawa, T., Giannakopoulos, S. et al. (2002) Largest subunits of the human SWI/SNF chromatin-remodelling complex promote transcriptional activation by steroid hormone receptors. Journal of Biological Chemistry 277, 41674–41685.

Jenster, G., Spencer, T.H., Burcin, M. et al. (1997) Steroid receptor induction of gene transcription: a two step model. Proceedings of the National Academy of Sciences, USA 94, 7879–7884.

Joliot, V., Demma, M. and Prywes, R. (1995) Interaction with RAP74 subunit of TFIIF is required for transcriptional activation by serum response factor. Nature 373, 632–635.

Jones, K. (1997) Taking a new TAK on Tat transactivation. Genes and Development 11, 2593–2599.

Kim, Y. K., Bourgeois, C. F., Isel, C. et al. (2002) Phosphorylation of the RNA polymerase II carboxyl-terminal domain by CDK9 is directly responsible for human immunodeficiency virus type 1 Tat-activated transcriptional elongation. Molecular and Cellular Biology 22, 4622–4637.

King, I. F. G. and Kingston, R. E. (2001) Specifying transcription. Nature 414, 858–861.

King, R.J.B. and Manwaring, W.I.P. (1974) Steroid Cell Interactions. Butterworths.

Kinzler, M., Braus, G.H., Georgiev, O. et al. (1994) Functional differences between mammalian transcriptional activation domains at the yeast GAL1 promoter. EMBO Journal 13, 641–645.

Kuras, L. and Struhl, K. (1999) Binding of TBP to promoters in vivo is stimulated by activators and requires Pol II holoenzyme. Nature 399, 609–613.

Kuras, L., Kosa, P., Mencia, M. and Struhl, K. (2000) TAF-containing and TAF-independent forms of transcriptionally active TBP in vivo. Science 288, 1244–1248.

Latchman, D. S. (2002) Gene regulation – a Eukaryotic Perspective, 4th edn. p. 323, Nelson Thornes Publishers.

Li, X-Y., Bhaumik, S. R. and Green, M. R. (2000) Distinct classes of yeast promoters revealed by differential TAF recruitment. Science 288, 1242–1244.

Li, X-Y., Virbasius, A., Zhu, X. and Green, M. R. (1999) Enhancement of TBP binding by activators and general transcription factors. Nature 399, 605–609.

Lis, J. and Wu, C. (1993) Protein traffic on the heat shock promoter: parking, stalling and trucking along. Cell 74, 1–4.

Lohr, D. (1997) Nucleosome transactions on the promoters of the yeast GAL and PHO genes. Journal of Biological Chemistry 272, 26795–26798.

Lomvardas, S. and Thanos, D. (2002) Modifying gene expression programs by altering core promoter chromatin architecture. Cell 110, 261–271.

Malik, S. and Roeder, R. G. (2000) Transcriptional regulation through mediator-like coactivators in yeast and metazoan cells. Trends in Biochemical Sciences 25, 277–283.

McKenna, N. J. and O'Malley, B. W. (2002) Combinational control of gene expression by nuclear receptors and coregulators. Cell 108, 465–474.

Mermod, N., O'Neil, E.A., Kelley, T.J. and Tjian, R. (1989) The proline-rich transcriptional activator of CTF/NF-1 is distinct from the replication and DNA binding domain. Cell 58, 741–753.

Mitchell, P.J. and Tjian, R. (1989) Transcriptional regulation in mammalian cells by sequence specific DNA binding proteins. Science 245, 371–378.

Myers, L. C. and Kornberg, R. D. (2000) Mediator of transcriptional regulation. Annual Reviews of Biochemistry 69, 729–749.

Nakajima, T., Fukamizu, A., Takahashi, J. et al. (1996) The signal-dependent coactivator CBP is a nuclear target for $PP90_{RSK}$. Cell 86, 465–474.

Ogryzko, V.V., Schiltz, L., Russanova, V. et al. (1996) The transcriptional coactivators p300 and CBP are histone acetyltransferases. Cell 87, 953–959.

Ozer, J., Moore, P.A., Bolden, A.H. et al. (1994) Molecular cloning of the small (α) sub-unit of human TFIIA reveals functions critical for activated transcription. Genes and Development 8, 2324.

Ptashne, M. (1988) How eukaryotic transcriptional activators work. Nature 335, 683–689.

Ptashne, M. and Gann, A. (1997) Transcriptional activation by recruitment. Nature 386, 569–577.

Radhakrishnan, I., Perez-Alvarado, G.C., Parker, D. et al. (1997) Solution structure of the KIX domain of CBP bound to the transactivation domain of CREB a model for activator: co-activator interactions. Cell 91, 741–752.

Roberts, S.G.E. and Green, M.R. (1994) Activator-induced conformational change in general transcription factor TFIIB. Nature 371, 717–720.

Rosenfeld, M. G. and Glass, C. K. (2001) Coregulator codes of transcriptional regulation by nuclear receptors. Journal of Biological Chemistry 276, 36865–36868.

Schumacher, A. and Magnuson, T. (1997) Murine polycomb- and trithorax-group genes regulate homeotic pathways and beyond. Trends in Genetics 13, 167–170.

Seipel, K., Georgiev, O. and Schaffner, W. (1992) Different activation domains stimulate transcription from the remote ('enhancer') and proximal ('promoter') positions. EMBO Journal 11, 4961–4968.

Shaywitz, A. J. and Greenberg, M. E. (1999) CREB: A stimulus-induced transcription factor activated by a diverse array of extracellular signals. Annual Reviews of Biochemistry 68, 821–861.

Shen, W-C. and Green, M.R. (1997) Yeast TAF$_{II}$145 functions as a core promoter selectivity factor, not a general co-activator. Cell 90, 615–624.

Shikama, N., Lyon, J. and La Thangue, N.B. (1997) The p300/CBP family: integrating signals with transcription factors and chromatin. Trends in Cell Biology 7, 230–236.

Simon, J.A. and Tamkun, J.W. (2002) Programming off and on states in chromatin: mechanisms of polycomb and trithorax group complexes. Current Opinion in Genetics and Development 12, 210–218.

Spencer, C.A. and Groudine, M. (1990) Transcriptional elongation and eukaryotic gene regulation. Oncogene 5, 777–785.

Struhl, K. and Moqtaderi, Z. (1998) The TAFs in the HAT. Cell 94, 1–4.

Svaren, J. and Horz, W. (1997) Transcription factors vs nucleosomes: regulation of the PH05 promoter in yeast. Trends in Biochemical Sciences 22, 93–97.

Triezenberg, S.J. (1995) Structure and function of transcriptional activation domains. Current Opinion in Genetics and Development 5, 190–196.

Tsukiyama, T. and Wu, C. (1997) Chromatin remodelling and transcription. Current Opinion in Genetics and Development 7, 182–191.

Tsukiyama, T., Becker, P.B. and Wu, C. (1994) ATP-dependant nucleosome disruption at a heat-shock promoter mediated by DNA binding of GAGA transcription factor. Nature 367, 525–532.

Uesugi, M., Nyanguile, O., La, H. et al. (1997) Induced α-helix in the VP16 activation domain upon binding to a human TAF. Science 277, 1310–1313.

Uptain, S. M., Kane, C. M. and Chamberlin, M. J. (1997) Basic mechanisms of transcript elongation and its regulation. Annual Reviews of Biochemistry 66, 117–172.

Verrijzer, C. P. (2001) Transcription factor IID – not so basal after all. Science 293, 2010–2011.

Wang, E.H. and Tjian, R. (1994) Promoter selective defect in cell cycle mutant ts13 rescued by hTAF$_{II}$250 in vitro. Science 263, 811–814.

Wilkins, R.C. and Lis, J.T. (1997) Dynamics of potentiation and activation: GAGA factor and its role in heat shock gene regulation. Nucleic Acids Research 25, 3963–3968.

Wong, J., Shi, Y-B. and Wolffe, A.P. (1997) Determinants of chromatin disruption and transcriptional regulation instigated by the thyroid hormone receptor: hormone regulated chromatin disruption is not sufficient for transcriptional activation. EMBO Journal 16, 3158–3171.

Wu, C. (1985) An exonuclease protection assay reveals heat shock element and TATA-box binding proteins in crude nuclear extracts. Nature 317, 84–87.

Xiao, H., Pearson, A., Coulombe, B. *et al.* (1994) Binding of basal transcription factor TFIIH to the acidic activation domains of VP16 and p53. Molecular and Cellular Biology 14, 7013–7024.

Yankulov, K., Blau, J., Purton, T. *et al.* (1994) Transcriptional elongation by RNA polymerase II is stimulated by transactivators. Cell 77, 748–759.

CHAPTER 6

REPRESSION OF GENE EXPRESSION BY TRANSCRIPTION FACTORS

6.1 REPRESSION OF TRANSCRIPTION

Although the majority of transcription factors that have so far been described act in a positive manner, a number of cases have now been reported in which a transcription factor exerts an inhibitory effect on transcription initiation. This effect can occur by indirect repression, in which the repressor interferes with the action of an activating factor so preventing it stimulating transcription (Fig. 6.1a–d). Alternatively, it can occur via direct repression in which the factor reduces the activity of the basal transcriptional complex (Fig. 6.1e). These two mechanisms will be discussed in sections 6.2 and 6.3 respectively. Other targets for transcriptional repression, such as the alteration of chromatin structure or the inhibition of transcriptional elongation, are discussed in section 6.4 (for reviews of transcriptional repression see Hanna-Rose and Hansen, 1996; Latchman, 1996; Maldonado *et al.*, 1999; Courey and Jia, 2001).

6.2 INDIRECT REPRESSION

Several mechanisms exist by which an inhibitor can interfere with the action of an activator and these will be discussed in turn.

6.2.1 INHIBITION OF ACTIVATOR BINDING BY MASKING OF ITS DNA BINDING SITE

One means by which repression can occur is by the masking of the DNA binding site for the factor so preventing it binding to the DNA and activating transcription. By preventing the binding of the positively acting factor, the negatively acting factor effectively inhibits gene activation. This masking of the binding site can be achieved simply by the negatively acting factor binding to the same site as the positively acting factor but failing to activate transcription (Fig. 6.1a). This is seen for example in the case of the Sp3 factor, a factor

Figure 6.1

Potential mechanisms by which a transcription factor can repress gene expression. This can occur by the repressor (R) binding to DNA and preventing an activator (A) from binding and activating gene expression (a), by the repressor interacting with the activator in solution and preventing its DNA binding (b), by the repressor binding to DNA with the activator and neutralizing its ability to activate gene expression (c), by the repressor promoting degradation of the activator (d) or by direct repression by an inhibitory transcription factor (e).

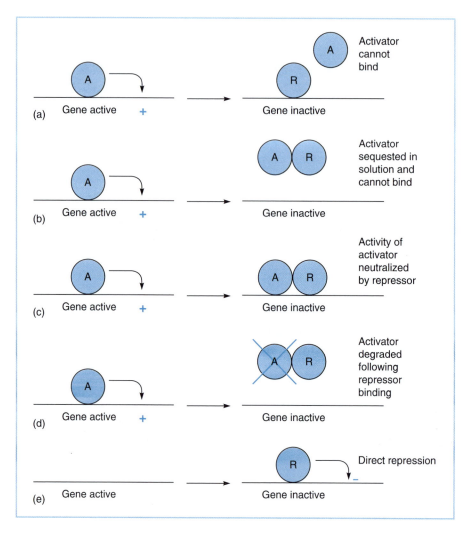

related to the Sp1 factor. Thus the Sp3 factor binds to the same Sp1 binding site as Sp1 itself (see Chapter 1, section 1.3.2) but unlike Sp1 it cannot activate transcription. It therefore blocks the Sp1 binding site, preventing Sp1 binding and activating transcription (for review see Lania *et al.*, 1997).

A similar example is seen in the case of the homeobox proteins, discussed in Chapter 4 (section 4.2), the proteins engrailed (eng), fushi-tarazu (Ftz), paired (prd) and zerknult (zen) can all bind to the sequence TCAATTAAAT (Hoey and Levine, 1988). When plasmids expressing each of these genes are co-transfected with a target promoter carrying multiple copies of this binding site, the Ftz, prd and zen proteins can activate transcription of the target promoter (Jaynes and O'Farrell, 1988; Han *et al.*, 1989). In contrast, the eng protein has no effect on the transcription of such a promoter. It does, however, interfere with the ability of the activating proteins to induce transcription, presumably by blocking the binding of the activating factor. Thus, for example, while Ftz can stimulate the target promoter when co-transfected with it, it cannot do so in the presence of eng since eng prevents binding of Ftz to its binding site (Jaynes and O'Farrell, 1988). Hence, the expression of Ftz alone in a cell would activate particular genes whereas its expression in a cell also expressing engrailed would not have any effect (Fig. 6.2).

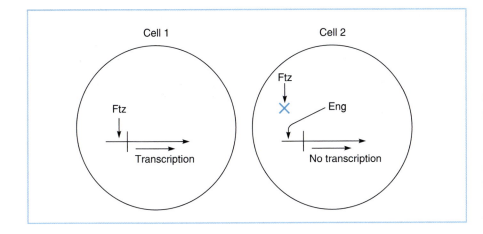

Figure 6.2
Blockage of gene induction by Ftz in cells expressing the engrailed (Eng) protein which binds to the same sequence as Ftz but does not activate transcription.

Hence, both Sp3 and eng act purely as transcriptional repressors by blocking binding of activators to their binding sites. Interestingly, the glucocorticoid receptor can also repress transcription of specific genes in this way, even though, as discussed in Chapter 4 (section 4.4), it acts primarily as an activator of target gene expression. Thus, treatment with glucocorticoid and activation of the glucocorticoid receptor inhibits, for example, expression of the genes encoding bovine prolactin and human pro-opiomelanocortin. The inhibitory effect observed in these cases is mediated by binding to DNA of the identical receptor/hormone complex which activates glucocorticoid-inducible genes (see Chapter 4, section 4.4 and Chapter 8, section 8.2.2) and these genes are therefore repressed by glucocorticoid. However, the DNA sequence element to which the complex binds when mediating its negative effect (nGRE) is distinct from the glucocorticoid response element (GRE) to which it binds when inducing gene expression, although the two are related (Fig. 6.3).

Figure 6.3
Relationship of the sites in DNA which mediate gene activation or repression by binding the glucocorticoid receptor. Note that the sites are related but distinct.

Binding site for positive regulation	**R G R A C A N N N T G T Y C Y**
Binding site for negative regulation	**A T Y A C A N N N T G A T C W**

This has led to the suggestion that the sequence difference causes the receptor/hormone complex to bind to the nGRE in a configuration in which its activation domain cannot interact with other transcription factors to activate transcription as occurs following binding to the positive element (Fig. 6.4). In agreement with this idea, the glucocorticoid receptor has been shown to bind to the nGRE in the POMC gene as a trimer rather than the dimer form which binds to the GRE and stimulates transcription (for review see Latchman, 2001). The receptor bound in this configuration to the negative element apparently acts by preventing binding of a positive acting factor to this or an adjacent site, thereby preventing gene induction. In agreement with this idea, the nGRE in the human glycoprotein hormone alpha subunit gene, which overlaps a cyclic AMP response element (CRE), is only able to inhibit gene expression when the CRE is left intact. Hence, it is likely that receptor bound at the negative element prevents binding of a transcriptional activator to the CRE and thereby inhibits gene expression (Fig. 6.5).

Figure 6.4

Consequences of glucocorticoid receptor binding to the DNA binding sites which mediate gene activation (GRE) or repression (nGRE). Note that the receptor is likely to bind in a different configuration to the two different sequences resulting in its ability to activate transcription only following binding to the GRE.

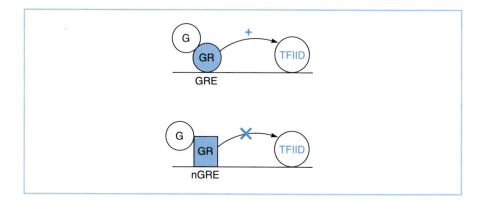

Figure 6.5

Inhibition of gene expression by glucocorticoid receptor binding to an nGRE is likely to be mediated by preventing the binding of a positively acting activator protein (A) to a site adjacent to or overlapping the nGRE.

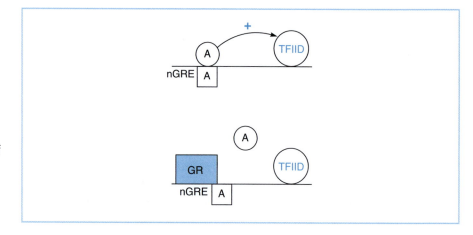

Hence the inhibition of DNA binding by a specific activator via masking of its binding site is a major method of transcriptional inhibition. This can involve either factors such as Sp3 or eng, which function only as transcriptional inhibitors, as well as factors such as the glucocorticoid receptor which either repress via this mechanism or activate transcription depending on the nature of their DNA binding site in a specific target gene.

6.2.2 INHIBITION OF ACTIVATOR BINDING BY FORMATION OF A NON-DNA BINDING COMPLEX

As well as preventing activator binding to DNA via masking its binding site, an inhibitor can also inhibit transcription via the formation of a non-DNA binding complex with an activating factor (see Fig. 6.1b). Thus, as discussed in Chapter 7 (section 7.2.1), the MyoD transcription factor is specifically expressed in skeletal muscle cells and plays a key role in activating skeletal muscle genes. Its expression can be induced in the 10T½ fibroblast cell line by treatment with 5-azacytidine and this converts the fibroblast cells into muscle cell precursors, known as myoblasts. However, activation of muscle-specific genes and the production of differentiated myotubes require these cells to be incubated in the absence of serum (Fig. 6.6). Paradoxically, however, MyoD levels do not change in this transition from myoblast to myotubes and yet MyoD-dependent muscle specific genes are activated.

The explanation of this paradox was provided by the identification of the Id protein (Benezra et al., 1990) which, like MyoD, contains a helix-loop-helix motif but lacks the basic domain mediating DNA binding (see Chapter 4, section 4.5 for a discussion of these motifs). Because the helix-loop-helix motif mediates dimerization of proteins containing it, Id can dimerize with other helix-loop-helix proteins such as MyoD and inhibit their DNA binding since the resulting heterodimer lacks the necessary pair of DNA binding motifs (Fig. 6.7). When 10T½-derived myoblasts are induced to form myotubes, Id levels decline indicating that this second stage of myogenesis is mediated by a decline in the inhibitory protein rather than an increase in the activator, MyoD.

The role of inhibitory helix-loop-helix proteins is not confined to myogenesis. Thus, the level of Id also declines during differentiation of early embryonic cells and erythroid cells. Moreover, inactivation of Id proteins in knock out mice results in abnormal development of the brain and the blood vessels providing direct evidence for the key role of proteins of this type in development (Lyden et al., 1999; for review see Carmeliet, 1999). Similarly, the product of the emc gene, which regulates neurogenesis in Drosophila, also contains a helix-loop-helix motif and lacks a basic DNA binding domain (for review

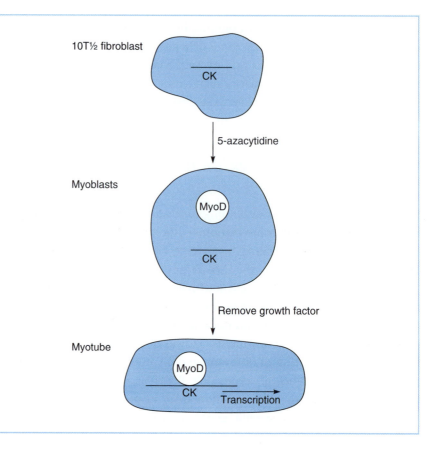

Figure 6.6

Differentiation of 10T½ cells into myoblasts by 5-azacytidine and then into myotubes by removal of growth factors. Note that the MyoD-dependent induction of genes encoding terminal differentiation markers such as creatine kinase (CK), which occurs in myotubes, occurs without an increase in MyoD concentration.

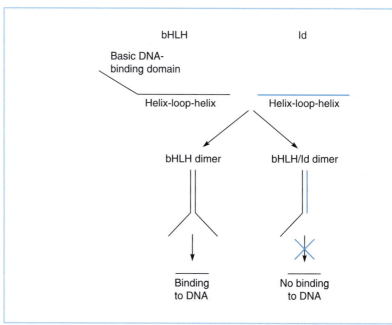

Figure 6.7

Dimerization of functional basic helix-loop-helix proteins (bHLH) with Id. Note that while Id can dimerize with other proteins via the helix-loop-helix domain, it lacks the basic DNA domain and hence the Id-containing heterodimer cannot bind to DNA.

see Jones, 1990. Hence, this form of repression is not confined to mammalian cells.

6.2.3 QUENCHING OF AN ACTIVATOR

The cases of repression described so far all involve the inhibition of DNA binding either by blocking the binding site for a factor (see Fig. 6.1a) or by forming a non-DNA binding protein–protein complex (see Fig. 6.1b). Since DNA binding is a necessary prerequisite for gene activation, this constitutes an effective form of repression. In addition, however, inhibition of transcription can also be achieved by interfering with transcriptional activation by a DNA bound factor in a phenomenon known as quenching (see Fig. 6.1c).

A simple example of this type is seen in the case of the negatively acting yeast factor GAL80 which inhibits gene activation by the positively acting GAL4 protein. This is achieved by the binding of GAL80 to DNA-bound GAL4, such binding occurring via the thirty amino acids located at the extreme C terminus of the GAL4 molecule. As these amino acids are located close to the GAL4 activation domain, the binding of GAL80 to GAL4 masks the GAL4 activation domain and hence inhibits the activation of gene expression by GAL4. In response to treatment with galactose, GAL80 dissociates from GAL4 allowing GAL4 to fulfill its function of activating galactose inducible genes. Hence this system provides an elegant means of modulating gene expression in response to galactose with the activating GAL4 factor being bound to DNA both prior to and after galactose addition but being able to activate gene expression only following the galactose-induced dissociation of the quenching GAL80 factor (Fig. 6.8). Similar cases involving repression of p53 by MDM2 and E2F by Rb-1 both of which involve, at least in part, masking of the activation domain are discussed in Chapter 9 (sections 9.4.2 and 9.4.3).

A related example of quenching, in which the inhibitory factor binds to a DNA sequence adjacent to the quenched factor rather than only to the factor itself, is seen in the case of the c-*myc* promoter. Thus an inhibitory transcription factor myc-PRF binds to a site adjacent to that occupied by an activating factor myc-CF1 and interferes with its ability to activate c-*myc* gene transcription (Kakkis et al., 1989). Hence quenching can occur either by an inhibitory factor binding to the positively acting factor (Fig. 6.9a) or by the inhibitory factor binding to DNA adjacent to the positive factor (Fig. 6.9b). In both cases, however, this effect involves the inhibitor interfering with the ability of the activator's activation domain to stimulate transcription.

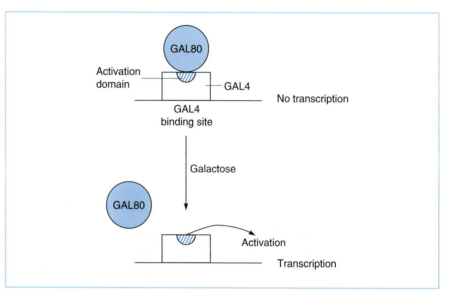

Figure 6.8
Galactose activates gene expression by removing the GAL80 protein from DNA-bound GAL4 protein, unmasking the activation domain on GAL4.

Figure 6.9
The ability of a bound activator (A) to stimulate transcription via its activation domain (hatched) can be inhibited by quenching of the activation domain by inhibitory factors (R), which either bind to the activator without binding to DNA (a) or which bind to a site adjacent to the activator (b).

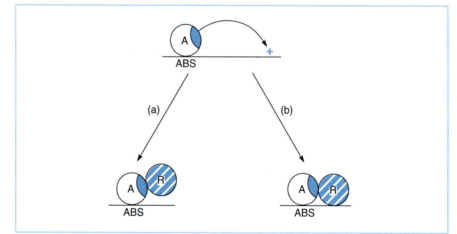

6.2.4 DEGRADATION OF THE ACTIVATOR

As well as interfering functionally with the action of the activator by preventing its DNA binding or quenching its activation domain, an indirect repressor can act by targeting an activator for degradation (see Fig. 6.1d). This is seen in the case of MDM2, which as well as quenching the activation domain of p53 (see above), also targets it for degradation so using multiple mechanisms to prevent p53 activating transcription (see Chapter 9, section 9.4.2).

In the case of p53, MDM2 is likely to do this by stimulating the recognition of p53 by protease enzymes which are present in the cell (Fig. 6.10a). However, the AEBP1 transcription factor, which regulates adipocyte differen-

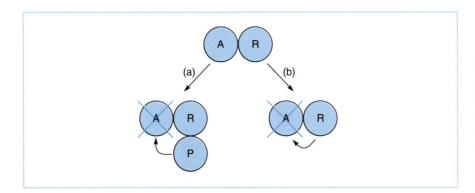

Figure 6.10
A repressor (R) can promote the degradation of an activator (A) either (a) indirectly by making it a target for a protease (P) or (b) directly by itself degrading the activator.

tiation, actually itself has the ability to degrade other proteins and it therefore likely acts directly by degrading activators to which it binds (Fig. 6.10b). Hence this factor combines the ability to bind to DNA with the ability to degrade other factors with which it comes into contact (He *et al.*, 1995).

Thus the degradation of activators mediated directly or indirectly by inhibiting factors appears to be an important mechanism of transcriptional repression. Interestingly, however, it is also possible for the reverse to occur with a factor activating transcription because it directs the degradation of a repressor. Thus in *Drosophila* the PHYL and SINA factors cause the degradation of the TTK88 transcription repressor hence producing activation of transcription (Li *et al.*, 1997; Tang *et al.*, 1997).

6.3 DIRECT REPRESSION

6.3.1 MECHANISMS OF TRANSCRIPTIONAL REPRESSION

In the cases described so far, a negative factor exerts its effect by neutralizing the action of a positively acting factor by preventing either its DNA binding (see Fig. 6.1a and b), inhibiting its activation of transcription following such binding (see Fig. 6.1c), or promoting its degradation (see Fig. 6.1d). In other cases, however, the inhibitory effect of a particular factor can be observed in the absence of any activating factors. This indicates that these inhibitory factors inhibit transcription directly by interacting with the basal transcriptional complex to reduce its activity (see Fig. 6.1e). Several factors of this type have now been shown to bind to specific DNA binding sites within their target genes and reduce the activity of the basal transcriptional complex (Fig. 6.11a). This effect is evidently similar in nature but opposite in effect to the stimulation of the basal complex by the binding of activating factors to specific DNA sequences in the promoter. Alternatively, a direct repressor may

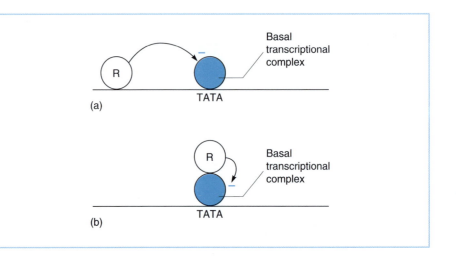

Figure 6.11

An inhibitory factor (R) can reduce the activity of the basal transcriptional complex either by binding to DNA and then interacting with the complex (a) or by binding directly to the complex by protein–protein interaction (b).

actually join the basal transcriptional complex via a protein–protein interaction, without binding to DNA and then inhibit transcription (Fig. 6.11b). These two mechanisms are discussed in the next two sections.

6.3.2 DIRECT REPRESSION BY DNA BINDING TRANSCRIPTION FACTORS

One factor capable of inhibiting the basal initiation complex following binding to the DNA is the *Drosophila* eve protein which is a member of the homeobox family discussed in Chapter 4 (section 4.2) and can act, for example, to repress the gene encoding the Ubx protein which is also a member of the homeobox family.

Thus, if the Ubx promoter linked to a marker gene is added to a suitable cell-free extract, transcription of the marker gene driven by the promoter can be observed. Addition of the purified protein even-skipped (eve) to this extract inhibits Ubx promoter activity, however, and this inhibition is dependent upon binding sites for the eve protein within the Ubx promoter. Such findings parallel the ability of a vector expressing eve to repress the Ubx promoter following co-transfection into cultured cells and the genetic evidence which originally led to the definition of eve as a repressor of Ubx (Fig. 6.12). This case thus represents an interesting example of the transcription of the gene encoding one homeobox transcription factor (Ubx) being repressed by another (eve).

Interestingly, the number of such directly inhibitory factors is growing steadily and now includes some which were previously thought to function only in an indirect manner. Thus, the MDM2 factor which was thought to

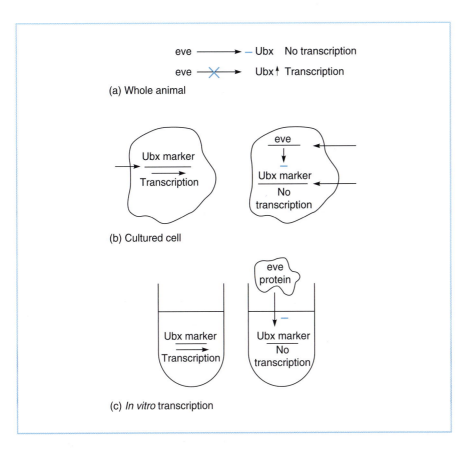

(a) Whole animal

(b) Cultured cell

(c) *In vitro* transcription

Figure 6.12

Inhibitory effect of the eve protein on expression of the Ubx gene. This inhibitory effect can be observed in the whole animal where mutation of the eve gene enhances Ubx expression (top panel); in cultured cells where introduction of a plasmid expressing the eve gene represses a co-transfected Ubx promoter driving a marker gene (middle panel) and in a test tube *in vitro* transcription system where addition of purified eve protein represses transcription of a marker gene driven by the Ubx promoter (bottom panel).

function solely by masking the activation domain of p53 has now been shown to function as a direct repressor of transcription as well as targeting p53 for degradation (see Chapter 9, section 9.4.2), while the Rb-1 protein which was originally thought to function solely by inhibiting E2F is now known to act also as a direct repressor (see Chapter 9, section 9.4.3).

An interesting example of a directly acting repressor is provided by the thyroid hormone receptor which is a member of the nuclear receptor family discussed in Chapter 4 (section 4.4). Thus, this receptor can bind to its response element (TRE) in the absence of thyroid hormone and inhibit gene expression. This effect is not due to the receptor preventing other positive factors from binding but involves a direct inhibitory effect of the receptor on transcription which requires a specific domain at the C-terminus of the molecule. In the presence of thyroid hormone the receptor undergoes a conformational change which exposes its activation domain and converts it from a repressor to an activator (Fig. 6.13). Hence, in this case, gene activation or repression can be mediated from the same DNA binding site with the effect depending on the presence or absence of the hormone.

Figure 6.13

In the absence of thyroid hormone (T) the thyroid hormone receptor inhibits gene expression via a discrete inhibitory domain (hatched box). Binding of thyroid hormone (T) exposes the activation domain of the receptor (solid box) and allows it to activate transcription.

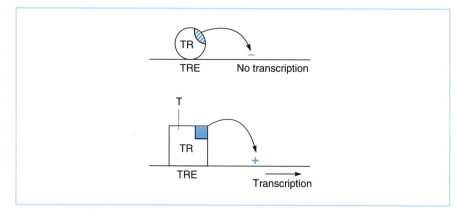

Although the thyroid hormone receptor can therefore act as either a transcriptional activator or repressor, the mechanism differs from that observed with the glucocorticoid receptor (which is also a member of the nuclear receptor family) and, as discussed in section 6.2.1 can also act as either an activator or a repressor. Thus, in the case of the glucocorticoid receptor, both activation and repression are dependent upon activation of the receptor by glucocorticoid and it is the nature of the binding site that determines whether activation or repression is observed. Moreover, repression is indirect being achieved by preventing an activator from binding. In contrast, in the case of the thyroid receptor, the activation/repression decision is controlled by thyroid hormone and inhibition of gene expression involves direct repression (for review see Latchman, 2001).

Interestingly, in addition to the thyroid binding form of the thyroid hormone receptor, alternative splicing generates another form (alpha 2) lacking a part of the hormone binding domain and therefore unable to bind hormone (Koenig *et al.*, 1989; Fig. 6.14a). Both the alpha 2 form and the hormone binding alpha 1 form can bind to DNA, however, binding of alpha 2 to the thyroid response element (TRE) sequence prevents binding of alpha 1 and thereby prevents gene induction in response to thyroid hormone (Fig. 6.14b). As discussed in Chapter 9 (section 9.3.2), a similar non-hormone binding form of the thyroid hormone receptor is encoded by the v-*erbA* oncogene which produces cancer by inhibiting the expression of thyroid hormone responsive genes involved in erythroid differentiation.

Inhibitory factors of the directly acting type, such as the thyroid hormone receptor or the eve factor, generally contain a small domain which can confer the ability to repress gene expression upon the DNA binding domain of another factor when the two are artificially linked (see for example, Han and Manley, 1993; Lillycrop *et al.*, 1994) (Fig. 6.15). Hence these directly

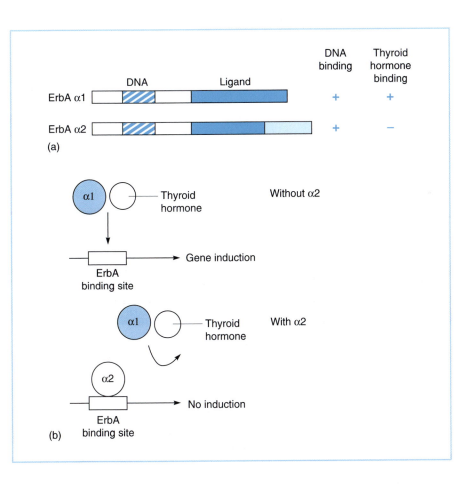

Figure 6.14
(a) Relationship of the ErbA alpha 1 and alpha 2 proteins. Note that only the alpha 1 protein has a functional thyroid hormone binding domain. (b) Inhibition of ErbA alpha 1 binding and of gene activation in the presence of the alpha 2 protein.

Figure 6.15
A specific region (CD) of the eve factor, which is distinct from the DNA-binding domain (B), acts as a transferable inhibitory domain. Thus its deletion from the eve protein results in a loss of the ability to repress transcription while its linkage to the DNA binding domain of GAL4 generates a functional repressor.

repressing factors contain specific inhibitory domains paralleling the existence of specific activation domains in activating transcription factors.

Interestingly, the inhibitory domain in the human Wilms tumour anti-oncogene product and those from several *Drosophila* inhibitory factors, including eve, appear to share the common features of proline richness and an absence of charged residues (Han and Manley, 1993) suggesting that these factors have a common inhibitory domain. However, other inhibitory domains, such as those in the mammalian factors Oct-2 (Lillycrop *et al.*, 1994) and E4BP4 (Cowell and Hurst, 1994), are distinct both from this common domain and from each other indicating that, as with activation domains, several types of inhibitory domain may exist.

By analogy with activation domains, inhibitory domains are likely to inhibit either the assembly of the basal transcriptional complex or reduce its activity and/or stability after it has assembled. In agreement with this idea, the inhibitory domain of the Kruppel repressor in *Drosophila* has been shown to interact with a component of the basal transcriptional complex, TFIIEβ (Sauer *et al.*, 1995). Interestingly, Kruppel can also act as an activator by interacting with TFIIB to stimulate its activity. This interaction with TFIIB is seen in the monomeric Kruppel factor which hence acts as an activator, whereas the Kruppel dimer which forms at high concentrations inhibits transcription by interacting with TFIIEβ. Hence Kruppel can act as activator or repressor depending on its concentration in the cell which results in its being present as an activating monomer or an inhibitory dimer (Fig. 6.16).

Although repressors may therefore interact directly with the basal transcriptional complex, in many cases they do so via a non-DNA binding co-repressor molecule which then actually represses transcription (Fig. 6.17). Such non-DNA binding co-repressors have been observed in a range of organisms from yeast to humans (for reviews see Knoepfler and Eisenman, 1999; Smith and Johnson, 2000; Chinnadurai, 2002). Thus, the inhibitory effect of the thyroid hormone receptor discussed above involves co-repressor mole-

Figure 6.16

The Kruppel factor (K) when present as a monomer can interact with TFIIB to stimulate transcription. At high concentrations when it forms a dimer it interacts with TFIIEβ to repress transcription.

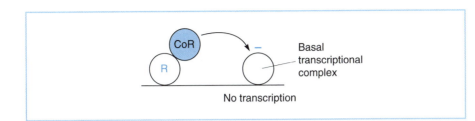

Figure 6.17
A DNA binding repressor (R) can recruit a non-DNA binding co-repressor (CoR) which then represses transcription.

cules such as N-CoR (nuclear receptor co-repressor) which bind to the receptor in the absence of hormone and produce its inhibitory effect on transcription (for reviews see Glass and Rosenfeld, 2000; Rosenfeld and Glass, 2001).

Interestingly, studies on mice lacking N-CoR have shown multiple defects in the development of numerous organs and cell types (Jepsen *et al.*, 2000; Hermanson *et al.*, 2002). Hence, co-repressors play critical roles in the regulation of gene expression with N-CoR, for example, being involved both in responses to thyroid hormone and in embryonic development.

In the case of the thyroid hormone receptor, following treatment with thyroid hormone, the conformation of the receptor changes and the co-repressor can no longer bind. This conformational change allows co-activator molecules such as CBP to bind and induce transcriptional activation. Hence, in this case, treatment with thyroid hormone causes a conformational change in the receptor resulting in the removal of co-repressors and the binding of co-activators (Fig. 6.18).

An interesting example of such a co-activator, co-repressor interchange has recently been described in the case of the LIM homeodomain transcription factors (Ostendorff *et al.*, 2002). Thus, these factors bind both the RLIM co-repressor molecule and the CLIM co-activator molecule. In a novel mechan-

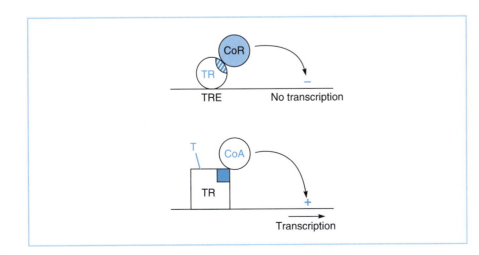

Figure 6.18
In the absence of thyroid hormone, the inhibitory domain (hatched) of the thyroid hormone receptor (TR) bound to its response element (TRE) can recruit a co-repressor (CoR) which then inhibits transcription. In the presence of thyroid hormone (T), the conformation of the receptor changes exposing its activation domain (solid), allowing recruitment of co-activator molecules (CoA), thereby producing activation of transcription in response to thyroid hormone.

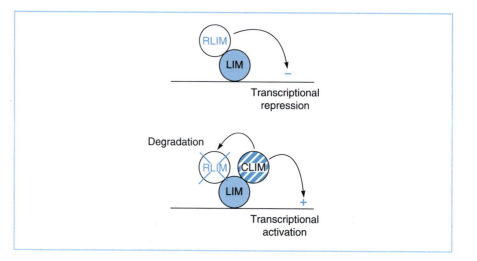

Figure 6.19

The LIM homeodomain protein binds the RLIM co-repressor producing transcriptional repression. However, following binding of the CLIM activator protein, CLIM digests the RLIM protein, allowing transcriptional activation to occur.

ism, the CLIM co-activator promotes degradation of RLIM, removing the repressor and allowing activation to occur (Fig. 6.19).

It is clear therefore that a number of factors can inhibit transcription by binding to upstream DNA sequences and inhibiting the activity of the basal transcriptional complex either directly or via co-repressor molecules. The binding of such factors is likely to be of vital importance in producing the inhibitory effect of many of the silencer elements which were described in Chapter 1 (section 1.3.5), the silencer element in the chicken lysozyme gene for example, having been shown to act by binding the inhibitory thyroid hormone receptor.

6.3.3 DIRECT REPRESSION BY FACTORS BINDING TO THE BASAL TRANSCRIPTIONAL COMPLEX

As well as interfering with the basal complex by binding to distinct DNA binding sites (see Fig. 6.11a), it is also possible for inhibitory factors to bind to the complex itself by protein–protein interaction and thereby interfere with its activity or assembly (Fig. 6.11b) (for review see Maldonado *et al.*, 1999). An example of this is provided by the Dr1 protein which inhibits the assembly of the basal transcriptional complex by binding to TBP and preventing TFIIB from binding (Fig. 6.20) (Inostroza *et al.*, 1992). As the recruitment of TFIIB to the promoter by interaction with the TBP component of TFIID is an essential step in the assembly of the basal transcriptional complex, this effectively inhibits transcription (see Chapter 3, section 3.5.1).

As noted in Chapter 3 (section 3.6), TBP is a component of the initiation complexes of all three polymerases and in each case acts by recruiting other

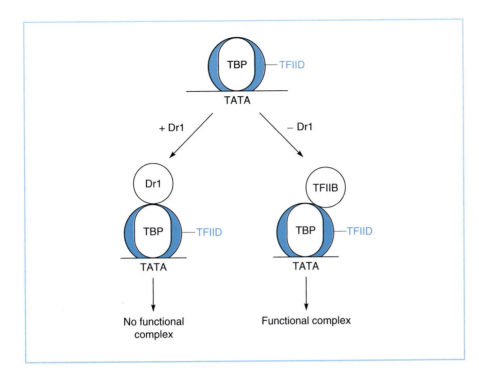

Figure 6.20

The Dr1 inhibitory factor can interact with the TBP component of TFIID, thereby preventing it binding TFIIB and thus inhibiting the assembly of the basal transcriptional complex.

factors to the promoter. It has been shown (White *et al.*, 1994) that Dr1 can inhibit this ability of TBP to recruit other factors within the RNA polymerase II and III initiation complexes but not within the RNA polymerase I complex. It therefore inhibits transcription by RNA polymerase II and III but not by RNA polymerase I. Thus Dr1 may play a critical role in regulating the balance of transcriptional activity between the ribosomal genes which are the only genes transcribed by RNA polymerase I and all the other genes in the cell.

As well as this potential role for Dr1, there is evidence that it can also alter the balance between transcription of different types of promoter by RNA polymerase II. Thus, although Dr1 inhibits transcription from TATA box-containing promoters, it actually stimulates transcription from polymerase II promoters lacking a TATA box and containing an initiator element and an associated downstream promoter element (Willy *et al.*, 2000) (see Chapter 1, section 1.3.2 for discussion of promoters with or without a TATA box). Hence, Dr1 may switch transcription between different genes transcribed by RNA polymerase II but containing different core promoter elements.

In addition to Dr1, other factors which bind to TBP and inhibit the assembly of the RNA polymerase II basal complex have been described and are likely to be important in controlling the rate of transcription (Chitikila *et al.*, 2002; for review see Maldonado *et al.*, 1999). Thus, for example, the Mot1

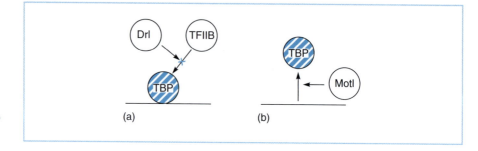

(a)   (b)

factor also targets TBP but rather than preventing TFIIB binding, it displaces TBP from the DNA, thereby inhibiting transcription (Fig. 6.21).

As well as interacting with activating factors (see Chapter 5, section 5.3.3), the TFIIA factor also appears to be able to bind to TBP preventing these inhibitors from binding, thus preventing Mot1 from inhibiting TBP binding to DNA or allowing the recruitment of TFIIB in the presence of Dr1 (Fig. 6.22). This indicates that the activity of inhibitory molecules which act by interacting with the basal transcriptional complex can be regulated by activating factors. Moreover, it illustrates that the balance between transcriptional activators and repressors is of central importance in the control of transcription with directly acting repressors playing a key role either by binding to upstream DNA sequences or by joining the basal transcriptional complex.

6.4 OTHER TARGETS FOR TRANSCRIPTIONAL REPRESSORS

6.4.1 MODULATION OF CHROMATIN STRUCTURE

Evidently, in the same way as an activating factor can activate transcription, by opening up the chromatin (see Chapter 5, section 5.5.1), an inhibitory factor can produce repression by directing a more tightly packed chromatin structure (for reviews see Tyler and Kadonaga, 1999; Courey and Jia, 2001) (Fig. 6.23). This could occur either via ATP-dependent remodelling of nucleosomes or via altering the modification of histones (see Chapter 1, section 1.2).

An example of a factor which acts in this way is the polycomb repressor of *Drosophila* which normally represses inappropriate expression of several homeotic genes by modulating their chromatin structure so that activating molecules cannot bind. When this factor is inactive, inappropriate expression of these genes in the wrong cell type is observed leading to dramatic transformations in the nature of specific parts of the body (for review see Jacobs and van Lohuizen, 2002; Simon and Tamkun, 2002; Orlando, 2003). By

Figure 6.22

Binding of TFIIA to TBP prevents inhibitory molecules (I) from binding but still allows the binding of TFIIB and thereby promotes the assembly of the basal transcriptional complex.

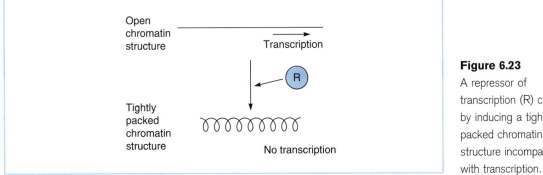

Figure 6.23

A repressor of transcription (R) can act by inducing a tightly packed chromatin structure incompatible with transcription.

directing the tight packing of specific genes and thereby preventing transcription, the polycomb factor evidently has the opposite effect to that of the GAGA/trithorax factor which, as discussed in section 5.5.1, directs an open chromatin structure allowing activator binding (Fig. 6.24) (for reviews see Schumacher and Magnusson, 1997; Simon and Tamkun, 2002).

Interestingly, both the polycomb and trithorax factors have been shown to affect the function of insulator elements (see Chapter 1, section 1.3.5) with mutation of polycomb preventing the blocking effects of insulators on enhancer function, while mutation of trithorax enhances the blocking effects of insulators. Hence, insulator elements represent one target for the antagonistic effects of polycomb and trithorax factors (Gerasimova and Corces, 1998).

It has recently been shown that the polycomb factor forms part of a multiprotein complex that has histone methyltransferase activity (Czermin *et al.*, 2002; Müller *et al.*, 2002). As discussed in Chapter 1 (section 1.2.3) the methylation of histones can produce a tightly packed chromatin structure and hence this is one mechanism by which polycomb can produce a closed chromatin structure.

As well as targeting histones at the level of methylation, transcriptional co-repressors such as the NuRD and SIN3 complexes can also produce a more tightly packed chromatin structure by acting as histone deacetylases, removing the acetyl groups which promote chromatin opening (for reviews see Ahringer, 2000; Ng and Bird, 2000) (see Chapter 1, section 1.2.3). This effect evidently parallels the ability of transcriptional co-activators to produce a more open chromatin structure by acetylating histones (see Chapter 5, section 5.5.1).

Figure 6.24

The GAGA/trithorax factor induces an open chromatin structure. Its inactivation by mutation produces an inappropriate tightly packed chromatin structure of its target genes producing a failure of gene transcription. In contrast, the polycomb factor (PC) induces a tightly packed chromatin structure. Its inactivation by mutation produces inappropriate chromatin opening and transcription.

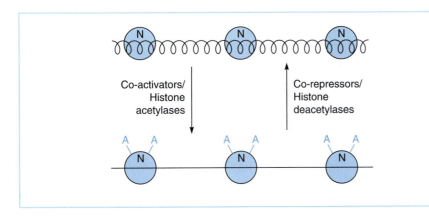

Figure 6.25

The balance between tightly packed chromatin (wavy line) and open chromatin (solid line) can be controlled by the balance between co-activating molecules which acetylate histones and co-repressors which deacetylate histones in the nucleosome (N).

Hence the state of histone acetylation and the structure of chromatin can be controlled by the balance of deacetylases and acetylases which are bound to the DNA (Fig. 6.25). In most cases the acetylating and deacetylating factors will be respectively co-activators and co-repressors and will be brought to the DNA via interactions with distinct activating and inhibiting transcription factors respectively. In the case of the thyroid hormone receptor, however, both types of factors bind to the same molecule. Thus, the N-CoR co-repressor, which binds to the thyroid hormone receptor prior to exposure to hormone, also binds other co-repressor complexes such as mRPD3 and SIN3 which have histone deacetylase activity. Conversely, following hormone binding, these factors dissociate and are replaced by co-activators such as CBP which acetylate histones and allow the receptor to activate transcription (Fig. 6.26).

Hence, transcriptional repressors can modulate chromatin structure, acting at least in part by altering the histone code enhancing methylation and reducing acetylation.

Figure 6.26

The inhibitory domain (hatched semi-circle) of the thyroid hormone receptor binds a co-repressor (Co-R) complex which deacetylates histones inducing a closed chromatin structure (wavy line). Binding of ligand results in the release of the co-repressor and binding of co-activator (Co-A) molecules to the exposed activation domain (solid square). The co-activators have histone acetyltransferase activity and produce a more open chromatin structure (solid line) compatible with transcription.

6.4.2 INHIBITION OF TRANSCRIPTION ELONGATION

Just as activators can stimulate transcriptional elongation, as well as transcriptional initiation (see Chapter 5, section 5.5.2) repressors can inhibit transcription by blocking transcriptional elongation. Thus, the zebrafish Foggy protein acts by interacting with the non-phosphorylated form of RNA polymerase and prevents it from catalysing transcriptional elongation. When the polymerase is phosphorylated on its C-terminal domain (see Chapter 3, section 3.5.1), it is no longer inhibited by Foggy and transcriptional elongation proceeds (Guo *et al.*, 2000) (Fig. 6.27). Importantly, when Foggy is mutated so that it cannot block transcriptional elongation, the development of the zebrafish nervous system is severely disrupted. This indicates that the correct regulation of transcriptional elongation by proteins such as Foggy is necessary for normal development.

Inhibition of transcriptional elongation is also produced by the von Hippel–Lindau protein (VHL). However, this factor targets the phosphorylated form of RNA polymerase. Thus, VHL forms part of a complex which adds the small protein ubiquitin to the large subunit of RNA polymerase II (see Chapter 8, section 8.4.5 for further details of this protein modification). This ubiquitination occurs only for the phosphorylated form of RNA polymerase II and targets it for degradation, thereby blocking transcriptional elongation (Kuznetsova *et al.*, 2003), (Figure 6.28). As with Foggy, the action of VHL is of critical importance for normal cell function. Thus, as discussed in Chapter 9 (section 9.4.4), VHL is an anti-oncogene and cancers result when the function of VHL is disrupted by mutations.

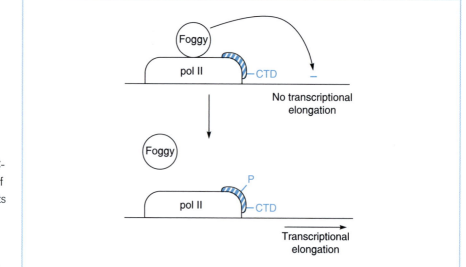

Figure 6.27

Phosphorylation of the C-terminal domain (CTD) of RNA polymerase prevents binding of the Foggy protein which would otherwise inhibit transcriptional elongation.

Figure 6.28

The von Hippel-Lindau gene product (VHL) adds ubiquitin (Ubi) to the phosphorylated C-terminal domain (CTD) of RNA polymerase. This promotes degradation of the polymerase inhibiting transcriptional elongation.

Hence, transcriptional elongation can be repressed both by Foggy which inhibits the non-phosphorylated form of RNA polymerase II and by VHL which targets the phosphorylated form of the polymerase for degradation. As with transcriptional initiation, the correct regulation of these inhibitory processes and their activity relative to processes which stimulate transcriptional elongation (see Chapter 5, section 5.5.2) is likely to be critical for the correct regulation of cellular function.

6.5 CONCLUSIONS

In this chapter, we have discussed how the repression of transcription can be produced by the neutralization of a positively acting factor, or by direct repression of the basal transcriptional complex as well as by alteration of chromatin structure or the inhibition of transcriptional elongation. These properties offer ample scope for gene regulation in different cell types or in different tissues. Thus in addition to the simple activation of gene expression by a positively acting factor present in only one cell type, the effect of a positively acting factor present in several different cell types can be affected by the presence or absence of a negatively acting factor which is active in only one cell type and which inhibits its activity. Similarly, a single factor may act either positively or negatively depending on the gene involved (as in the case of the glucocorticoid receptor) or depending on whether a specific hormone is present (as in the case of the thyroid hormone receptor).

Interestingly, two positive factors can also repress one another if they compete for the same co-factor. Thus, glucocorticoid hormones have been known for some time to be a potent inhibitor of the induction of the collagenase gene by phorbol esters resulting in their having an anti-inflammatory effect. This inhibition is mediated by the glucocorticoid receptor which inhibits the activity of the Jun and Fos proteins that normally activate the collagenase gene via the AP-1 sites in its promoter (for discussion of Fos, Jun and AP-1 see

Chapter 9, section 9.3.1). This effectively inhibits collagenase gene activation. Unlike the examples of repression by the glucocorticoid receptor discussed in section 6.2.1, however, the collagenase promoter does not contain any binding sites for the receptor adjacent to the AP-1 sites, nor does the receptor apparently bind to the collagenase promoter.

Interestingly, however, like the glucocorticoid receptor, the Fos/Jun complex requires the CBP protein as a co-activator to activate transcription. Hence, the glucocorticoid receptor may compete with Fos/Jun for limited quantities of the CBP co-activator which are present in the cell resulting in a failure of Fos/Jun to activate in the presence of activated glucocorticoid receptor (Kamei *et al.*, 1996) (Fig. 6.29). Clearly, such competition between Fos/Jun and the glucocorticoid receptor for limited quantities of CBP will also result in inhibition of glucocorticoid-dependent genes in response to hormone in the presence of high concentrations of Fos and Jun and this is indeed observed (Fig. 6.29).

Hence, mutual transrepression of two different activating proteins can be achieved by competition for a co-activator (Fig. 6.29). Moreover, this mutual repression illustrates how different cellular signalling pathways, which are activated respectively by phorbol esters and glucocorticoid hormones, can interact with one another resulting in cross talk between the pathways (for review see Janknecht and Hunter, 1996).

Hence, as well as being able to activate gene expression, the members of the steroid/thyroid hormone receptor family also illustrate three mechanisms by which repression of gene expression can be achieved namely, the direct inhibition of transcription (see section 6.3.2), the neutralization of a positive

Figure 6.29

Mutual transrepression by Fos/Jun and the glucocorticoid receptor. Competition between Fos/Jun and the glucocorticoid receptor for the CBP co-activator inhibits the expression of genes containing binding sites for either Fos/Jun (AP-1 sites) or for the glucocorticoid receptor (GRE).

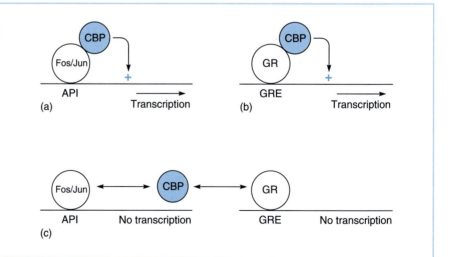

factor either by preventing its binding to DNA by masking of its site (see section 6.2.1), or by competing for a co-activator (see above). It should be noted, however, that these three examples differ in that the glucocorticoid receptor needs to be activated by steroid before it can inhibit gene expression by binding to an nGRE or competing for CBP whereas the thyroid hormone receptor directly inhibits by binding to its response element in the absence of hormone.

Hence, numerous methods of transcriptional repression exist paralleling the different mechanisms of transcriptional activation discussed in Chapter 5. Ultimately, however, as with transcriptional activation, all such potential mechanisms involving the inhibition of gene expression in response to specific stimuli or in specific cell types are dependent upon mechanisms that control the synthesis or activity of specific transcription factors in different cell types or in response to specific stimuli. These mechanisms are discussed in the next two chapters.

REFERENCES

Ahringer, J. (2000) NuRD and SIN3 histone deacetylase complexes in development. Trends in Genetics 16, 351–356.

Benezra, R., Davis, R. L., Lockshon, D. *et al.* (1990) The protein Id: a negative regulator of helix-loop-helix DNA binding proteins. Cell 61, 49–59.

Carmeliet, P. (1999) Controlling the cellular brakes. Nature 401, 657–658.

Chinnadurai, G. (2002) CtBP, an unconventional transcriptional corepressor in development and oncogenesis. Molecular Cell 9, 213–224.

Chitikila, C., Huisinga, K. L., Irvin, J. D. *et al.* (2002) Interplay of TBP inhibitors in global transcriptional control. Molecular Cell 10, 871–882.

Courey, A. J. and Jia, S. (2001) Transcriptional repression: the long and the short of it. Genes and Development 15, 2786–2796.

Cowell, I.G. and Hurst, H.C. (1994) Transcriptional repression by the human bZIP factor E4BP4: Definition of a minimal repressor domain. Nucleic Acids Research 22, 59–65.

Czermin, B., Melfi, R., McCabe, D. *et al.* (2002) *Drosophila* enhancer of Zeste/ESC complexes have a histone H3 methyltransferase activity that marks chromosomal polycomb sites. Cell 111, 185–196.

Gerasimova, T. I. and Corces, V. G. (1998) Polycomb and trithorax group proteins mediate the function of a chromatin insulator. Cell 92, 511–521.

Glass, C. K. and Rosenfeld, M. G. (2000) The coregulator exchange in transcriptional functions of nuclear receptors. Genes and Development 14, 121–141.

Guo, S., Yamaguchi, Y., Schillbach, S. et al. (2000) A regulator of transcriptional elongation controls vertebrate neuronal development. Nature 408, 366–369.

Han, K. and Manley, J.L. (1993) Transcriptional repression by the Drosophila even skipped protein: definition of a minimal repressor domain. Genes and Development 7, 491–503.

Han, K., Levine, M. S. and Manley, J. L. (1989) Synergistic activation and repression of transcription by Drosophila homeobox proteins. Cell 56, 573–583.

Hanna-Rose, W. and Hansen, U. (1996) Active repression mechanisms of eukaryotic transcriptional repressors. Trends in Genetics 12, 229–234.

He, G-P., Muise, A., Li, A.W. and Ro, H-S. (1995) A eukaryotic transcriptional repressor with carboxypeptidase activity. Nature 378, 92–96.

Hermanson, O., Jepsen, K. and Rosenfeld, M. G. (2002) N-CoR controls differentiation of neural stem cells into astrocytes. Nature 419, 934–939.

Hoey, T. and Levine, M. (1988) Divergent homeo box proteins recognise similar DNA sequences in Drosophila. Nature 332, 858–861.

Inostroza, J.A., Mermelstein, F.H., Ha, I. et al. (1992) Dr1, a TATA-binding protein-associated phosphoprotein and inhibitor of class II gene transcription. Cell 70, 477–489.

Jacobs, J. J. and van Lohuizen, M. (2002) Polycomb repression: from cellular memory to cellular proliferation and cancer. Biochimica et Biophysica Acta 1602, 151–161.

Janknecht, R. and Hunter, T. (1996) A growing co-activator network. Nature 383, 22–23.

Jaynes, J. B. and O'Farrell, P. H. (1988) Activation and repression of transcription by homeodomain-containing proteins that bind a common site. Nature 336, 744–749.

Jepsen, K., Hermanson, O., Onami, T. M. et al. (2000) Combinatorial roles of the nuclear receptor corepressor in transcription and development. Cell 102, 753–763.

Jones, N. (1990) Transcriptional regulation by dimerization: two sides to an incestuous relationship. Cell 61, 9–11.

Kakkis, E., Riggs, K.J., Gillespie, W. and Calame, K. (1989) A transcriptional repressor of C-myc. Nature 339, 718–721.

Kamei, Y, Heinzel, T., Torchia, J. et al. (1996) A CBP integrator complex mediates transcriptional activation and AP-1 inhibition by nuclear receptors. Cell 85, 403–414.

Knoepfler, P. S. and Eisenman, R. N. (1999) Sin meets NuRD and other tails of repression. Cell 99, 447–450.

Koenig, R. G., Lazar, M. A., Hoden, R. A. et al. (1989) Inhibition of thyroid hormone action by a non hormone binding c-erb A protein generated by alternative RNA splicing. Nature 337, 659–661.

Kuznetsova, A. V., Meller, J., Schnell, P. O. *et al.* (2003) von Hippel-Lindau protein binds hyperphosphorylated large subunit of RNA polymerase II through a proline hydroxylation motif and targets it for ubiquitination. Proceedings of the National Academy of Sciences USA 100, 2706–2711.

Lania, L., Majello, B. and DeLuca, P. (1997) Transcriptional regulation by the Sp family proteins. International Journal of Biochemistry and Cell Biology 29, 1313–1323.

Latchman, D.S. (1996) Inhibitory factors. International Journal of Biochemistry and Cell Biology 28, 965–974.

Latchman, D. S. (2001) Transcription Factors: Bound to activate or repress. Trends in Biochemical Sciences 26, 211–213.

Li, S., Li, Y., Carthew, R.W. and Lai, Z-C. (1997) Photoreceptor cell differentiation requires regulated proteolysis of the transcriptional repressor tramtrack. Cell 90, 469–478.

Lillycrop, K.A., Dawson, S.J., Estridge, J.K. *et al.* (1994) Repression of a herpes simplex virus immediate-early promoter by the Oct-2 transcription factor is dependent upon an inhibitory region at the N-terminus of the protein. Molecular and Cellular Biology 14, 7633–7642.

Lyden, D., Young, A. Z., Zagzag, D. *et al.* (1999) Id1 and Id3 are required for neurogenesis, angiogenesis and vascularization of tumour xenografts. Nature 401, 670–677.

Maldonado, E., Hampsey, M. and Reinberg, D. (1999) Repression: targeting the heart of the matter. Cell 99, 455–458.

Müller, J., Hart, C. M., Francis, N. J. *et al.* (2002) Histone methyltransferase activity of a *Drosophila* polycomb group repressor complex. Cell 111, 197–208.

Ng, H-H. and Bird, A. (2000) Histone deacetylases: silencers for hire. Trends in Biochemical Sciences 25, 121–126.

Orlando, V. (2003) Polycomb, Epigenomes and control of cell identity. Cell 112, 599–606.

Ostendorff, H. P., Peirano, R. I., Peters, M. A. *et al.* (2002) Ubiquitination-dependent cofactor exchange on LIM homeodomain transcription factors. Nature 416, 99–103.

Rosenfeld, M. G. and Glass, C. K. (2001) Coregulator codes of transcriptional regulation by nuclear receptors. Journal of Biological Chemistry 276, 36865–36868.

Sauer, F., Fondell, J.D., Ohkuma, Y. *et al.* (1995) Control of transcription by Kruppel through interactions with TFIIB and TFIIEβ. Nature 375, 162–164.

Schumacher, A. and Magnusson, T. (1997) Murine polycomb and trithorax group genes regulate homeotic pathways and beyond. Trends in Genetics 13, 167–170.

Simon, J. A. and Tamkun, J. W. (2002) Programming off and on states in chromatin: mechanisms of polycomb and trithorax group complexes. Current Opinion in Genetics and Development 12, 210–218.

Smith, R. L. and Johnson, A. D. (2000) Turning genes off by Ssn6-Tup1: a conserved system of transcriptional repression in eukaryotes. Trends in Biochemical Sciences 25, 325–330.

Tang, A.H., Neufeld, T.P., Kwan, E. and Rubin, G.M. (1997) PHYL acts to down regulate TTK88, a transcriptional repressor of neuronal cell fates by a SINA-dependent mechanism. Cell, 90, 459–467.

Tyler, J. K. and Kadonaga, J. T. (1999) The 'dark side' of chromatin remodelling: repressive effects on transcription. Cell 99, 443–446.

White, R.J., Khoo, B.C-E., Inostroza, J.A. *et al.* (1994) Differential regulation of RNA polymerases I, II and III by the TBP-binding repressor Dr1. Science 266, 448–450.

Willy, P. J., Kobayashi, R. and Kadonaga, J. T. (2000) A basal transcription factor that activates or represses transcription. Science 290, 982–984.

CHAPTER 7

REGULATION OF TRANSCRIPTION FACTOR SYNTHESIS

7.1 TRANSCRIPTION FACTOR REGULATION

Transcription factors play a central role in a number of biological processes, producing, for example the induction of specific genes in response to particular stimuli, as well as controlling the cell type-specific or developmentally regulated expression of other genes. The ability to bind to DNA (Chapter 4) and influence the rate of transcription either positively (Chapter 5) or negatively (Chapter 6) are clearly features of many transcription factors that regulate gene expression in response to specific stimuli or in specific cell types. Most importantly, however, such factors must also have their activity regulated such that they only become active in the appropriate cell type or in response to the appropriate stimulus, thereby producing the desired pattern of gene expression.

Two basic mechanisms by which the action of transcription factors can be regulated have been described. These involve either controlling the synthesis of the transcription factor so that it is made only when necessary (Fig. 7.1a) or alternatively, regulating the activity of the factor so that pre-existing protein becomes activated when required (Fig. 7.1b). This chapter considers the regulation of transcription factor synthesis while Chapter 8 considers the regulation of transcription factor activity.

7.2 REGULATED SYNTHESIS OF TRANSCRIPTION FACTORS

Regulating the synthesis of transcription factors such that they are only made when the genes that they regulate are to be activated is an obvious mechanism of ensuring that specific genes become activated only at the appropriate time and place. This mechanism is widely used therefore, particularly for transcription factors that regulate the expression of cell type-specific or developmentally regulated genes. Specific examples of the regulated synthesis of particular transcription factors which illustrate the role of this mechanism

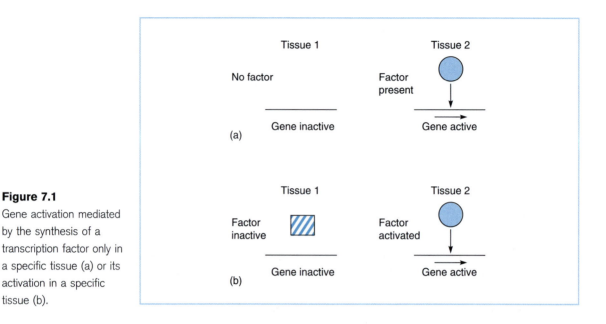

Figure 7.1

Gene activation mediated by the synthesis of a transcription factor only in a specific tissue (a) or its activation in a specific tissue (b).

in regulating cell type-specific or developmental gene expression are discussed in the following sections.

7.2.1 THE MyoD TRANSCRIPTION FACTOR

Probably the most novel approach to the cloning of the gene encoding a transcription factor was taken by Davis *et al.* (1987), who isolated cDNA clones encoding MyoD, a factor which plays a critical role in skeletal muscle-specific gene regulation. They used an embryonic muscle fibroblast cell line known as C3H 10T½. Although these cells do not exhibit any differentiated characteristics, they can be induced to differentiate into myoblast cells expressing a number of muscle-lineage genes upon treatment with 5-azacytidine (Constantinides *et al.*, 1977). This agent is a cytidine analogue having a nitrogen instead of a carbon atom at position 5 on the pyrimidine ring and is incorporated into DNA instead of cytidine. Unlike cytidine, however, it cannot be methylated at this position and hence its incorporation results in demethylation of DNA. As methylation of DNA at C residues is thought to play a critical role in transcriptional silencing of gene expression (for review see Latchman, 2002), this artificial demethylation can result in the expression of particular genes which were previously silent.

In the case of 10T½ cells therefore this demethylation was thought to result in the expression of previously silent regulatory loci that are necessary for differentiation into muscle myoblasts. Several experiments also suggested that the activation of only one key regulatory locus might be involved. Thus 5-azacytidine induces myoblasts at very high frequency consistent with only the demethylation of one gene being required while DNA prepared from differentiated cells can also induce differentiation in untreated cells at a frequency consistent with the transfer of only one activated locus.

Hence differentiation is thought to occur via the activation of one regulatory locus (gene X in Fig. 7.2) whose expression in turn switches on the expression of genes encoding muscle lineage markers which is observed in the differentiated 10T½ cells and thereby induces their differentiation. This suggested that the regulatory locus might encode a transcription factor which switched on muscle-specific gene expression.

To isolate the gene encoding this factor, Davis *et al.* (1987) reasoned that it would continue to be expressed in the myoblast cells but would evidently not be expressed in the undifferentiated cells. They therefore prepared RNA from the differentiated cells and removed from it by subtractive hybridization all the RNAs which were also expressed in the undifferentiated cells. After various further manipulations to exclude RNAs characteristic of terminal muscle differentiation such as myosin and others induced non-specifically in all cells by 5-azacytidine, the enriched probe was used to screen a cDNA library prepared from differentiated 10T½ cells.

This procedure (Fig. 7.3) resulted in the isolation of three clones, MyoA, MyoD and MyoH whose expression was specifically activated when 10T½ cells

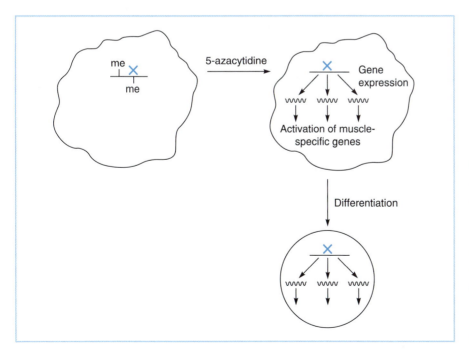

Figure 7.2

Model for differentiation of 10T½ cells in response to 5-azacytidine. Activation of a master locus (x) by demethylation allows its product to activate the expression of muscle-specific genes thereby producing differentiation.

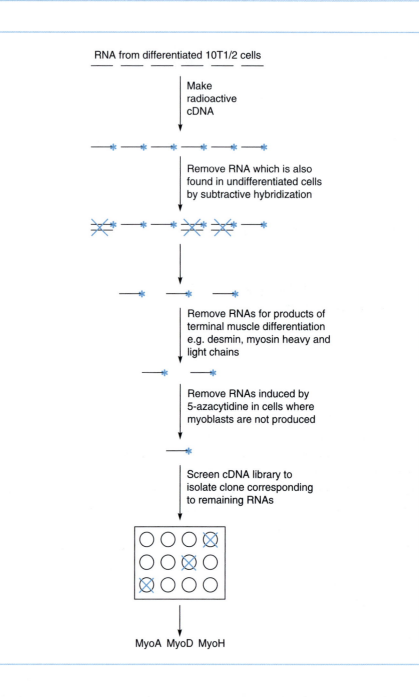

Figure 7.3

Strategy for isolating the master regulatory locus expressed in 10T½ cells after but not before treatment with 5-azacytidine. Subtractive hybridization was used to isolate all RNA molecules that are present in 10T½ cells only following treatment with 5-azacytidine. After removal of RNAs for terminal differentiation products of muscle and RNAs induced in non-muscle-producing cells by 5-azacytidine, the remaining RNAs were used to screen a cDNA library. Three candidates for the master regulatory locus MyoA, MyoD and MyoH were isolated in this way.

were induced to form myoblasts with 5-azacytidine. When each of these genes was artificially expressed in 10T½ cells, MyoA and MyoH had no effect. However, artificial expression of MyoD was able to convert undifferentiated 10T½ cells into myoblasts (Fig. 7.4). Hence expression of MyoD alone can induce differentiation of 10T½ cells into muscle cells and it is the induction of

Figure 7.4

Test of each of the putative master regulatory loci MyoA, MyoD and MyoH. Each of the genes was introduced into 10T½ cells and tested for the ability to induce the cells to differentiate into muscle cells. Note that while MyoA and MyoH have no effect, introduction of MyoD results in the production of muscle cells which contain the muscle protein myosin. The differentiated muscle cells induced by MyoD cease to divide on differentiation resulting in less cells being detectable by staining with crystal violet compared to the MyoA and MyoH treated cells which continue to proliferate. Hence only MyoD has the capacity to cause 10T½ cells to differentiate into non-proliferating muscle cells producing myosin, identifying it as a master regulatory locus for muscle differentiation.

MyoD expression by 5-azacytidine that is responsible for the ability of this compound to induce muscle differentiation.

The differentiated 10T½ cells produced by artificial expression of MyoD, like those induced by 5-azacytidine express a variety of muscle lineage markers and indeed also switch on both MyoA and MyoH as well as the endogenous MyoD gene itself. This suggests that MyoD is a transcription factor that switches on genes expressed in muscle cells. In agreement with this, MyoD was shown to bind to a region of the creatine kinase gene upstream enhancer which was known to be necessary for its muscle specific gene activity.

Moreover, it has been shown that MyoD can actually bind to its binding sites within target genes when they are in the tightly packed chromatin structure characteristic of genes that are inactive in a particular lineage (Gerber *et al.*, 1997). This binding results in the remodelling of the chromatin to a more open form and is then followed by enhanced transcription stimulated by MyoD (Fig. 7.5). This alteration in chromatin structure is likely to be dependent on the ability of MyoD to interact with the p300 co-activator protein (Puri *et al.*, 1997) (see Chapter 5, section 5.4.3). Like CBP, p300 has histone acetyltransferase activity and is therefore able to alter chromatin to the more open structure associated with acetylated histones (see Chapter 1, section 1.2.3).

Hence MyoD is capable of activating transcription by two distinct means, namely the remodelling of chromatin and the direct stimulation of enhanced

Figure 7.5

MyoD binding to its binding site (solid box) both converts the chromatin structure from a closed (wavy line) to a more open (solid line) structure compatible with transcription and also directly enhances the rate of transcription (arrow).

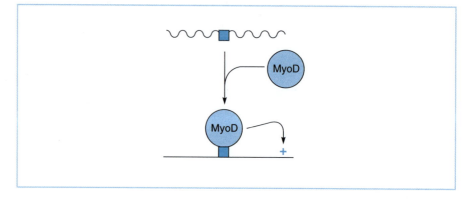

transcription (see Chapter 5, for a discussion of the mechanisms of transcriptional activation). This is particularly important since it allows enhanced synthesis of MyoD to induce the development of myogenic cells from non-differentiated precursors, in which the genes that must be switched on are in an inactive closed chromatin structure which is inaccessible to many transcriptional activators.

Interestingly, as well as stimulating muscle-specific genes, MyoD also promotes differentiation by modulating gene expression so as to inhibit cellular proliferation, thereby producing the non-dividing phenotype characteristic of muscle cells. Thus MyoD has been shown to activate the gene encoding the p21 inhibitor of cyclin-dependent kinases (Halevy *et al.*, 1995). This results in the inhibition of these kinases whose activity is necessary for cell division (see Chapter 9, section 9.4.2). In addition, MyoD can also repress the promoter of the c-*fos* gene whose protein product is important for cellular proliferation (see Chapter 9, section 9.3.1) indicating that MyoD can also act by repressing genes whose products are not required in non-dividing muscle cells (Trouche *et al.*, 1993).

Like gene activation by MyoD, repression of the c-*fos* promoter is dependent on DNA binding which, in this case, prevents the binding of a positively acting factor to a site known as the serum response element which overlaps the MyoD binding site in the c-*fos* promoter (Fig. 7.6). Obviously, in contrast to its binding to the creatine kinase enhancer, MyoD must bind to its binding site in the c-*fos* promoter in a form which cannot activate transcription. Hence, like the glucocorticoid receptor, MyoD can have different effects on gene expression depending on the nature of its binding site (see Chapter 6, section 6.2.1 for discussion of the mechanism of transcriptional repression by the glucocorticoid receptor). In both cases, however, DNA binding by MyoD is dependent upon a basic region of the protein which binds directly to the DNA and an adjacent region which can form a helix-loop-helix structure and is

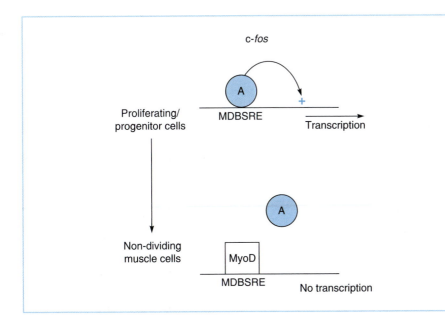

c-fos

Proliferating/
progenitor cells

MDBSRE

Transcription

Non-dividing
muscle cells

MyoD

MDBSRE   No transcription

Figure 7.6
MyoD binds to its binding
site (MDBS) in the c-*fos*
promoter in a
configuration which does
not activate transcription
and prevents binding of
an activating factor (A) to
the overlapping serum
response element (SRE).
This therefore results in
the repression of c-*fos*
transcription in MyoD-
expressing muscle cells.

essential for dimerization of MyoD (see Chapter 4, section 4.5 for further discussion of these motifs).

Hence, synthesis of MyoD results in the production of the skeletal muscle phenotype by activating and repressing the expression of specific target genes. As expected in view of the critical role that MyoD plays in the development of muscle cells, the MyoD mRNA is present in skeletal muscle tissue taken from a variety of different sites in the body but is absent in all other tissues including cardiac muscle (Davis *et al.*, 1987; Fig. 7.7). The MyoD mRNA and protein therefore accumulate only in a specific cell type where it is required and the activation of the MyoD gene during myogenesis is likely to be of central importance in switching on the expression of muscle-specific genes. In turn, this suggests that other developmentally regulated transcription factors will be involved in switching on MyoD expression during myogenesis. In agreement with this, the paired-type homeobox factor Pax 3 (Chapter 4, section 4.2.7) has been shown to activate MyoD expression and myogenic differentiation in a variety of non-muscle cell types (for review see Rawls and Olson, 1997) while the classical homeobox factor MSX1 can repress the transcription of the MyoD gene (Woloshin *et al.*, 1995).

Thus, MyoD is a transcription factor whose regulated synthesis results in the activation of muscle-specific gene expression and the production of skeletal muscle cells. Interestingly, the observation that the introduction of MyoD into cells switches on the endogenous MyoD gene (see above) suggests that a positive feedback loop normally regulates MyoD expression so that once the

Figure 7.7

Northern blotting experiment to detect the mRNAs encoding MyoA, MyoD and MyoH in different muscle and non-muscle tissues. Note that the MyoD mRNA is present only in skeletal muscle as expected in view of its ability to produce muscle differentiation, whereas the MyoA and MyoH mRNAs are more widely distributed. nb, indicates new born, ad, indicates adult, rRNA indicates the ribosomal RNA control used to show that all samples contain intact RNA.

gene is initially expressed, expression is maintained producing commitment to the myogenic lineage (Fig. 7.8). This is of importance since MyoD appears to be essential for the repair of damaged muscle in adult animals indicating that its expression must be maintained throughout life (Megeney *et al.*, 1996).

MyoD therefore offers a classic example of the role of transcription factor synthesis in regulating cell type-specific gene expression. It should be noted, however, that, as discussed in Chapter 6 (section 6.2.2), the activity of MyoD is

Figure 7.8

Ability of MyoD protein to activate expression of its own gene creating a positive feedback loop which ensures that, following an initial stimulus, the MyoD protein is continuously produced and hence maintains myoblast differentiation.

also regulated by its interaction with the Id inhibitor protein. Hence, MyoD is regulated both by regulating its synthesis and by regulating its activity.

7.2.2 HOMEOBOX TRANSCRIPTION FACTORS

In addition to its role in controlling cell type-specific gene expression, regulation of transcription factor synthesis is also widely used in the control of developmentally-regulated gene expression. Thus numerous studies of the *Drosophila* homeobox transcription factors discussed in Chapter 4 (section 4.2), using both immunofluorescence with specific antibodies and *in situ* hybridization, have revealed highly specific expression patterns for individual factors and the mRNAs which encode them indicating that their role in regulating gene expression in development is dependent, at least in part, on the regulation of their synthesis (Fig. 7.9).

Moreover, such regulated synthesis of specific transcription factors can specifically determine the nature of the cell types that are produced during development. Thus, the LIM homeobox factors Lhx3 and Lhx4 are expressed transiently in motor neurons whose axons extend ventrally (v-MN) but not in those which extend dorsally (d-MN). In knock out mice lacking both Lhx3 and Lhx4, cells which should become v-MN cells, instead become d-MN cells (Fig. 7.10). In contrast, misexpression of Lhx3 is sufficient to convert d-MN cells into v-MN cells (Sharma *et al.*, 1998). Hence, the regulated synthesis of these two homeodomain proteins results in cells which express them becoming one type of motor neuron (v-MN), whereas the cells which do not express them, form a different type of motor neuron (d-MN).

Figure 7.9

Localization of the Ftz protein in the *Drosophila* blastoderm embryo using a fluorescent antibody which reacts specifically with the protein. The anterior end of the embryo is to the left and the dorsal surface to the top of the photograph. Note the precise pattern of seven stripes of Ftz-expressing cells around the embryo.

Figure 7.10

The homeobox
transcription factors Lhx3
and Lhx4 are expressed
in motor neurons whose
axons project ventrally
(vMN) but not in those
which project dorsally
(dMN). Inactivation of
Lhx3 and Lhx4 in knock
out mice converts vMN
cells into dMN cells
whereas artificial
expression of Lhx3 in
dMN cells converts them
into vMN cells.

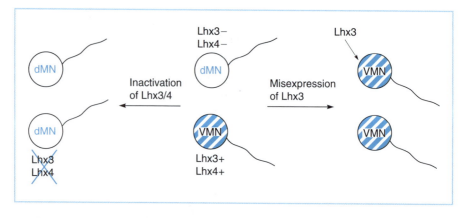

Figure 7.10

The homeobox
transcription factors Lhx3
and Lhx4 are expressed
in motor neurons whose
axons project ventrally
(vMN) but not in those
which project dorsally
(dMN). Inactivation of
Lhx3 and Lhx4 in knock
out mice converts vMN
cells into dMN cells
whereas artificial
expression of Lhx3 in
dMN cells converts them
into vMN cells.

In an even more dramatic example of this effect, the Pitx2 homeobox
factor is expressed only on the left side of the developing embryo in the
mouse or chicken. Expressing Pitx2 on the right side of the embryo affects
the normal pattern of asymmetry between the left and right sides of the
embryo (Logan *et al.*, 1998; Piedra *et al.*, 1998), indicating that the appropriate
regulation of its synthesis is required for the embryo to develop distinct left
and right sides.

Hence, the regulated expression of homeobox factors is essential for their
role in regulating gene transcription and cell fate in development. As discussed in Chapter 4 (section 4.2), the mouse homeobox genes are found in
clusters containing a number of different genes (Fig. 7.11). Most interestingly,
in both *Drosophila* and mammals, the position of a gene within a cluster is
related to its expression pattern during embryogenesis. Thus, in the mouse
Hoxb cluster all the genes are expressed in the developing central nervous

Figure 7.11

Hoxb gene cluster on
mouse chromosome 11.
Note that in moving from
the left to the right of the
mouse complex, the
genes are expressed
progressively earlier in
development, have a
more anterior boundary of
expression and a greater
responsiveness to retinoic
acid.

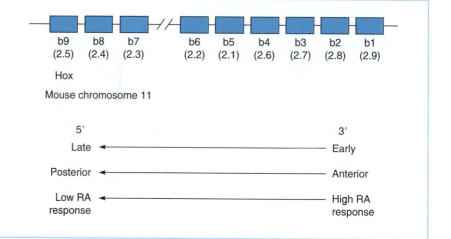

system of the embryo. However, in moving from the 5' to the 3' end of the cluster (i.e. from Hoxb-9 (2.5) to Hoxb-1 (2.9) in Fig. 7.11) each successive gene is expressed earlier in development and also displays a more anterior boundary of expression within the central nervous system (Figs 7.12 and 7.13). Similar expression patterns have also been observed in *Drosophila* where each successive gene in the Bithorax and Antennapedia clusters is expressed more

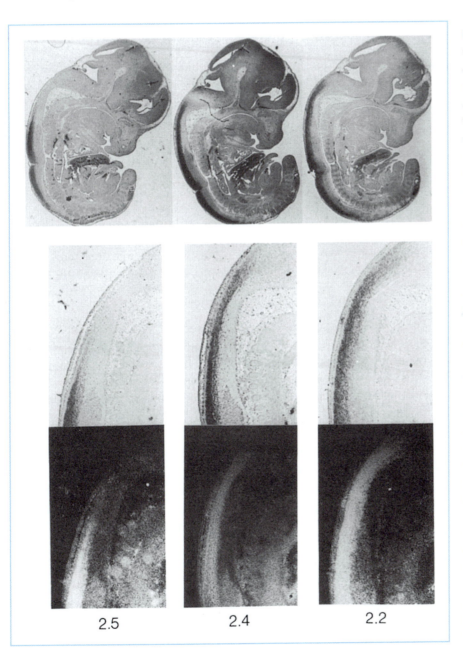

2.5   2.4   2.2

Figure 7.12

Comparison of the expression pattern of the Hoxb-9 (2.5), b-8 (2.4) and b-6 (2.2) genes in the 12.5-day mouse embryo. The top panel shows *in situ* hybridization with the appropriate gene probe to a section of the entire embryo while the middle row shows a high power view of the region in which the anterior limit of gene expression occurs. In these panels, which show the sections in bright field, hybridization of the probe and therefore gene expression is indicated by the dark areas. In the lower panel, which shows the same area in dark field, hybridization is indicated by the bright areas. Note the progressively more anterior boundary of expression of Hoxb-6 (2.2) compared to Hoxb-8 (2.4) and to Hoxb-9 (2.5) and compare with their positions in the Hoxb (Hox 2) complex in Figure 7.11.

Figure 7.13

Summary of the anterior boundary of expression of the genes in the Hoxb (2) complex indicated on a section of a 12.5-day mouse embryo and compared to the position of the gene in the Hoxb (2) cluster. Note the progressively more anterior boundary of expression from the 5' to the 3' end of the Hoxb (2) cluster.

anteriorly and affects progressively more anterior segments when it is mutated. Indeed, studies in which regulatory elements from the invertebrate *Amphioxus* were tested in mouse and chick embryos have indicated that the elements regulating homeobox gene expression have been highly conserved in evolution with the *Amphioxus* elements functioning in these very different species (Manzanares *et al.*, 2000).

In the case of the mouse genes a possible molecular mechanism for the differential expression pattern across a cluster is provided by the finding that genes in the 3' half of the Hoxb cluster are activated in cultured cells by treatment with low levels of retinoic acid, whereas genes in the 5' half of the cluster require much higher levels of retinoic acid for their activation (for review see Conlon, 1995; Tabin, 1995). Considerable evidence exists that retinoic acid can act as a morphogen in vertebrate development and it has been suggested that a gradient of retinoic acid concentration may exist across the developing embryo (reviewed by Conlon, 1995; Tabin, 1995). Hence, the observed difference in expression of the Hoxb genes could be controlled by a retinoic acid gradient (Fig. 7.14). In turn, the Hoxb genes like their *Drosophila* counterparts would switch on other genes required in cells at

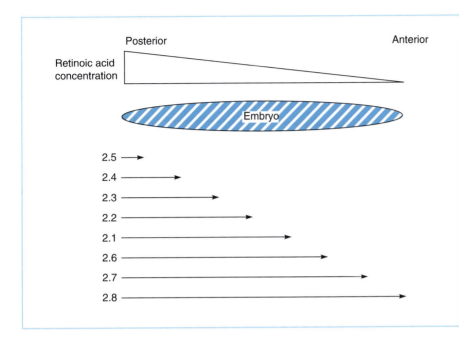

Figure 7.14

Model for the progressively more anterior expression of the genes in the Hoxb (2) cluster in which expression is controlled by a posterior to anterior gradient in retinoic acid concentration and the increasing sensitivity to induction by retinoic acid which occurs from the 5' to the 3' end of the cluster. Thus because genes at the 3' end of the cluster are inducible by very low levels of retinoic acid they will be expressed in anterior points of the embryo where the retinoic acid level will be too low to induce the genes at the 5' end of the cluster which require a much higher level of retinoic acid to be activated.

particular positions in the embryo accounting for the morphogenetic effects of retinoic acid.

Retinoic acid functions by binding to and activating specific receptors which are members of the steroid-thyroid hormone receptor super family and which in turn bind to specific sequences within retinoic acid responsive genes activating their expression (see Chapter 4, section 4.4, and Chapter 8, section 8.2.2). Hence, the activation of regulatory genes and the initiation of a regulatory cascade can be achieved by the activation of specific receptors-transcription factors by an inducing stimulus.

This illustrates therefore how the synthesis of one set of transcription factors (the homeobox proteins) can be regulated by the activation of another set of transcription factors (the retinoic acid receptors). In agreement with this idea, the treatment of mouse embryos with retinoic acid results, for example, in changes in the expression pattern of the Hoxb-1 gene which contains a retinoic acid response element in its 3' regulatory region. Moreover, the inactivation of this element so that it no longer binds the retinoic acid receptors, abolishes expression of Hoxb-1 in the neuroectoderm of the early embryo, providing direct evidence that the retinoic acid response element is necessary to produce the expression pattern of this gene observed in the developing embryo (Marshall *et al.*, 1994; for review see Stern and Foley, 1998).

Interestingly, the regulation of Hox gene expression by such DNA response elements located adjacent to the individual genes appears to interact with other regulatory processes which operate over the whole gene cluster. Thus, in experiments where individual Hox genes (with their adjacent control elements) were moved to a different position within the gene cluster, their pattern of expression was altered so that they behaved similarly to genes normally located at that position in the cluster for example, in terms of the time at which they were switched on during development (van der Hoeven *et al.*, 1996) (Fig. 7.15).

In the case of the HoxD cluster, this effect appears to involve the order of the genes relative to a distant enhancer element located at least 100 000 bases away. Thus, in this cluster the first gene, HoxD13 is expressed most anteriorly and at the highest level with each successive gene being exposed at lower levels and more posteriorly. If Hox D13 is deleted, the next gene in the cluster HoxD12 is now expressed in the manner typical of Hox D13 even though it remains in its normal position (Kmita *et al.*, 2002; for review see Zeller and Deschamps, 2002). In this case therefore, the genes appear to compete to interact with the distant enhancer element so that the closest gene is expressed in a particular pattern and so on (Fig. 7.16). This effect is evidently reminiscent of the locus control region (LCR) in the β-globin gene cluster (see Chapter 1, section 1.3.4) where the globin genes were expressed in a specific order in development which is determined by their position relative to the LCR.

The specific pattern of expression of individual homeobox genes, which is determined by their position in the cluster is absolutely critical to their function. Indeed, it appears that it is the different patterns of regulation rather

Figure 7.15

Each gene in a Hox cluster has its own specific expression pattern (panel i). Moving a particular gene to a new position in the cluster results in it having the expression pattern of the gene which is normally located at that position (panel ii).

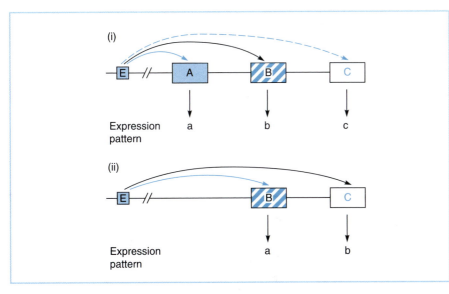

Figure 7.16

In the HoxD cluster, the expression of the genes is affected by their order relative to a distant enhancer element (E) (panel i). Deletion of gene A results in gene B being the closest to the enhancer. It is therefore expressed in the normal pattern for gene A even though its physical location is unchanged (panel ii).

than the different proteins that they encode, that determine the different roles of specific genes in a cluster. Thus, if an individual gene in a cluster is deleted, the other genes in the cluster cannot substitute for it and an abnormal animal results (Fig. 7.17). However, if the deleted gene is replaced by a further copy of another gene in the cluster, then a normal animal results (for review see Duboule, 2000) (Fig. 7.17). This occurs because the expression of the inserted gene is now determined by its position in the cluster and it is therefore expressed in the manner characteristic of the deleted gene. Hence, the products of different genes in a cluster can functionally substitute for one another but only if they are expressed in the appropriate pattern, as determined by their position in the cluster. This illustrates the critical role of the regulated synthesis of transcription factors in allowing them to produce their functional effects.

The manner in which the regulated synthesis of multiple homeobox factors can regulate the production of several different cell types has been analysed in detail in the ventral neural tube. In this case, the system is regulated by a gradient in the concentration of a protein signalling molecule known as sonic hedgehog (Shh) rather than via a retinoic acid gradient. The expression of several homeobox factors (Dbx1, Dbx2, Irx3 and Pax6) is repressed by Shh, but their sensitivity to such repression differs so that Pax6, for example, is expressed at higher Shh concentrations than Irx3 and so on (Fig. 7.18). In contrast, two other homeobox genes are activated by Shh but their sensitivity differs so that Nkx6.1 is expressed at lower levels of Shh than Nkx2.2 (Fig. 7.18).

Figure 7.17

The normal expression pattern of Hox genes results in the production of a normal animal (panel i). Inactivation of a specific gene results in a mutant animal being produced (panel ii). However, if the mutant gene is replaced with a further copy of another gene in the cluster, this gene is expressed in the same way as the deleted gene and a normal animal results (panel iii).

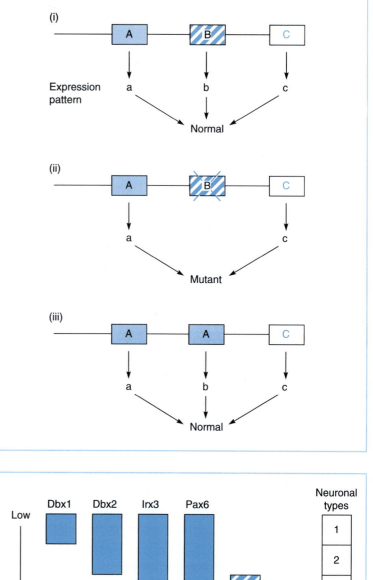

Figure 7.18

In the ventral neural tube, a gradient of sonic hedgehog regulates the expression of several homeobox genes. Dbx1, Dbx2, Irx3 and Pax6 are repressed by sonic hedgehog but differ in their sensitivity to repression. Thus, Pax6, which is the least sensitive to repression, is expressed at higher sonic hedgehog concentrations than Irx3 and so on. Conversely, Nkx6.1 and Nkx2.2 are activated by sonic hedgehog with Nkx6.1 being activated at lower concentrations than Nkx2.2. Together these effects create a homeodomain code in which each region has a different pattern of expression of the six genes and hence different neuronal types (1–5) form at each point.

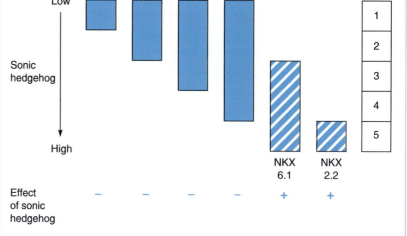

The different expression patterns of these genes, therefore, convert the gradient of Shh expression into a homeobox code in which each region has a unique pattern of expression of the different homeobox genes. In turn, this results in five different neuronal types forming at different positions in the ventral neural tube (Fig. 7.18) (Briscoe *et al.*, 2000). Hence, in this case, the precise combination of specific homeobox genes expressed in each position controls the precise cell type that is formed (for review see Marquardt and Pfaff, 2001).

In our discussion so far, it has been assumed that a homeobox factor is either present in a particular cell or is entirely absent. In fact, however, a further level of complexity exists since many homeobox factors are not expressed in a simple on/off manner but rather show a concentration gradient ranging from high levels in one part of the embryo via intermediate levels to low levels in another part. For example, in *Drosophila*, the bicoid protein (bcd) whose absence leads to the development of a fly without head and thoracic structures is found at high levels in the anterior part of the embryo and declines progressively posteriorly, being absent in the posterior one third of the embryo (Fig. 7.19).

Most interestingly, genes that are activated in response to bicoid contain binding sites in their promoters which have either high affinity or low affinity for the bicoid protein. If these sites are linked to a marker gene, it can be demonstrated that genes with low affinity binding sites are only activated at high concentrations of bicoid and are therefore expressed only at the extreme anterior end of the embryo. In contrast, genes that have higher affinity binding

Figure 7.19

The gradient in Bicoid concentration from the anterior to the posterior point of the embryo results in bicoid-dependent genes with only low affinity binding sites for the protein being active only at the extreme anterior part of the embryo, whereas genes with high affinity binding sites are active more posteriorly. Note that, in addition to the different posterior boundaries in the expression of genes with high and low affinity binding sites, genes with high affinity binding sites will be expressed at a higher level than genes with low affinity binding sites at any point in the embryo.

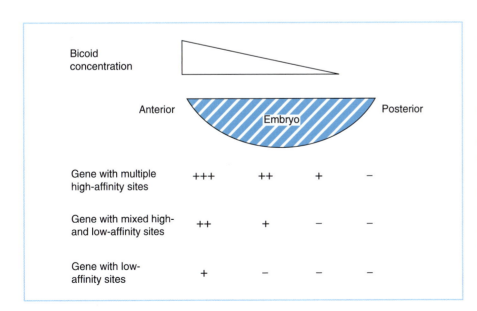

Bicoid concentration				
	Anterior	Embryo		Posterior
Gene with multiple high-affinity sites	+++	++	+	–
Gene with mixed high- and low-affinity sites	++	+	–	–
Gene with low-affinity sites	+	–	–	–

sites are active at much lower protein concentrations and will be active both at the anterior end and more posteriorly. Moreover, the greater the number of higher affinity sites the greater the level of gene expression that will occur at any particular point in the gradient (Driever *et al.*, 1989; Fig. 7.19).

The gradient in bicoid expression can be translated therefore into the differential expression of various bicoid-dependent genes along the anterior part of the embryo. Each cell in the anterior region will be able to 'sense' its position within the embryo and respond by activating specific genes. One of the genes activated by bicoid is the homeobox-containing segmentation gene hunchback. In turn, this protein regulates the expression of the gap genes, Kruppel and giant (Struhl *et al.*, 1992). All four of these proteins then act on the eve gene with bicoid and hunchback activating its expression while Kruppel and giant repress it. The concentration gradients of these four factors thus result in the spatial localization of eve gene expression in a defined region of the embryo where it exerts its inhibitory effects on gene expression (Small *et al.*, 1991; Fig. 7.20). Hence, the gradient in bicoid gene expression results in changes in the expression of other genes encoding regulatory proteins leading to the activation of regulatory networks involving the controlled synthesis of multiple transcription factors.

The bicoid factor therefore has all the properties of a morphogen whose concentration gradient determines position in the anterior part of the embryo. This idea is strongly supported by the results of genetic experiments in which the bicoid gradient was artificially manipulated, cells containing artificially increased levels of bicoid assuming a phenotype characteristic of more anterior cells which normally contain the new level of bicoid and vice versa (Driever and Nusslein-Volhard, 1988).

The anterior to posterior gradient in bicoid levels is required to produce the opposite posterior to anterior gradient in the level of another protein, caudal. However, the caudal mRNA is equally distributed throughout the embryo indicating that the bicoid gradient does not regulate transcription of the caudal gene. Rather, the bicoid protein binds to the caudal mRNA and represses its translation into protein so that caudal protein is not produced when bicoid levels are high (reviewed by Carr, 1996; Chan and Struhl, 1997). As well as providing further evidence for the key role of the bicoid factor, this finding also shows that homeodomain proteins can bind to RNA as well as to DNA and that they may therefore act at the post-transcriptional level as well as at transcription.

The bicoid case clearly illustrates therefore how the regulated synthesis of an individual factor, resulting in a gradient in its concentration, can alter the expression of a regulatory network of other genes and ultimately control the differentiation of specific cells during development.

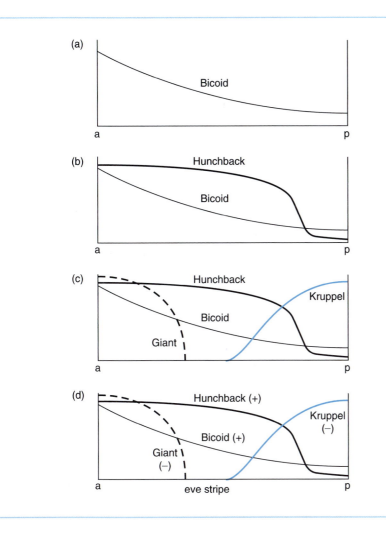

Figure 7.20
Model illustrating how the concentration gradients of the activators bicoid and hunchback and the repressors Kruppel and giant produce a stripe of eve gene expression. Note that the bicoid gradient (a) affects hunchback expression (b) which in turn affects giant and kruppel expression (c). Eve gene activation (+) by hunchback and bicoid and its repression (-) by giant and Kruppel then produces a specific stripe or region of the embryo in which eve is expressed (d).

7.3 MECHANISMS REGULATING THE SYNTHESIS OF TRANSCRIPTION FACTORS

The cases discussed in the previous section illustrate therefore that where a factor must be active in a particular cell type or at a specific point in development, this is frequently achieved by the factor being present only in the particular cells where it is required. Clearly, such regulated synthesis of a specific transcription factor could be achieved by any of the methods that are normally used to regulate the production of individual proteins, such as the regulation of gene transcription, RNA splicing or translation of the mRNA (Fig. 7.21, for review of the levels at which gene regulation can occur see

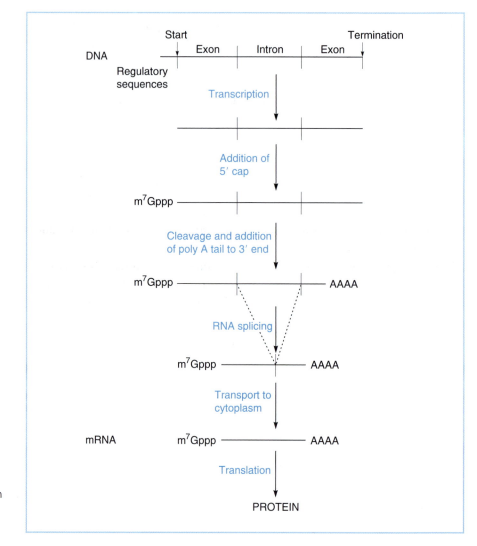

Figure 7.21
Potential regulatory stages in the expression of a gene encoding a transcription factor.

Latchman, 2002). Several of these mechanisms of gene regulation are utilized in the case of individual transcription factors and these will be discussed in turn.

7.3.1 REGULATION OF TRANSCRIPTION

As discussed above, a number of cases where the cell type-specific expression of a transcription factor is paralleled by the presence of its corresponding mRNA in the same cell type have now been described. In turn this cell type-specific expression of the transcription factor mRNA is likely to result from the regulated transcription of the gene encoding the transcription factor.

Unfortunately, the low abundance of many transcription factors has precluded the direct demonstration of the regulated transcription of the genes that encode them. This has been achieved, however, in the case of the CCAAT box binding factor C/EBP which regulates the transcription of several different liver-specific genes such as transthyretin and alpha-1 anti-trypsin. Thus by using nuclear run on assays to measure directly transcription of the gene encoding C/EBP, Xanthopoulos *et al.* (1989) were able to show that this gene is transcribed at high levels only in the liver, paralleling the presence of C/EBP itself and the mRNA encoding it at high levels only in this tissue (Fig. 7.22). Hence the regulated transcription of the C/EBP gene in turn controls the production of the corresponding protein which, in turn, directly controls the liver specific transcription of other genes such as alpha-1 anti-trypsin and transthyretin.

Interestingly, as well as being used to regulate the relative amounts of a particular factor produced by different tissues, transcriptional control can also be used to regulate factor levels within a specific cell type. Thus the levels of the liver-specific transcription factor DBP are highest in rat hepatocytes in the afternoon and evening, with the protein being undetectable in the morning. This fluctuation is produced by regulated transcription of the gene encoding DBP, which is highest in the early evening and undetectable in the morning, whereas the C/EBP gene is transcribed at equal levels at all times. In turn the alterations in DBP level produced in this way produce similar diurnal fluctuations in the transcription of the albumin gene which is dependent on DBP for its transcription (Wuarin and Schibler, 1990).

Although regulated transcription of the genes encoding the transcription factors themselves is likely therefore to constitute an important means of regulating their synthesis it is clear that this process simply sets the problem of gene regulation one stage further back. Thus it will be necessary to have

Figure 7.22

Nuclear run on assay of transcription in the nuclei of kidney and liver. Values indicate the degree of transcription of each gene in the two tissues. Note the enhanced transcription in the liver of the gene encoding the transcription factor C/EBP as well as of the genes encoding the liver-specific proteins transthyretin (TTR) and alpha-1 anti-trypsin (alpha 1AT). The positive control transfer RNA gene is, as expected, transcribed at equal levels in both tissues while the negative control, pBR322 bacterial plasmid does not detect any transcription.

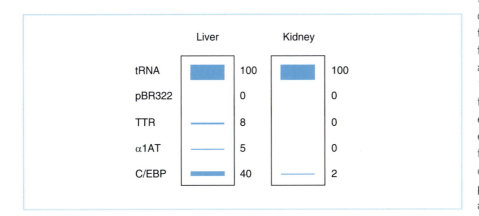

some means of regulating the specific transcription of the gene encoding the transcription factor itself which, in turn, may require other transcription factors that are synthesized or are active only in that specific cell type. It is not surprising therefore that the synthesis of transcription factors is often modulated by post-transcriptional control mechanisms not requiring additional transcription factors. These mechanisms will now be discussed.

7.3.2 REGULATION OF RNA SPLICING

Numerous examples have now been described in eukaryotes where a single RNA species transcribed from a particular gene can be spliced in two or more different ways to yield different mRNAs encoding proteins with different properties (for review see Latchman, 2002). This process is also used in several cases of genes encoding specific transcription factors, for example, in the case of the *era*-1 gene which encodes a transcription factor that mediates the induction of gene expression in early embryonic cells in response to retinoic acid. In this case two alternatively spliced mRNAs are produced, one of which encodes the active form of the molecule, while the other produces a protein lacking the homeobox region. As the homeobox mediates DNA binding by the intact protein (see Chapter 4, section 4.2.3), this truncated form of the protein is incapable of binding to DNA and activating gene expression (Larosa and Gudas, 1988). A similar use of alternative splicing to create mRNAs encoding proteins with and without the homeobox has also been reported for the Hoxb-6 (2.2) gene (Shen *et al.*, 1991).

Hence in these cases where one of the two proteins encoded by the alternatively spliced mRNAs is inactive, alternative splicing can be used in the same way as the regulation of transcription in order to control the amount of functional protein that is produced.

Interestingly, however, unlike transcriptional regulation, alternative splicing can also be used to regulate the relative production of two distinct functional forms of a transcription factor which have different properties. This is seen in the case of the Pax8 factor which is a member of the Pax family (see Chapter 4, section 4.2.7). In this case, alternative splicing results in the insertion of a single serine residue in the recognition helix of the paired domain which is critical for DNA binding (Fig. 7.23). This alters the DNA binding properties of the factor so that it recognizes different DNA sequences to the form of Pax8 which lacks this residue (Kozmik *et al.*, 1997). Hence alternative splicing can introduce a subtle, single amino acid change in a transcription factor which results in the existence of two forms of the factor with different DNA binding specificities.

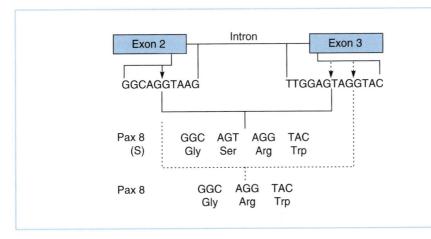

Figure 7.23

Alternative splicing in the Pax8 gene involving the use of different splice sites in exon 3 (dotted arrows) together with the same splice site in exon 2 (solid arrow) generates different forms of the protein with and without an additional serine residue and thus having different DNA binding specificities.

As well as affecting DNA binding specificity, alternative splicing can also produce forms of a transcription factor with distinct effects on transcription. This is seen in the case of the CREM factor which is related to the CREB factor discussed in Chapter 5 (section 5.4.3). Thus CREM resembles CREB in being phosphorylated following cyclic AMP treatment at a site located between two glutamine-rich activation domains. Like CREB, it can therefore bind to the CRE and activate transcription in response to cyclic AMP by binding the co-activator CBP (for reviews see de Cesare *et al.*, 1999; de Cesare and Sassoni-Corsi, 2000.)

Interestingly, however, alternative splicing produces distinct forms of the CREM factor that lack the activation domains, although they retain the leucine zipper and basic DNA binding domain (Fig. 7.24a) (see Chapter 4, section 4.5 for a discussion of these motifs). These forms can therefore bind to DNA but cannot activate transcription since they lack an activation domain. They therefore inhibit transcription by competing for binding to the CRE with the activating forms (Fig. 7.25) (see Chapter 6, section 6.2.1 for a discussion of indirect repression of this type). Since the proportion of the activating and inhibitory forms of CREM varies in different cell types, the level of transcription directed by a CRE following cyclic AMP treatment will be different in these cells depending on the precise balance between the activating and inhibitory forms.

As well as producing distinct forms with and without the activation domain, the CREM factor also undergoes alternative splicing in another manner. Thus two distinct exons in the CREM gene contain two distinct DNA binding domains. Alternative splicing results in the proteins that either do or do not contain the activation domains also having one or other of the DNA

Figure 7.24

The CREM protein contains two transcriptional activation domains (A), a region containing a site for cyclic AMP-induced phosphorylation (P) and two DNA-binding domains containing a basic domain and leucine zipper (BD/LZ). After transcription from the P$_1$ promoter, alternative splicing can result in forms with or without the activation domains (a) or having either of the DNA-binding domains (b). In addition, cyclic AMP-inducible transcription from the P$_2$ promoter can produce forms (inducible cyclic AMP early repressors: ICERs) containing only one or other of the DNA-binding domains but lacking the activation domains and the phosphorylated region (c). Arrows indicate the transcriptional start sites used in each case.

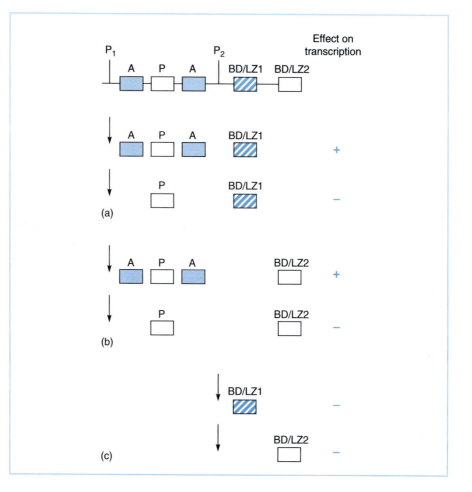

binding domains (Fig. 7.24b). As the relative usage of the two DNA binding domains is different in different cell types, this effect is likely to have biological significance but its precise role is at present unclear.

The different forms of the CREM factor which have been discussed so far are all produced by alternative splicing of a single RNA transcript whose rate of production is unaffected by cyclic AMP. The ability of the CREB factor and the activating forms of CREM to switch on gene expression is then stimulated post-translationally by their phosphorylation following cyclic AMP treatment, hence allowing them to switch on gene expression in response to cyclic AMP (see Chapter 5, section 5.4.3).

In contrast to such post-translational regulation, the CREM gene also contains a promoter which is activated in response to cyclic AMP. This promoter produces transcripts encoding short proteins which contain one or other of the DNA binding domains and the phosphorylated region but lack the activa-

Figure 7.25
Gene activation by the
activating forms of the
CREM protein (A) can be
inhibited by forms (I)
which contain the DNA-
binding domain (hatched
shading) but lack the
activation domain (solid
shading). They therefore
bind to the CRE and
prevent binding by the
activating forms.

tion domain (Fig. 7.24c). These proteins can therefore bind to the cyclic AMP response element and repress transcriptional activation by the activating forms, exactly as described above for the alternatively spliced forms lacking the activation domain. These forms are therefore known as ICERs (inducible cyclic AMP early repressors). As they are inducible by cyclic AMP, these forms are likely to play a key role in making the cyclic AMP response self-limiting. Thus following cyclic AMP treatment CREB and CREM will become phosphorylated and will then activate the expression of promoters containing a CRE including that which produces the ICERs. The ICERs produced in this manner will then bind to the CRE and switch off the inducible genes by preventing the binding of CREB and CREM (Fig. 7.26) thereby making the cyclic AMP response a transient one.

The regulation of cyclic AMP inducible transcription by the CREB and CREM factors is therefore extraordinarily complex with both alternative splicing and the use of two different promoters in the CREM gene. It illustrates therefore how the combination of transcriptional and post-transcriptional control of synthesis can be used to produce multiple forms of transcription factors with different functional roles.

Alternative splicing can also occur in factors which contain a specific inhibitory domain and which can therefore function as direct repressors interfering with the activity of the basal transcriptional complex (see Chapter 6, section 6.3.2). Thus, although it is transcribed in B cells and not in most other cell types, the gene encoding the Oct-2 transcription factor (which is a member of the POU family, discussed in Chapter 4, section 4.2.6) is also transcribed in neuronal cells. In neuronal cells, the Oct-2 RNA is spliced so that the protein it encodes does not contain the C-terminal activation domain which allows it to activate transcription. It does, however, retain the N-terminal inhibitory domain discussed in Chapter 6 (section 6.3.2) as well as the DNA binding domain and can therefore act as a direct inhibitor of gene expression (Lillycrop et al., 1994). In contrast, in B cells, alternative splicing produces an mRNA which encodes a protein containing both the inhibitory domain and the stronger activation domain and which therefore activates transcription (Fig. 7.27). Hence in this case alternative splicing produces

Figure 7.26

(a) In the absence of cyclic AMP, the CREM gene is transcribed from the P_1 promoter. However, neither the CREM produced in this way nor the CREB protein can activate transcription until they are activated post-translationally. (b) Following cyclic AMP treatment, the CREB and CREM proteins become activated post-translationally by phosphorylation. They therefore activate the cyclic AMP inducible genes (cAI) which contain a cyclic AMP response element (CRE) in their promoters. In addition, they also activate the P_2 promoter of CREM which also contains a CRE. (c) The ICERs (inducible cyclic AMP early repressors) produced by the CREM P_2 promoter bind to the CREs and prevent activation by CREB and CREM thereby repressing transcription.

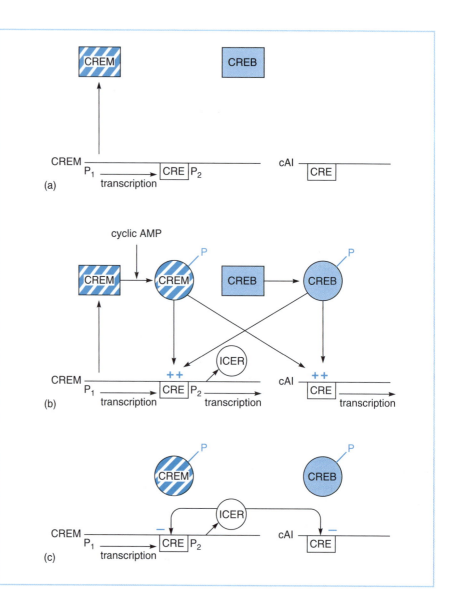

different forms of a factor in different cell types which have opposite effects on the activity of their target promoters.

Such alternative splicing is also seen in the case of another transcription factor containing an inhibitory domain, namely the thyroid hormone receptor. Thus as discussed in Chapter 6 (section 6.3.2), alternative splicing produces two forms of the receptor, one of which lacks the ligand binding domain and therefore cannot bind thyroid hormone (see Fig. 6.14). Although it cannot therefore respond to thyroid hormone, this alpha 2 form of the protein still contains the DNA binding domain and can therefore

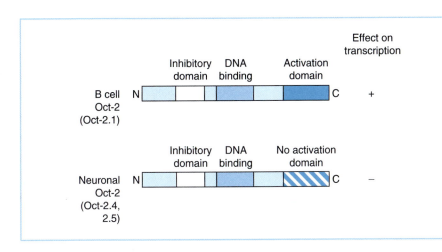

Figure 7.27

In B lymphocytes the predominant form of Oct-2 (Oct 2.1) contains the C-terminal activation domain as well as the DNA binding domain and an inhibitory domain. As the activation domain overcomes the effect of the inhibitory domain, this form is able to activate transcription. In contrast the predominant neuronal forms of Oct-2 (Oct 2.4 and 2.5) contain different C-terminal regions and lack the activation domain. As they retain both the inhibitory domain and the DNA binding domain, however, they can bind to specific DNA-binding sites and inhibit gene expression.

bind to the specific binding site for the receptor in hormone-responsive genes. By doing so, it acts as a dominant repressor of gene activation mediated by the normal receptor in response to hormone binding. Hence these two alternatively spliced forms of the transcription factor, which are made in different amounts in different tissues, mediate opposing effects on thyroid hormone-dependent gene expression.

As well as affecting the actual properties of a transcription factor, regulation of splicing can also be used to determine how much of the protein accumulates. This is seen in the case of the Haclp protein which is a member of the basic-leucine zipper transcription factor family discussed in Chapter 4 (section 4.5). This factor accumulates at an increased level in the presence of unfolded proteins in the cell and then activates the expression of genes which assist other proteins to fold properly. This increased accumulation of Haclp is controlled by a splicing event which removes an intron from the Haclp transcript. When this intron is present, the RNA forms a folded structure which cannot be translated to produce Haclp protein. When the intron is removed by splicing, this folded structure no longer forms and the Haclp mRNA is translated (Rüegsegger and Leber, 2001) (Fig. 7.28). Hence, in this case the regulation of splicing alters the amount of the transcription factor produced rather than its activity.

The examples of regulated splicing discussed above thus illustrate the potential of this process in controlling the level of functional transcription factor which is produced or in generating different forms of a particular transcription factor which, because of differences in the regions that mediate DNA binding or transcriptional activation, have different properties that result in differences in their effects on gene expression.

Figure 7.28

Regulated splicing of the RNA encoding Haclp results in its enhanced synthesis in response to the presence of unfolded proteins in the cell. In the absence of unfolded proteins, the intron is not removed from the RNA and base pairing between the first exon and the intron prevents the RNA from being translated into protein. In the presence of unfolded protein, the intron is removed and the unfolded mRNA is translated to produce functional Haclp protein.

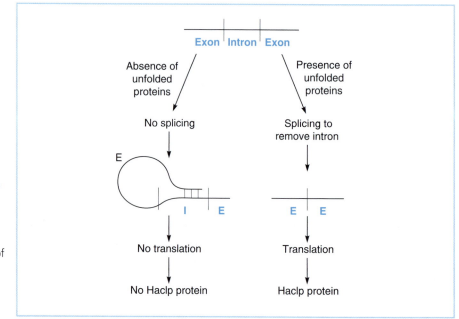

7.3.3 REGULATION OF TRANSLATION

The final stage in the expression of a gene is the translation of its corresponding mRNA into protein. In theory therefore, the regulation of synthesis of a particular transcription factor could be achieved by producing its mRNA in all cell types but translating it into active protein only in the particular cell type where it was required. However, the observed parallels between the cell type-specific expression of a particular transcription factor and the cell type-specific expression of its corresponding mRNA discussed above (section 7.2) indicate that this cannot be the case for the majority of transcription factors. Nonetheless, this mechanism is used to control the synthesis of at least one transcription factor in yeast.

Thus the yeast GCN4 transcription factor controls the activation of several genes in response to amino acid starvation and the factor itself is synthesized in increased amounts following such starvation, allowing it to mediate this effect. This increased synthesis of GCN4 following amino acid starvation is mediated via increased translation of pre-existing GCN4 mRNA (for reviews see Hinnebusch, 1997; Morris and Geballe, 2000). This translational regulation is dependent upon short sequences within the 5' untranslated region of the GCN4 mRNA, upstream of the start point for translation of the GCN4 protein.

Most interestingly, such sequences are capable of being translated to produce short peptides of two or three amino acids (Fig. 7.29). Under conditions

Figure 7.29

Presence of short open reading frames capable of producing small peptides in the 5' untranslated region of the yeast GCN4 RNA. Translation of the RNA to produce these small proteins suppresses translation of the GCN4 protein. The position of the methionine residue beginning each of the small peptides is indicated together with the number of additional amino acids incorporated before a stop codon is reached.

when amino acids are plentiful, these short peptides are synthesized and the ribosome fails to reinitiate at the start point for GCN4 production resulting in this protein not being synthesized. Following amino acid starvation, however, the production of the small peptides is suppressed and the production of GCN4 is correspondingly enhanced. Hence this mechanism ensures that GCN4 is synthesized only in response to amino acid starvation and then activates the genes encoding the enzymes required for the biosynthetic pathways necessary to make good this deficiency.

Interestingly, the use of distinct translational start sites is also seen in the case of the C/EBP transcription factors expressed in the mammalian liver. In this case, however, the two start sites of translation result in two different forms of the C/EBP proteins. The longer form contains an activation domain as well as a basic DNA binding domain and leucine zipper. The other is produced by translational initiation from a downstream start site and therefore lacks the activation domain, although it retains both the basic domain and the leucine zipper (Fig. 7.30). This shorter protein can bind to the same sites as the longer form and since it cannot activate transcription, acts as an inhibitor of gene activation by the longer form (Descombes and Schibler, 1991).

Figure 7.30

The use of different translational initiation codons (vertical arrows) in the mRNA encoding the C/EBP transcription factors produces the longer LAP (liver activator protein) form of the protein which possesses an activation domain and the shorter LIP (liver inhibitor protein) form of the protein which lacks this domain and therefore inhibits gene activation by LAP.

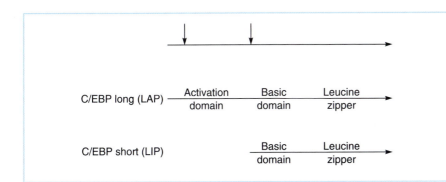

Interestingly, the balance between the long and short forms of C/EBP is controlled by the level in the cell of factors required for the translation of all mRNAs. Thus, when a low level of these translation factors is present in the cell, the upstream start site of translation is used preferentially and the full length protein predominates. In contrast, when higher levels of the translation factors are present, the shorter form of C/EBP is produced in increasing amounts (Calkhoven *et al.*, 2000). Moreover, it has been shown that the shorter form of C/EBP promotes cellular proliferation, whereas the longer form promotes growth arrest and terminal differentiation. Hence, in this case the regulated translation of a transcription factor produces two distinct forms with opposite effects on cellular proliferation and differentiation (Fig. 7.31).

Figure 7.31

The level of translation initiation factors controls the balance between the activating long form of C/EBP which induces growth arrest and differentiation and the inhibitory short form which induces cellular proliferation.

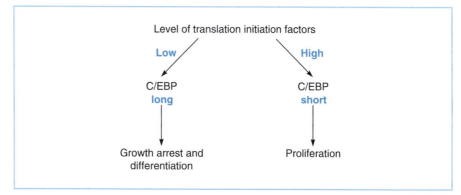

As with the regulation of splicing, the regulation of translation can therefore be used to control the amount of an active factor that is produced as well as to regulate the balance between two functionally antagonistic factors encoded by the same gene.

7.4 CONCLUSIONS

Regulating the synthesis of a transcription factor constitutes a metabolically inexpensive way of controlling its activity. Thus in situations where the activity of a particular factor is not required, no energy is expended on making it in an inactive form. Such regulation probably takes place predominantly at the level of transcription so that no energy is expended on the production of an RNA, its splicing, transport, etc. However, even in cases where regulation occurs at later stages such as splicing or translation, the system is relatively efficient in terms of energy usage, since the step in gene expression that requires the most energy is the final one of translation.

In view of its metabolic efficiency, it is not surprising therefore that the regulation of their synthesis is widely used to control the activity of the factors which mediate cell type-specific gene regulation where differences in the activity of a given factor in different cell types are maintained for long periods of time. Similarly, alternative splicing or use of different translational initiation codons is used to produce different forms of the same factor which often have antagonistic effects on gene expression.

The regulation of factor activity by regulating its synthesis, does suffer, however, from the defect that a change in the level of activity of a factor which is controlled purely by a change in its actual amount can take some time to occur. Thus in response to a signal which induces new transcription of the gene encoding a particular factor, it is necessary to go through all the stages illustrated in Figure 7.21, before the production of active factor which is capable of activating the expression of other genes in response to the inducing signal. It is not surprising therefore that although some factors such as GCN4 which mediate inducible gene expression are regulated by the regulation of their synthesis, the majority of such factors are regulated by post-translational mechanisms which activate pre-existing transcription factor protein in response to the inducing signal. Thus, although mechanisms of this type are metabolically expensive in that they require the synthesis of the factor in situations where it is not required, they have the necessary rapid response time required for the regulation of inducible gene expression. Moreover, unlike transcriptional regulation, they constitute an independent method of gene regulation rather than requiring the activation of other transcription factors in order to activate the transcription of the gene encoding the factor itself. The regulation of transcription factor activity and the manner in which it is achieved is discussed in the next chapter.

REFERENCES

Briscoe, J., Pierani, A., Jessell, T. M. and Ericson, J. (2000) A homeodomain protein code specifies progenitor cell identity and neuronal fate in the ventral neural tube. Cell 101, 435–445.

Calkhoven, C. F., Müller, C. and Leutz, A. (2000) Translational control of C/EBPa and C/EBPb isoform expression. Genes and Development 14, 1920–1932.

Carr, K. (1996) RNA bound to silence. Nature 379, 676.

Chan, S.K. and Struhl, G. (1997) Sequence specific RNA binding by bicoid. Nature 388, 634.

Conlon, R.A. (1995) Retinoic acid and pattern formation in vertebrates. Trends in Genetics 11, 314–319.

Constantinides, P.G., Jones, P.A. and Gevers, W. (1977) Functional striated muscle cells from non-myoblast precursors following 5-azacytidine treatment. Nature 267, 364–366.

Davis, H.L., Weintraub, H. and Lassar, A.B. (1987) Expression of a single transfected cDNA converts fibroblasts to myoblasts. Cell 51, 987–1000.

De Cesare, D., Fimia, G. M. and Sassoni-Corsi, P. (1999) Signalling routes to CREM and CREB: plasticity in transcriptional activation. Trends in Biochemical Sciences 24, 281–285.

De Cesare, D. and Sassoni-Corsi, P. (2000) Transcriptional regulation by cyclic AMP-responsive factors. Progress in Nucleic Acids Research and Molecular Biology 64, 343–369.

Descombes, P. and Schibler, U. (1991) A liver transcriptional activator protein LAP and a transcriptional inhibitory protein LIP are translated from the same mRNA. Cell 67, 569–580.

Driever, W. and Nusslein-Volhard, C. (1988) The bicoid protein determines position in the Drosophila embryo in a concentration-dependent manner. Cell 54, 95–104.

Driever, W., Thoma, G. and Nusslein-Volhard, C. (1989) Determination of spatial domains of zygotic gene expression in the Drosophila embryo by the affinity of binding sites for the bicoid morphogen. Nature 340, 363–367.

Duboule, D. (2000) A Hox by any other name. Nature 403, 607–610.

Gerber, A.N., Klesert, T.R., Bergstrom, D.A. and Tapscott, S.J. (1997) Two domains of MyoD mediate transcriptional activation of genes in repressive chromatin: a mechanism for lineage determination in myogenesis. Genes and Development 11, 436–450.

Halevy, O., Novitch, B.G., Spicer, D.B. et al. (1995) Correlation of terminal cell cycle arrest of skeletal muscle with induction of p21 by MyoD. Science 267, 1018–1021.

Hinnebusch, A.G. (1997) Translational regulation of GCN4. Journal of Biological Chemistry 272, 21661–21664.

Kmita, M., Fraudeau, N., Hérault, Y. and Duboule, D. (2002) Serial deletions and duplications suggest a mechanism for the collinearity of Hoxd genes in limbs. Nature 420, 145–150.

Kozmik, Z., Czerny, T. and Busslinger, M. (1997) Alternatively spliced insertions in the paired domain restrict the DNA sequence specificity of Pax6 and Pax8. EMBO Journal 16, 6793–6803.

Larosa, G.J. and Gudas, L.J. (1988) Early retinoic acid-induced F9 teratocarcinoma stem cell gene ERA-1: alternative splicing creates transcripts for a homeobox-containing protein and one lacking the homeobox. Molecular and Cellular Biology 8, 3906–917.

Latchman, D.S. (2002) Gene Regulation: A eukaryotic perspective. Fourth Edition, Nelson Thornes, Cheltenham, UK.

Lillycrop, K.A., Dawson, S.J., Estridge, J.K. *et al.* (1994) Repression of a herpes simplex virus immediate-early promoter by the Oct-2 transcription factor is dependent upon an inhibitory region at the N-terminus of the protein. Molecular and Cellular Biology 14, 7633–7642.

Logan, M., Pagán-Westphal, S. M., Smith, D. M. *et al.* (1998) The transcription factor Pitx2 mediates situs-specific morphogenesis in response to left-right asymmetric signals. Cell 94, 307–317.

Manzanares, M., Wada, H., Itasaki, N. *et al.* (2000) Conservation and elaboration of *Hox* gene regulation during evolution of the vertebrate head. Nature 408, 854–857.

Marquardt, T. and Pfaff, S. L. (2001) Cracking the transcriptional code for cell specification in the neural tube. Cell 106, 651–654.

Marshall, H., Studer, M., Popperl, H. *et al.* (1994) A conserved retinoic acid response element required for early expression of the homeobox gene Hoxb-1. Nature 370, 567–571.

Megeney, L.A., Kabler, B., Garrett, K. *et al.* (1996) MyoD is required for myogenic stem cell function in adult skeletal muscle. Genes and Development 10, 1173–1183.

Morris, D. R. and Geballe, A. P. (2000) Upstream open reading frames as regulators in mRNA translation. Molecular and Cellular Biology 20, 8635–8642.

Piedra, M. E., Icardo, J. M., Albajar, M. *et al.* (1998) *Pitx2* participates in the late phase of the pathway controlling left-right asymmetry. Cell 94, 319–324.

Puri, P.L., Avantaggiati, M.L., Balsone, C. *et al.* (1997) p300 is required for MyoD-dependent cell cycle arrest and muscle-specific gene transcription. EMBO Journal 6, 369–383.

Rawls, A. and Olson, E.N. (1997) MyoD meets its maker. Cell 89, 5–8.

Rüegsegger, U. and Leber, J. H. (2001) Block of *HAC1* mRNA translation by long-range base pairing is released by cytoplasmic splicing upon induction of the unfolded protein response. Cell 107, 103–114.

Sharma, K., Sheng, H. Z., Lettieri, K. *et al.* (1998) LIM homeodomain factors Lhx3 and Lhx4 assign subtype identities for motor neurones. Cell 95, 817–828.

Shen, W.-F., Detmer, K., Simonitch-Easton, T. *et al.* (1991) Alternative splicing of the Hox 2.2 homeobox gene in human hematopoetic cells and murine embryonic and adult tissues. Nucleic Acids Research 19, 539–545.

Small, S., Krant, R., Hoey, T. *et al.* (1991) Transcriptional regulation of a pair-rule stripe in *Drosophila*. Genes and Development 5, 827–839.

Stern, C. D. and Foley, A. C. (1998) Molecular dissection of *Hox* gene induction and maintenance in the hindbrain. Cell 94, 143–145.

Struhl, G., Johnston, P. and Lawrence, P.A. (1992) Control of *Drosophila* body pattern by the hunchback morphogen gradient. Cell 69, 237–249.

Tabin, C. (1995) The initiation of the limb bud: growth factors, Hox genes and retinoids. Cell 80, 671–674.

Trouche, D., Grigoriev, M., Lenormard, J.C. *et al.* (1993) Repression of c-fos promoter by MyoD on muscle cell differentiation. Nature 363, 79–82.

van der Hoeven, F., Zakany, J. and Duboule, D. (1996) Gene transpositions in the HoxD complex reveal a hierarchy of regulatory controls. Cell 85, 1025–1035.

Woloshin, P., Song, K., Degnin, C. *et al.* (1995) MSXI inhibits MyoD expression in fibroblast X 10T½ cell hybrids. Cell 82, 611–620.

Wuarin, J. and Schibler, U. (1990) Expression of the liver-enriched transcriptional activator protein DBP follows a stringent circadian rhythm. Cell 63, 1257–1269.

Xanthopoulos, K.G., Mirkovitch, J., Decker, T. *et al.* (1989) Cell-specific transcriptional control of the mouse DNA binding protein mC/EBP. Proceedings of the National Academy of Sciences, USA, 86, 4117–4121.

Zeller, R. and Deschamps, J. (2002) First come, first served. Nature 420, 138–139.

CHAPTER 8

REGULATION OF TRANSCRIPTION FACTOR ACTIVITY

8.1 EVIDENCE FOR THE REGULATED ACTIVITY OF TRANSCRIPTION FACTORS

In a number of cases, it has been shown that a particular transcription factor pre-exists in an inactive form prior to its activation and the consequent switching on of the genes that depend on it for their activity. Thus, as discussed in Chapter 5 (section 5.5.1) and in section 8.3.1 of this chapter, the activation of heat-inducible genes by elevated temperature is dependent on the activity of the heat shock transcription factor (HSF). However, this induction can be achieved in the presence of the protein synthesis inhibitor cycloheximide (Zimarino and Wu, 1987; for review see Morimoto, 1998). Hence this process cannot be dependent on the synthesis of HSF in response to heat but rather must depend on the heat-induced activation of pre-existing inactive HSF (for further details see section 8.3.1). Similarly, as discussed in Chapter 6 (section 6.2.3), the yeast GAL4 transcription factor pre-exists in cells prior to galactose treatment, which activates it by causing the dissociation of the inhibitory GAL80 protein.

Although for the reasons discussed in Chapter 7 (section 7.4) the activation of pre-existing transcription factors is predominantly used to modulate transcription factors involved in controlling inducible rather than cell type-specific gene expression, it has also been reported for factors involved in regulating cell type-specific gene expression. Thus, the transcription factor NFκB (which is a heterodimer of two subunits p50 and p65) plays an important role in the B cell-specific expression of the immunoglobulin κ gene (for reviews see Foo and Nolan, 1999; Karin and Ben-Neriah, 2000; Perkins, 2000; Dixit and Mak, 2002). However, both subunits of NFκB are expressed in a wide variety of cell types and the factor is present in an inactive form both in pre-B cells and in a wide variety of other cell types such as T cells and HeLa cells which do not express the immunoglobulin genes. This pre-existing form of NFκB can be activated by treatment of pre-B cells with substances such as lipopolysaccharides. As in the case of HSF, this activation can take place in the presence of inhibitors of protein synthesis, indicating that it does not require *de novo* synthesis of NFκB protein. These treatments therefore activate pre-existing NFκB and thus result in the activation of the immunoglobulin κ gene in pre-B cells which do not normally express it.

Interestingly, the inactive form of NFκB is widely distributed in different cell types and can be activated in both T cells and HeLa cells by treatment with phorbol ester. Although in these cases NFκB activation does not result in immunoglobulin light chain gene expression since the gene has not rearranged and is tightly packed within inactive chromatin, it does play a role in gene regulation. Thus the activation of NFκB by agents which activate T cells, results in the active transcription factor inducing increased expression of cellular genes such as that encoding the interleukin-2 α receptor and is also responsible for the increased activity of the human immunodeficiency virus promoter in activated T cells. NFκB therefore plays a role not only in B-cell specific gene activity but also in gene activity specific to activated T cells. Indeed further work has suggested an additional role for NFκB in bone development indicating that it plays a key role in a number of different cell types (for review see Abu-Amer and Tondravi, 1997; Dixit and Mak, 2002).

The process in which pre-existing NFκB becomes activated both during B-cell differentiation and by agents such as phorbol esters which activate T cells therefore allows NFκB to play a dual role both in B-cell specific gene expression and in the expression of particular genes in response to T-cell activation by various agents (Fig. 8.1). This effect would otherwise require a complex pattern of regulation in which NFκB was synthesized both in response to B-cell maturation and to agents which activate T cells.

Figure 8.1

Activation of NFκB during B-cell differentiation or by agents such as PMA which activate T cells allows it to activate expression of the immunoglobulin κ chain gene in B cells and the interleukin 2 receptor gene in activated T cells.

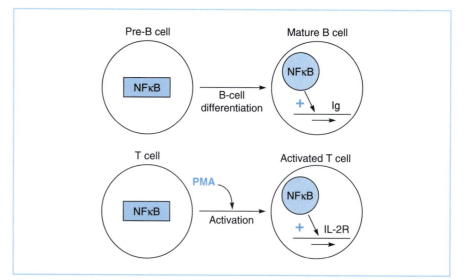

Hence modulating the activity of a transcription factor represents a rapid and flexible means of activating a particular factor. Moreover, unlike transcriptional control, such mechanisms allow a direct linkage between the inducing stimulus and the activation of the factor rather than requiring the regulated activity of other transcription factors which, in turn, activate transcription of the gene encoding the regulated factor. Hence they represent a highly efficient means of allowing specific cellular signalling pathways to produce changes in cellular transcription factor activity and hence affect gene expression (for reviews see Barolo and Posakony, 2002; Brivanlou and Darnell, 2002).

In the most extreme example of the linkage between signalling pathways and transcription factors, the signalling molecule and the transcription factor are identical. Thus, in response to microbial infection, mammalian neutrophils secrete the protein lactoferrin into the medium. It has been shown that the lactoferrin protein can be taken up by other cells of the immune system. It then enters the nucleus of the cells and binds to specific DNA sequences, activating genes whose protein products are required for the cells to neutralize the microbial infection (He and Furmanski, 1995). Hence in this case, the signalling factor and the transcription factor are the same protein (for discussion see Baeuerle, 1995). In most cases, however, the signalling molecule acts indirectly to produce a change in the activity of a distinct transcription factor which pre-existed within the cell in an inactive form prior to exposure to the signal. Four basic means by which such mechanisms can regulate factor activity have been described (Fig. 8.2) and these will be discussed in turn.

8.2 REGULATION BY PROTEIN–LIGAND BINDING

8.2.1 EXAMPLES OF REGULATION BY LIGAND BINDING

As discussed above, one of the principal advantages of regulating the activity of a factor in response to an inducing stimulus is that it allows a direct interaction between the inducing stimulus and the activation of the factor, ensuring a rapid response. The simplest method for this is for an inducing ligand to bind to the transcription factor and alter its structure so that it becomes activated (Fig. 8.2a).

An example of this effect is seen in the case of the ACE1 factor which mediates the induction of the yeast metallothionein gene in response to copper. In this case, the transcription factor undergoes a major conformational change in the presence of copper which converts it to an active form that is able to bind to its appropriate binding sites in the metallothionein gene promoter and activate transcription (Fig. 8.3; for review see Thiele, 1992).

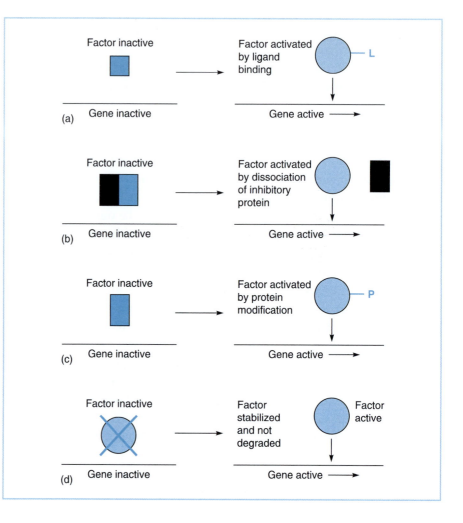

Figure 8.2

Methods of activating a transcription factor in response to an inducing stimulus. This can occur by a ligand-mediated conformational change (a), by removal of an inhibitory protein (b), by a modification to the protein such as phosphorylation (c) or by stabilizing the factor so that it is not degraded (d).

A similar example in mammalian cells involves the DREAM transcription factor which represses the transcription of the dynorphin gene and thereby enhances the response to painful stimuli (for review see Costigan and Woolf, 2002). The activity of this factor is directly modulated by the level of calcium, which binds directly to the DREAM protein and reduces its ability to bind to its binding site in the dynorphin gene. Hence, the repression of the gene by DREAM is relieved and the dynorphin gene is transcribed (for review see Mandel and Goodman, 1999) (Fig. 8.4).

8.2.2 THE NUCLEAR RECEPTORS

A more complex example of regulation by ligand binding to a transcription factor is provided by the steroid hormone receptors. These receptors are members of the nuclear receptor or steroid-thyroid receptor family discussed

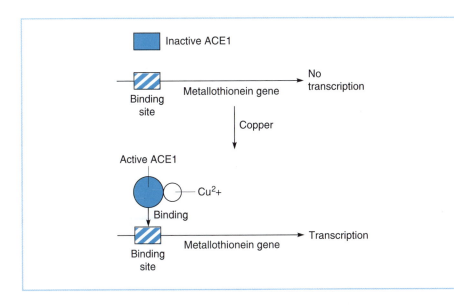

Figure 8.3

Activation of the ACE1 factor in response to copper results in transcription of the metallothionein gene.

Figure 8.4

In the absence of calcium, the DREAM repressor binds to its response element (DRE) in the dynorphin gene and represses its transcription. When calcium is present, it binds to the DREAM factor and changes its conformation so that it does not bind to the DRE. This relieves the repression and allows transcription of the dynorphin gene.

in Chapter 4 (section 4.4) and mediate gene activation in response to steroids such as glucocorticoid or oestrogen (for review see Weatherman *et al.*, 1999).

Following identification of the steroid hormone receptors, it was very rapidly shown that the receptors were only found associated with DNA after hormone treatment. These early studies were subsequently confirmed by using DNAseI footprinting on whole chromatin to show that the receptor was only bound to the hormone response sequence following hormone treatment (Becker *et al.*, 1986). These studies were therefore consistent with a model in which the hormone induces a conformational change in the receptor activating its ability to bind to DNA and thereby activate transcription.

Subsequent studies have suggested that the situation is more complex however. Thus, although in the intact cell the receptor binds to DNA only in the presence of the hormone, purified receptor can bind to DNA *in vitro* in a band shift or footprinting assay regardless of whether hormone is present or not (Wilmann and Beato, 1986; Figs 8.5 and 8.6).

This discrepancy led to the suggestion that the receptor is inherently capable of binding to DNA but is prevented from doing so in the absence of steroid because it is anchored to another protein. The hormone acts to release it from this association and allow it to fulfill its inherent ability to bind to DNA. In agreement with this possibility, in the absence of hormone, the glucocorticoid receptor protein is found in the cytoplasm complexed to a 90 000 molecular weight heat-inducible protein (hsp90) in an 8S complex.

Figure 8.5

DNAseI footprint analysis of the binding of the glucocorticoid receptor to the glucocorticoid-inducible mouse mammary tumour virus long terminal repeat promoter (MMTV-LTR). In tracks I and II the DNAseI digestion has been carried out without any added receptor. In tracks 1–3, glucocorticoid receptor has been added prior to DNAseI digestion either alone (track 1+), with the glucocorticoid hormone corticosterone (track 2) or with the anti-hormone RU486 which inhibits steroid-induced activation of the receptor (track 3). Track 1 shows the result of adding receptor to the DNA in the absence of DNAseI addition in which some cleavage by endogenous nucleases (e) occurs, while track G is a marker track produced by cleaving the same DNA at each guanine residue. Minus signs indicate footprinted regions protected by receptor, plus signs are hypersensitive sites at which cleavage is increased by the presence of the receptor. The DNA fragment used and position of the radioactive label (diamond) are shown together with the distances upstream from the initiation site for transcription.

Note that the identical footprint is produced by the receptor either alone or in the presence of hormone or anti-hormone. Hence *in vitro* the receptor can bind to DNA in the absence of hormone.

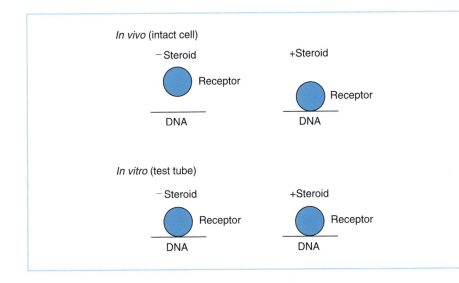

Figure 8.6

Comparison of steroid receptor binding to DNA in the presence or absence of hormone *in vivo* and *in vitro*. Note that while *in vivo* DNA binding can occur only in the presence of hormone, *in vitro*, it can occur in the presence or absence of hormone.

This complex is dissociated upon steroid treatment releasing the 4S receptor protein (for reviews see Pratt, 1997; Pratt and Toft, 1997). The released receptor is free to dimerize and move into the nucleus. Since these processes have been shown to be essential for DNA binding and transcriptional activation by steroid hormone receptors, dissociation of the receptor from hsp90 is essential if gene activation is to occur. In agreement with this antiglucocorticoids, which inhibit the positive action of glucocorticoids, have been shown to stabilize the 8S complex of hsp90 and the receptor.

Similar complexes with hsp90 have also been reported for the other steroid hormone receptors. Thus the activation of the different steroid receptors such as the glucocorticoid and oestrogen receptors by their specific hormones is likely to involve disruption of the protein–protein interaction with hsp90 (Fig. 8.7).

Most interestingly, the association of hsp90 with the glucocorticoid receptor occurs via the C-terminal region of the receptor, which also contains the steroid binding domain. It has been suggested therefore that by associating with the C terminal region of the receptor, hsp90 masks adjacent domains whose activity is necessary for gene activation by the receptor for example, those involved in receptor dimerization or subsequent DNA binding, thereby preventing DNA binding from occurring. Following steroid treatment, however, the steroid binds to the C terminus of the receptor displacing hsp90 and thereby unmasking these domains and allowing DNA binding to occur (Fig. 8.8). Hence, activation of the steroid receptors involves a ligand-induced conformational change which results in the dissociation of an inhibitory protein.

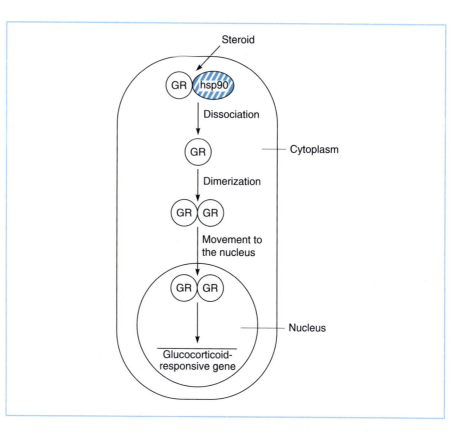

Figure 8.7

Activation of the glucocorticoid receptor (GR) by steroid involves dissociation of hsp90 allowing dimerization and movement to the nucleus.

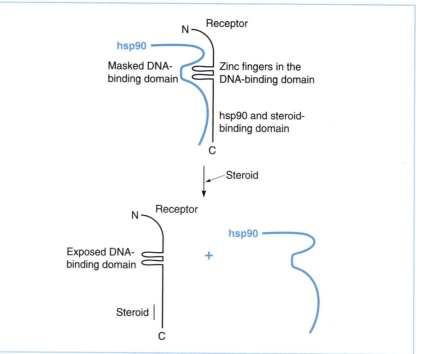

Figure 8.8

Interaction of hsp90 and the glucocorticoid receptor. hsp90 binds to the receptor via the C terminal region of the receptor which also binds steroid and may mask regions of the receptor necessary for dimerization or DNA binding. When steroid is added it binds to the receptor at the C terminus displacing hsp90 and exposing the masked regions.

In addition to the steroid-induced dissociation of the receptors from hsp90 it is clear that a second step following dissociation from hsp90 is also required for receptor activation. Thus in a cell-free system in which the progesterone receptor exists in a 4S form, free of bound hsp90, the addition of progesterone is still required for the activation of progesterone responsive genes. This indicates that the hormone has an additional effect on the receptor apart from dissociating it from hsp90. This effect involves the unmasking of a previously inactive transcriptional activation domain in the receptor allowing it to activate gene expression in a hormone-dependent manner following DNA binding. Thus, domain swopping experiments (see Chapter 2, section 2.4.1) have identified C-terminal regions in both the glucocorticoid and oestrogen receptors which, when linked to the DNA binding domain of another factor, can activate transcription only following hormone addition (see Fig. 4.29). These regions hence constitute hormone-dependent activation domains.

Moreover, in the case of the oestrogen receptor, it has been shown that the oestrogen antagonist 4-hydroxytamoxifen induces the receptor to bind to DNA (presumably by promoting dissociation from hsp90 and dimerization), but does not induce gene activation suggesting that it fails to activate the oestrogen-responsive transactivation domain. Hence the mechanism by which the steroid receptors are activated is now thought to involve both dissociation from hsp90 and a change in their transcriptional activation ability (Fig. 8.9a). This second step is likely to involve a change in the activation domain which allows it to bind co-activator proteins that are essential for transcriptional activation (see Chapter 5, section 5.4.3 for discussion of co-activator molecules).

Interestingly, other members of the nuclear receptor family which bind to substances that are related to steroids, such as retinoic acid or thyroid hormone, do not associate with hsp90 and are bound to DNA prior to exposure to ligand. Their activation by their appropriate ligand thus involves only the second stage discussed above, namely a ligand-induced structural change in their C-terminal activation domain, which is adjacent to the ligand binding domain, allowing it to bind co-activator molecules and activate transcription (Fig. 8.9b). Indeed, crystallographic studies of the ligand binding domain and the C-terminal activation domain of the retinoic acid receptors, both in the presence or absence of hormone, have provided direct evidence for this change. Thus, as illustrated in Plate 7, the activation domain is not closely associated with the ligand binding domain in the absence of ligand but is much more closely associated with it following ligand binding and forms a lid covering the ligand binding region (Renaud et al., 1996).

Although first defined in the retinoic acid receptors, a similar structural change occurs upon ligand binding in other members of the nuclear receptor

Figure 8.9

(a) Activation of the steroid receptors (SR) by treatment with steroid. As well as inducing dissociation of the receptor from hsp90, steroid treatment also increases the ability of the receptor to activate transcription following DNA binding by changing the structure of the activation domain (shaded) allowing it to bind co-activator proteins (CA) which stimulate transcription. (b) Activation of other members of the nuclear receptor family which bind non-steroids such as retinoic acid or thyroid hormone involves only the second of these stages.

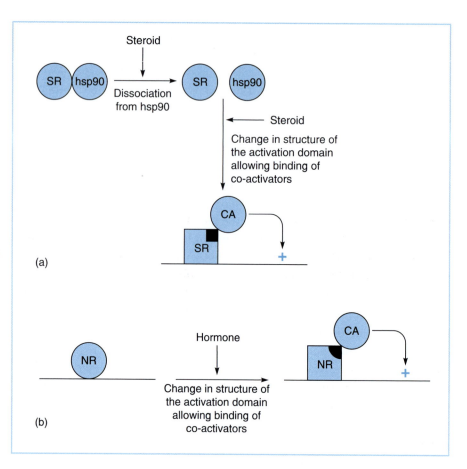

family including the glucocorticoid and oestrogen receptors and the thyroid hormone receptor (Wurtz *et al.*, 1996). Indeed, it has been shown that while oestrogen induces this realignment of the oestrogen receptor activation domain, the oestrogen antagonist raloxifene does not do so, thereby explaining its antagonistic action (Brzozowski *et al.*, 1997) (Fig. 8.10). In turn this ligand-induced structural change allows the activation domain to bind co-activator proteins, which bind to the receptors only after exposure to hormone and appear to play a key role in the ability of the receptors to activate transcription (see Fig. 8.9) (see Chapter 5, section 5.4.3 for a discussion of co-activator molecules).

Interestingly, in the case of receptors such as the thyroid hormone receptor, where DNA binding is observed even prior to hormone treatment, the receptor actually represses transcription prior to thyroid hormone treatment. As discussed in Chapter 6 (section 6.3.2), this is because in the absence of ligand, the receptor binds co-repressor molecules which are displaced by co-

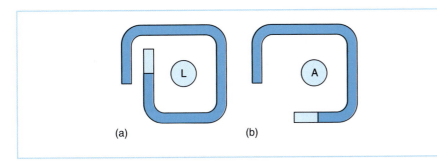

(a)   (b)

Figure 8.10
(a) The binding of the ligand (L) induces the realignment of the C-terminal activation domain of the nuclear receptors (light shading) so that it forms a lid over the ligand binding domain and the activation domain then stimulates transcription. (b) This realignment is not induced by binding of antagonists (A) which therefore do not stimulate transcriptional activation.

activators on hormone treatment. The importance of this conversion from repressor to activator is seen in the case of mutant forms of the thyroid hormone receptor which cannot undergo this conformational change because they do not bind thyroid hormone. This is observed not only in the v-*erbA* oncogene as discussed in Chapter 9 (section 9.3.2) but also in patients with generalized thyroid hormone resistance. Thus these patients have been shown to produce forms of the receptor which can repress gene expression but which cannot activate genes in response to thyroid hormone. Most interestingly, the presence of these dominant negative forms of the receptor results in impairment of physical and mental development which is much more severe than that observed if the receptor is absent completely (Baniahmad *et al.*, 1992).

Hence, in all the nuclear receptors, activation by ligand involves a structural change in the C-terminal activation domain which allows it to bind co-activators. In the steroid hormone receptors, this is preceded by an earlier step which involves the disruption of the receptor hsp90 association. Activation of these steroid receptors, therefore involves both the ligand-induced conformational changes seen in ACE1 and DREAM as well as the dissociation of an inhibitor protein and thus combines the mechanisms illustrated in Figure 8.2a and b.

8.3 REGULATION BY PROTEIN–PROTEIN INTERACTIONS

8.3.1 INHIBITION OF TRANSCRIPTION FACTOR ACTIVITY BY PROTEIN–PROTEIN INTERACTION

As described above, the glucocorticoid receptor is regulated by its interaction with hsp90 which prevents it binding to DNA and activating transcription in the absence of steroid hormone. A similar mechanism is used in the case of the NFκB factor which, as discussed above, only activates transcription in mature B cells or in other cell types following treatment with agents such as lipopolysaccharides or phorbol esters. In agreement with this, no active form of NFκB capable of binding to DNA can be detected in DNA mobility shift assays (see Chapter 2, section 2.2.1) using either cytoplasmic or nuclear extracts prepared from pre-B cells or non-B cell types. Interestingly, however, such activity can be detected in the cytoplasm but not the nucleus of such cells following denaturation and subsequent renaturation of the proteins in the extract. Hence NFκB exists in the cytoplasm of pre-B cells and other cell types in an inactive form which is complexed with another protein known as IκB that inhibits its activity (for reviews see Karin and Ben-Neriah, 2000; Perkins, 2000; Dixit and Mak, 2002). The release of NFκB from IκB by the denaturation/renaturation treatment therefore results in the appearance of active NFκB capable of binding to DNA (Fig. 8.11a).

These findings suggested therefore that treatments with substances such as lipopolysaccharides or phorbol esters do not activate NFκB by interacting directly with it in a manner analogous to the activation of the ACE1 factor by copper. Rather they are likely to produce the dissociation of NFκB from

Figure 8.11a

Regulation of NFκB. Panel (a) In pre-B cells NFκB is located in the cytoplasm in an inactive form which is complexed to IκB. DNA mobility band shift assays do not therefore detect active NFκB. If a cytoplasmic extract is first denatured and renatured, however, active NFκB will be released from IκB and will be detected in a subsequent band shift assay.

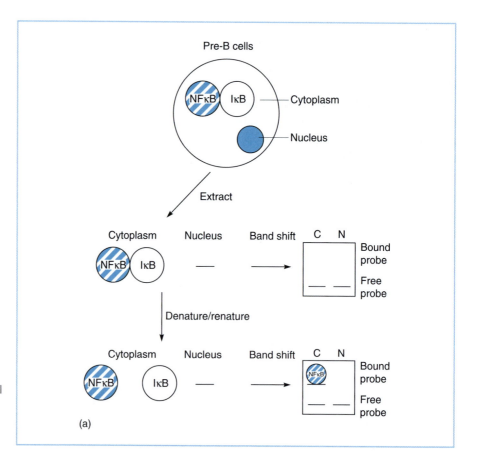

IκB resulting in its activation. In agreement with this idea, phorbol ester treatment of cells prior to their fractionation eliminated the latent NFκB activity in the cytoplasm and resulted in the appearance of active NFκB in the nucleus (Fig. 8.11b). These substances act therefore by releasing NFκB from IκB allowing it to move to the nucleus where it can bind to DNA and activate gene expression. Hence this constitutes an example of the activation of a factor by the dissociation of an inhibitory protein (see Fig. 8.2b).

Such a mechanism is used to regulate the activity of many different transcription factors. Thus apart from the NFκB/IκB and glucocorticoid receptor/hsp90 interactions, other examples of inhibitory interactions include those between DNA binding helix-loop-helix proteins and Id (Chapter 6, section 6.2.2) and p53 and the MDM2 protein (Chapter 9, section 9.4.2). Hence inhibitory interactions of this type are widely used to regulate the activity of specific transcription factors.

A highly complex example of such regulation by protein–protein interaction is seen in the case of the heat shock factor (HSF) which, as discussed in Chapter 5 (section 5.5.1) activates gene transcription in response to elevated

Figure 8.11b

Panel (b) In mature B cells, NFκB has been released from IκB and is present in the nucleus in an active DNA-binding form. It can therefore be detected in a DNA mobility shift assay without a denaturation, renaturation step which has no effect on the binding activity.

temperature. HSF achieves this effect by binding to its binding site in target genes, which is known as the heat shock element (HSE) (see Chapter 1, section 1.3.3). The amount of HSF bound to the HSE increases with the time of exposure to elevated temperature and with the extent of temperature elevation. Moreover, increased protein binding to the HSE is also observed following exposure to other agents which also induce the transcription of the heat shock genes, such as 2,4-dinitrophenol (Fig. 8.12). Thus, activation of the heat shock genes, mediated by the HSE is accompanied by the binding of a specific transcription factor to this DNA sequence.

As noted in section 8.1, this activation of HSF can occur in the absence of new HSF protein synthesis (for review see Morimoto, 1998). Thus, if cells are heat treated in the presence of cycloheximide, which is an inhibitor of protein synthesis, increased binding of HSF to the HSE is observed exactly as in cells treated in the absence of the drug (Zimarino and Wu, 1987). This indicates that the observed binding of HSF following heat shock does not require *de novo* protein synthesis. Rather, this factor must pre-exist in non-heat treated cells in an inactive form whose ability to bind to the HSE sequence in DNA is activated post-translationally by heat. In agreement with this, activation of HSF can also be observed following heat treatment of cell extracts *in vitro* when new protein synthesis would not be possible (Larson *et al.*, 1988).

Analysis of the activation process using *in vitro* systems from human cells (Larson *et al.*, 1988) has indicated that it is a two-stage process (Fig. 8.13). In

Figure 8.12

Detection of HSF binding to the HSE 91 bases upstream (-91) of the start site for transcription in the *Drosophila* hsp82 gene and protecting this region from digestion with exonuclease III. Note the increased binding of HSF with increasing time of exposure to heat shock or increased severity of heat shock. HSF binding is also induced by exposure to 2, 4-dinitrophenol (DNP) which is known to induce transcription of the heat shock genes.

Figure 8.13

Stages in the activation of HSF in mammalian and *Drosophila* cells. Initial activation of HSF to a DNA-binding form following elevated temperature is followed by its phosphorylation which converts it to a form capable of activating transcription.

the first stage, the HSF is activated to a form which can bind to DNA by an ATP-independent mechanism which is directly dependent on elevated temperature. Subsequently, this protein is further modified by phosphorylation allowing it to activate transcription. Interestingly, the second of these two stages appears to be disrupted in murine erythroleukaemia (MEL) cells in which heat shock results in increased binding of HSF to DNA but transcriptional activation of the heat shock genes is not observed (Hensold *et al.*, 1990).

The activation of HSF into a form capable of binding DNA involves its conversion from a monomeric to a trimeric form which can bind to the HSE (for review see Morimoto, 1998). The maintenance of the monomeric form of HSF prior to heat shock is dependent on a region at the C terminus of the molecule since when this region is deleted, HSF spontaneously trimerizes and can bind to DNA even in the absence of heat shock (Rabindran *et al.*, 1993). The C-terminal region contains a leucine zipper (see Chapter 4, section 4.5). As leucine zippers are known to be able to interact with one another, it is thought that this region acts by interaction with another leucine zipper located adjacent to the N-terminal DNA binding domain promoting intramolecular folding which masks the DNA binding domain. Following heat shock HSF unfolds, unmasking the DNA binding domain and allowing a DNA-binding trimer to form (Fig. 8.14).

Recent studies have also shown that the transition of HSF from monomer to trimer requires two specific cysteine residues within HSF. These cysteine residues are thought to promote this transition by forming disulphide

Figure 8.14

Prior to heat shock, HSF is present in a monomeric form in which the leucine zipper motifs (L) at the C-terminus and within the molecule promote intra-molecular folding which masks the N-terminal DNA binding domain (shaded) preventing binding to the HSE. Following heat shock, the protein unfolds and forms the DNA binding trimeric form. This form binds to the HSE and activates transcription following its subsequent phosphorylation.

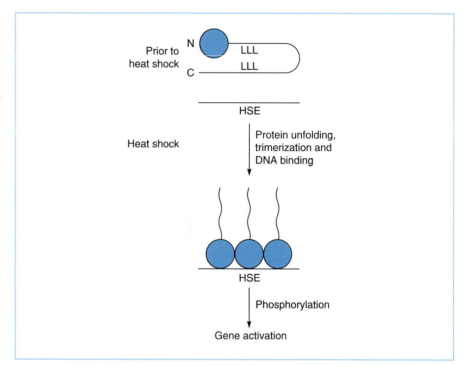

bonds with one another in response to heat or other stresses, although it is currently unclear whether these bonds form between the cysteines in one molecule of HSF or between different molecules in the trimer (Ahn and Thiele, 2003).

Interestingly, as with the glucocorticoid receptor (see section 8.2.2), the conversion of HSF from a monomer to a DNA-binding trimer involves the dissociation of hsp90 which binds to HSF in untreated cells and stabilizes it in the inactive form which cannot bind to DNA (Zou *et al.*, 1998). Interestingly, hsp90 acts as a so-called 'chaperone' protein, assisting the proper folding of other proteins. Evidently, following heat or other stress, the level of such unfolded proteins will increase. Hsp90 will therefore be 'called away' to deal with these unfolded proteins leaving HSF free to trimerize and bind to DNA (Fig. 8.15).

Hence, the response of HSF to stress involves both the loss of an inhibitory protein and changes in the HSF molecule itself. Together these changes promote the transition from an HSF monomer to a DNA binding trimer. However, this DNA binding by HSF is insufficient to produce transcriptional activation. This requires phosphorylation of HSF on serine 230 which allows the DNA bound form of HSF to activate transcription (see Figs. 8.14, 8.15) (Holmberg *et al.*, 2001).

Figure 8.15
Prior to heat shock or other stress, HSF is bound to hsp90 which stabilizes its inactive monomeric form. Following heat shock, hsp90 dissociates from HSF to fulfil its function of refolding other proteins which have unfolded due to the elevated temperature. This allows HSF to trimerize and bind to DNA. However, transcriptional activation requires subsequent phosphorylation of HSF.

This two-stage process, involving DNA binding induced by trimerization and transcriptional activation induced by serine phosphorylation, represents a common mechanism for the activation of HSF in higher eukaryotes such as *Drosophila* and mammals. In contrast, however, the *Saccharomyces cerevisiae* (budding yeast) HSF is activated by a much simpler mechanism. Thus, unlike *Drosophila* or mammalian HSF, the budding yeast protein lacks the C-terminal leucine zipper region, which promotes monomer formation, and therefore exists as a trimer prior to heat shock. As expected from this, HSF can be observed bound to the HSE even in non-heat shocked cells (Sorger *et al.*, 1987). HSF can activate transcription, however, only following heat treatment when the protein becomes phosphorylated. Interestingly, in *Schizosaccharomyces pombe* (fission yeast) HSF regulation follows the *Drosophila* and mammalian system with HSF becoming bound to DNA only following heat shock (Gallo *et al.*, 1991).

Hence in mammals, *Drosophila* and fission yeast, activation of HSF is more complex than in budding yeast, involving an initial stage activating the DNA

Figure 8.16

HSF activation in *Drosophila*, mammals and fission yeast compared to that in budding yeast. Note that in budding yeast HSF is already bound to DNA prior to heat shock and hence its activation by heat involves only the second of the two stages seen in other organisms, namely, its phosphorylation allowing it to activate transcription.

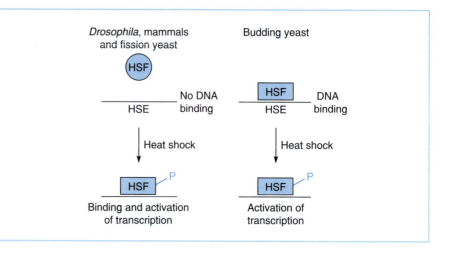

binding ability of HSF in response to heat as well as the stage, common to all organisms, in which the ability to activate transcription is stimulated by phosphorylation (Fig. 8.16). It thus combines regulation by protein–protein interaction (between HSF itself and HSF/hsp90) as well as regulation by phosphorylation which will be discussed more generally in section 8.4.

Interestingly, as well as being regulated by interacting with another transcription factor protein, it is also possible for a factor to be regulated by interaction with lipid within the cell. This is seen in the case of the Tubby factor which regulates the expression of genes involved in fat metabolism. It has been shown that the Tubby protein is anchored at the plasma membrane by interaction with a phospholipid $PI(4,5)P_2$. However, following activation of specific G-protein coupled receptors in the plasma membrane, the enzyme phospholipase C is activated. This enzyme then cleaves $PI(4,5)P_2$, releasing Tubby and allowing it to move to the nucleus and activate transcription (Fig. 8.17) (for review see Cantley, 2001).

This example is evidently similar to the glucocorticoid receptor/hsp90 and NFκB/IκB examples in that it involves the transcription factor moving from the cytoplasm to the nucleus but differs in that the activation process involves disruption of a protein–lipid interaction rather than a protein–protein interaction.

8.3.2 ACTIVATION OF TRANSCRIPTION FACTORS BY PROTEIN–PROTEIN INTERACTION

As well as inhibition, protein–protein interactions can actually stimulate the activity of a transcription factor. Thus, some transcription factors may be inactive alone and may need to complex with a second factor in order to be

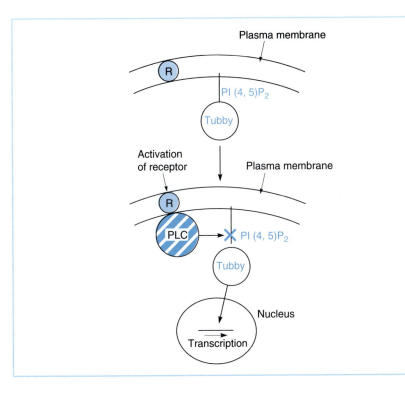

Figure 8.17
The Tubby transcription factor is anchored to the plasma membrane by binding to the phospholipid PI(4,5)P$_2$. Following activation of a membrane G-protein coupled receptor (R), phospholipase (PLC) is activated and cleaves PI(4,5)P$_2$. This releases Tubby, allowing it to move to the nucleus and activate gene expression.

active. This is seen in the case of the Fos protein which cannot bind to DNA without first forming a heterodimer with the Jun protein (see Chapter 4, section 4.5). A similar mechanism also operates in the case of the Myc factor which cannot bind to DNA except as a complex with the Max protein (see Chapter 9, section 9.3.3). Hence protein–protein interactions between transcription factors can result in either inhibition or stimulation of their activity. The need for Fos and Myc to interact with another factor prior to DNA binding arises from their inability to form a homodimer, coupled with the need for factors of this type to bind to DNA as dimers. Hence they need to form heterodimers with another factor prior to DNA binding (see Chapter 4, section 4.5 for further discussion).

8.3.3 ALTERATION OF TRANSCRIPTION FACTOR FUNCTION BY PROTEIN–PROTEIN INTERACTION

Even in the case of factors such as Jun which can form DNA-binding homodimers, the formation of heterodimers with another factor offers the potential to produce a dimer with properties distinct from those of either homodimer. Thus, the Jun homodimer can bind strongly to AP1 sites but only weakly to the cyclic AMP response element (CRE). In contrast a heterodimer of Jun and the CREB factor binds strongly to a CRE and more weakly to an AP1 site. Heterodimerization can therefore represent a means of producing multi-protein factors with unique properties different from that of either protein partner alone (for reviews see Jones, 1990; Lamb and McKnight, 1991).

Hence, as well as stimulating or inhibiting the activity of a particular factor, the interaction with another factor can also alter its properties, directing it to specific DNA binding sites to which it would not normally bind. Thus, as discussed in Chapter 4 (section 4.2.4) the *Drosophila* extradenticle protein changes the DNA binding specificity of the Ubx protein so that it binds to certain DNA binding sites with high affinity in the presence of extradenticle and with low affinity in its absence. Similarly, as described in Chapter 4 (section 4.2.4), the yeast $\alpha 2$ repressor factor forms heterodimers of different DNA binding specificities with the a1 or MCM1 transcription factors.

Although several examples of one transcription factor altering the DNA binding specificity of another have thus been defined, such protein–protein interactions can also change the specificity of a transcription factor in at least one other way. This is seen in the case of the *Drosophila* dorsal protein which is related to the mammalian NFκB factors. Thus this factor is capable of both activating and repressing specific genes. Such an ability is not due for example to the production of different forms by alternative splicing since both activation and repression take place in the same cell type. Rather it appears to depend on the existence of a DNA sequence (the ventral repression element or VRE) adjacent to the dorsal binding site in genes such as zen, which are repressed by dorsal, whereas the VRE sequence is absent in genes such as twist, which are activated by dorsal.

It has been shown that DSP1 (dorsal switch protein), a member of the HMG family of transcription factors (see Chapter 4, section 4.6), binds to the VRE and interacts with the dorsal protein changing it from an activator to a repressor. Hence in genes such as twist where DSP1 cannot bind, dorsal activates expression, whereas in genes such as zen which DSP1 can bind, dorsal represses expression (Fig. 8.18) (for review see Ip, 1995). It has been shown that DSP1 can interact with the basal transcriptional complex and disrupt the association of TFIIA with TBP (Kirov *et al.*, 1996). It therefore acts as an active transcriptional repressor interfering with the assembly of the basal transcriptional complex (see Chapter 6, section 6.3.2 for further discussion of this repression mechanism).

Interestingly, like DSP1, the *Drosophila* groucho protein can switch dorsal from activator to repressor indicating that multiple proteins can mediate this effect (Dubnicoff *et al.*, 1997). Moreover, a similar negative element to the VRE is associated with the NFκB binding site in the mammalian β-interferon

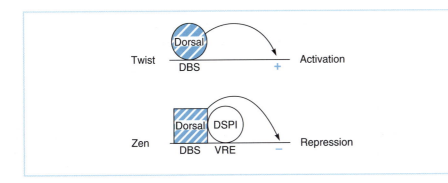

Figure 8.18

The interaction of DSPI bound at the ventral repression element (VRE) with the dorsal protein bound at its adjacent binding site (DBS) in the zen promoter results in dorsal acting as a repressor of transcription, whereas in the absence of binding sites for DSPI as in the twist promoter, it acts as an activator.

promoter and the *Drosophila* DSP1 protein can similarly switch NFκB from activator to repressor when DSP1 is artificially expressed in mammalian cells. This mechanism may thus not be confined to *Drosophila* and a mammalian homologue of DSP1 may regulate NFκB activity in a similar manner (for discussion see Thanos and Maniatis, 1995).

Protein–protein interactions between different factors can thus either stimulate or inhibit their activity or alter that activity either in terms of DNA binding specificity or even from activator to repressor. It is likely that the wide variety of protein–protein interactions and their diverse effects allow the relatively small number of transcription factors which exist to produce the complex patterns of gene expression which are required in normal development and differentiation.

8.4 REGULATION BY PROTEIN MODIFICATION

8.4.1 TRANSCRIPTION FACTOR MODIFICATION

Many transcription factors are modified extensively following translation, for example, by phosphorylation, particularly on serine or threonine residues (for reviews see Hill and Treisman, 1995; Treisman, 1996) or via the modification of lysine residues either by acetylation or by addition of the small protein ubiquitin (for review see Freiman and Tjian, 2003). Such modifications represent obvious targets for agents that induce gene activation. Thus, such agents could act by altering the activity of a modifying enzyme, such as a kinase. In turn this enzyme would modify the transcription factor, resulting in its activation and providing a simple and direct means of activating a particular factor in response to a specific signal (see Fig. 8.2c). The various modifications that have been shown to affect transcription factor activity will be discussed in turn.

8.4.2 PHOSPHORYLATION

Many cellular signalling pathways involve the activation of cascades of kinase enzymes which ultimately lead to the phosphorylation of specific transcription factors. The most direct example of such an effect of a signalling pathway on a transcription factor is seen in the case of gene activation by the interferons α and γ. Thus these molecules bind to cell surface receptors which are associated with factors having tyrosine kinase activity. The binding of interferon to the receptor stimulates the kinase activity and results in the phosphorylation of transcription factors known as STATs (signal transducers and activators of transcription). In turn this results in the dimerization of the STAT proteins allowing them to move to the nucleus where they bind to DNA and activate interferon-responsive genes (Fig. 8.19) (for reviews see Horvath, 2000; Ihle, 2001).

Another example of this type is provided by the CREB factor which mediates the induction of specific genes in response to cyclic AMP treatment. As discussed in Chapter 5 (section 5.4.3) CREB binds to DNA in its non-phosphorylated form but only activates transcription following phosphorylation by the protein kinase A enzyme which is activated by cyclic AMP. Hence, in this case, the activation of a specific enzyme by the inducing agent allows the transcription factor to activate transcription and hence results in the activation of cyclic AMP inducible genes.

Figure 8.19

Binding of interferon (IFN) to its receptor results in activation of an associated tyrosine kinase (TK) activity leading to phosphorylation of a STAT transcription factor allowing it to dimerize and move to the nucleus and stimulate interferon responsive genes (IRG).

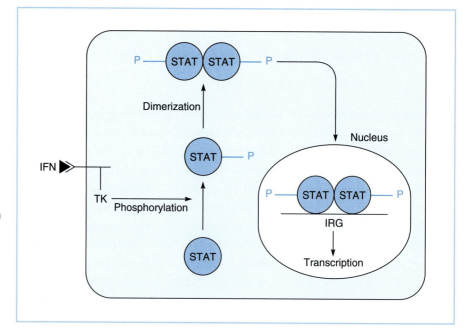

Similarly, the phosphorylation of the heat shock factor (HSF) following exposure of cells to elevated temperature increases the activity of its activation domain leading to increased transcription of heat-inducible genes (see section 8.3.1), while the ability of the retinoic acid receptor to stimulate transcription is enhanced by phosphorylation of its activation domain by the basal transcription factor TFIIH (see Chapter 3, section 3.5).

In contrast to these effects on transcriptional activation ability, phosphorylation of the serum response factor (SRF), which mediates the induction of several mammalian genes in response to growth factors or serum addition, increases its ability to bind to DNA rather than directly increasing the activity of its activation domain. Interestingly, SRF normally binds to DNA in association with an accessory protein p62^{TCF}. The ability of p62^{TCF} to associate with SRF is itself stimulated by phosphorylation.

Similarly, as discussed in Chapter 5 (section 5.4.3), phosphorylation of CREB on serine 133 by protein kinase A allows it to stimulate transcription because it allows it to associate with the CBP co-activator. Protein kinase A can also phosphorylate the equivalent serine residue in the CREM transcription factor which is closely related to CREB (see Chapter 7, section 7.3.2). As well as allowing it to activate its target genes, this phosphorylation also enhances the ability of CREM to bind to the DREAM repressor protein, discussed in section 8.2.1. As binding to CREM removes DREAM from its binding site in the dynorphin promoter, it provides an alternative means of activating this promoter, apart from direct calcium binding to DREAM (for review see Costigan and Woolf, 2002) (Fig. 8.20). Hence, the phosphorylation state of a transcription factor can control its ability to associate with other factors and

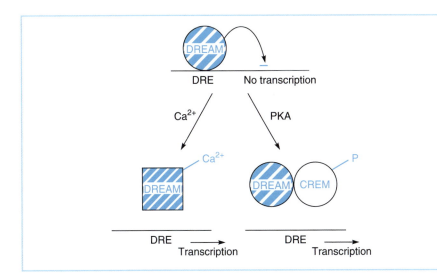

Figure 8.20

The DREAM repressor can be removed from its binding site (DRE) in the dynorphin promoter either by direct binding of calcium (compare Fig. 8.4) or by binding to DREAM of the CREM transcription factor following its phosphorylation by protein kinase A.

regulate their activity as well as its ability to enter the nucleus, bind to DNA or stimulate transcription.

The effect of phosphorylation on protein–protein interactions is also involved in the dissociation of NFκB and its associated inhibitory protein IκB which was discussed above (section 8.3.1). In this case, however, the target for phosphorylation is the inhibitory protein IκB rather than the potentially active transcription factor itself. Thus, following treatment with phorbol esters or other stimuli such as tumour necrosis factor or interleukin 1, IκB becomes phosphorylated. Such phosphorylation results in the dissociation of the NFκB/IκB complex and targets IκB for rapid degradation. This breakdown of the complex results in NFκB being free to move to the nucleus and activate transcription (Fig. 8.21) (for review see Karin and Ben-Neriah, 2000; Perkins, 2000). Hence in this case, as before, the inducing agent has a direct effect on the activity of a kinase enzyme but the resulting phosphorylation inactivates the IκB inhibitory transcription factor rather than stimulating an activating factor.

This example therefore involves a combination of two of the post-translational activation mechanisms we have discussed, namely protein modification (see Fig. 8.2c) and dissociation of an inhibitory protein (see Fig. 8.2b).

Figure 8.21

Activation of NFκB by dissociation of the inhibitory protein IκB, allowing NFκB to move to the nucleus and switch on gene expression. Note that dissociation of IκB from NFκB is caused by its phosphorylation (P) and degradation. NFκB is shown as a single factor for simplicity, although it normally exists as a heterodimer of two subunits p50 and p65.

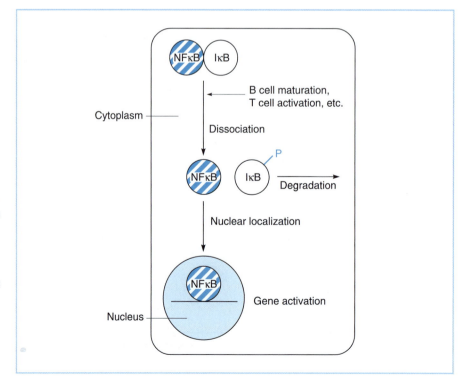

Moreover, as with the glucocorticoid receptor and its dissociation from hsp90 or the release of Tubby from PI(4.5)P$_2$ discussed in section 8.3.1, the net effect of the activation process is the movement of the activating factor from the cytoplasm to the nucleus where it can bind to DNA. Thus regulatory processes can activate a transcription factor by changing its localization in the cell as well as altering its inherent ability to bind to DNA or to activate transcription (for review see Vandromme *et al.*, 1996).

Clearly a key role in the regulation of the NFκB pathway will therefore be played by the enzymes which actually phosphorylate IκB in response to specific stimuli. Several IκB kinases have been identified and shown to be activated following treatment with substances which stimulate NFκB activity (for reviews see May and Ghosh, 1999; Israel, 2000). Hence, such stimuli act by activating the IκB kinase, resulting in phosphorylation of IκB leading to its degradation and thus activation of NFκB (Fig. 8.22a).

Figure 8.22
Regulation of NFκB activity by IκB can be modulated by stimuli which result in its phosphorylation and degradation leading to activation of NFκB (a) or by stimuli which enhance its synthesis thereby inactivating NFκB (b).

In contrast, other stimuli such as glucocorticoid hormone treatment can inhibit NFκB activity. Although this may involve a direct inhibitory interaction between the activated glucocorticoid receptor and the NFκB protein itself (Nissen and Yamamoto, 2000), it is also likely to involve the ability of glucocorticoid to induce enhanced IκB synthesis resulting in inhibition of NFκB (for review see Marx, 1995) (Fig. 8.22b). Hence the ability of IκB to interfere with NFκB is modulated both by processes which alter the activity of IκB by phosphorylating it (Fig. 8.22a) and by altering its rate of synthesis (Fig. 8.22b).

Interestingly, one form of IκB is actually induced by activated NFκB. Hence, following activation of NFκB, new IκB is synthesized and binds to NFκB. As this binding inhibits NFκB, a feedback loop is created which limits the effects of activating the NFκB pathway (Fig. 8.23) (for review see Ting and Endy, 2002).

Figure 8.23

Following the release of active NFκB from the inhibitory IκB protein, it can activate the gene encoding one form of IκB. This newly synthesized IκB can bind to active NFκB and inactivate it, thereby creating a negative feedback loop which limits NFκB activity.

In addition to its activation of NFκB, treatment with phorbol esters also results in the increased expression of several cellular genes which contain specific binding sites for the transcription factor AP-1. As discussed in Chapter 9 (section 9.3.1), this transcription factor in fact consists of a complex mixture of proteins including the proto-oncogene products Fos and Jun. Following treatment of cells with phorbol esters, the ability of Jun to bind to AP-1 sites in DNA is stimulated. This effect, together with the increased levels of Fos and Jun produced by phorbol ester treatment, results in the increased transcription of phorbol ester inducible genes. As with the activation of NFκB, phorbol esters appear to increase DNA binding of Jun by activating protein kinase C. Paradoxically, however, it has been shown (Boyle *et al.*, 1991) that the increased DNA binding ability of Jun following phorbol ester treatment is mediated by its dephosphorylation at three specific sites located immediately adjacent to the basic DNA binding domain indicating that protein kinase C acts by stimulating a phosphatase enzyme which in turn dephosphorylates Jun (Fig. 8.24).

Such an inhibitory effect of phosphorylation on the activity of a transcription factor is not unique to the Jun protein, a similar effect of phosphorylation in reducing DNA binding activity having also been observed in the Myb proto-oncogene protein discussed in Chapter 9 (section 9.3.4) (Luscher *et al.*, 1990). Moreover, DNA binding ability is not the only target for such inhibitory effects of phosphorylation. Thus phosphorylation of the bicoid protein reduces its ability to activate transcription without affecting its DNA binding activity, presumably by inhibiting the activity of its activation domain (Ronchi *et al.*, 1993). Similarly, phosphorylation of the Rb-1 anti-oncogene protein inhibits its ability to bind to the E2F transcription factor and inhibit its activity (see Chapter 9, section 9.4.3, for discussion of the Rb-1/E2F interaction).

As well as targeting factors themselves, phosphorylation has also been shown to modulate the activity of histone modifying enzymes which, in

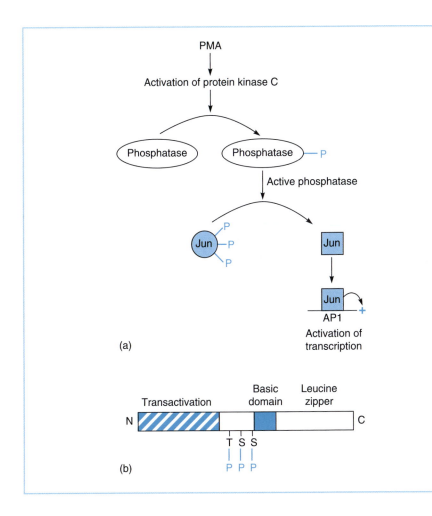

(a)

(b)

Figure 8.24
(a) Activation of Jun binding to DNA by dephosphorylation. The dephosphorylation of Jun protein following PMA treatment increases its ability to bind to AP-1 sites and activate PMA-responsive genes. This is likely to be mediated via the PMA-dependent activation of protein kinase C which, in turn, phosphorylates a phosphatase enzyme allowing it to dephosphorylate Jun. (b) Position in the Jun protein of the two serine (S) and one threonine (T) residues which are dephosphorylated in response to PMA. Note the close proximity to the basic domain (shaded) which mediates DNA binding. The positions of the transactivation domain and leucine zipper are also indicated.

turn, regulate chromatin structure. Thus, in the absence of calcium stimulation, the MEF2 transcription factor is bound to the promoters of muscle-specific genes. However, gene activation does not occur since histone deacetylase enzymes are bound to MEF2 and, as discussed in Chapter 1 (section 1.2.3), a lack of acetylated histones produces a tightly packed chromatin structure incompatible with transcription. However, in response to calcium, kinase enzymes are activated and phosphorylate the histone deacetylases. This phosphorylation results in the histone deacetylase enzymes being exported from the nucleus, allowing MEF-2 to fulfil its function and activate muscle-specific gene expression (Fig. 8.25) (for reviews see Stewart and Crabtree, 2000; McKinsey *et al.*, 2002).

This ability of calcium to activate a kinase which then phosphorylates a target protein is evidently in contrast to the direct binding of calcium to the DREAM transcription factor which was discussed in section 8.2.1

Figure 8.25

The ability of MEF-2 transcription factor to activate gene transcription can be blocked by histone deacetylase enzymes (HDA) which deacetylate histones and thereby block transcription. Following calcium treatment, the histone deacetylases are phosphorylated and exported from the nucleus, allowing MEF-2 to activate gene transcription.

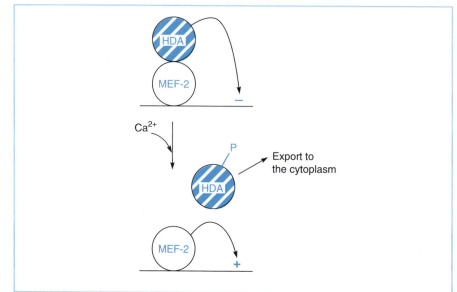

(compare Fig. 8.4 and Fig. 8.25) and illustrates the fact that a specific stimulus can use multiple mechanisms to activate transcription.

Hence, protein modification by phosphorylation can have a wide variety of effects on transcription factors, either stimulating or inhibiting their activity and acting via a direct effect on the ability of the factor to enter the nucleus, bind to DNA, associate with another protein or activate transcription or by an indirect effect affecting the activity of an inhibitory protein or a histone-modifying enzyme. The directness and rapidity of this means of transcription factor activation evidently renders it of particular importance in the response to cellular signalling pathways.

8.4.3 ACETYLATION

In view of the directness and rapidity of using post-translational modification as a means of modulating the activity of transcription factors, it is not surprising that other transcription factor modifications apart from phosphorylation, are used in this way.

In particular, acetylation of transcription factors, particularly on lysine residues has now been defined as an important means of regulating their activity. Thus, although acetylation was initially defined as a modification able to modulate histone activity (see Chapter 1, section 1.2.3), it has now been shown also to occur for transcription factors themselves (for review see Freiman and Tjian, 2003).

Thus, the addition of acetyl residues to the C-terminal domain of the p53 protein (see Chapter 9, section 9.4.2) increases the activity of p53 (Gu and Roeder, 1997; Luo *et al.*, 2000), although the precise manner in which acetylation enhances the ability of p53 to stimulate transcription is currently unclear (for review see Prives and Manley, 2001). This acetylation of p53 is carried out by the p300 co-activator molecule which, as described in Chapter 5 (section 5.4.3), associates with p53 as well as with a wide variety of other transcription factors. This finding indicates that as well as acetylating histones and thereby modifying chromatin structure (see Chapter 1, section 1.2.3), p300 and the related CBP co-activators may also use their acetyltransferase activity to acetylate specific transcription factors and thereby modify their activity (Fig. 8.26).

Hence, acetylation can modulate the activity of p53 by targeting its C terminus. However, the N terminus of p53 can be modified by phosphorylation and this reduces its ability to bind to the MDM2 inhibitory protein (see Chapter 9, section 9.4.2), thereby enhancing the stability of p53. Therefore, the activity of p53 can be modified by phosphorylation and by acetylation, indicating that different post-translational modifications can target the same transcription factor molecule.

Acetylation also occurs in the NFκB/IκB system which also involves regulated phosphorylation as discussed above (section 8.4.2). Thus, it has been shown that NFκB is acetylated and that this inhibits its interaction with IκB (Chen *et al.*, 2001). Hence, interaction of NFκB with IκB requires both deacetylated NFκB and dephosphorylated IκB (Fig. 8.27).

As well as targeting the same transcription factor (as in the case of p53) or two interacting transcription factors (as in the case of NFκB/IκB), there is

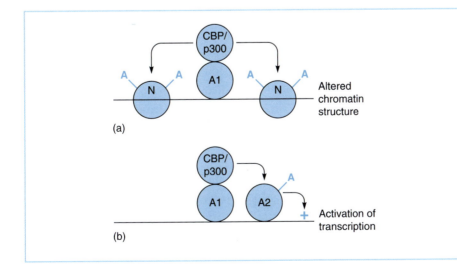

(a)

(b)

Figure 8.26

Possible mechanisms of action of CBP/p300. Following recruitment to DNA by an activating molecule (A1) the acetyl transferase activity of CBP/p300 may either (a) acetylate histones producing a more open chromatin structure or (b) acetylate another activating transcription factor (A2) allowing it to stimulate transcription.

Figure 8.27

Either acetylation of NFκB or phosphorylation of IκB can inhibit the NFκB/IκB interaction.

evidence that the phosphorylation and acetylation systems can interact with one another. Thus, for example, the ATF-2 transcription factor has been shown to have histone acetyltransferase activity and this activity is stimulated by ATF-2 phosphorylation (Kawasaki *et al.*, 2000).

8.4.4 METHYLATION

As with acetylation, methylation has been shown to play an important role in the modification of histones (see Chapter 1, section 1.2.3) and as described in Chapter 6 (section 6.4.1), the polycomb repressor complex contains an activity capable of methylating histones. However, as with acetylation, methylation has also been shown to occur for transcription factors. Thus, for example, the STAT-1 transcription factor is modified by the addition of methyl groups to specific arginine residues and this stimulates its DNA binding ability (Mowen *et al.*, 2001). As described in section 8.4.2, the activity of STAT-1 is also modified by phosphorylation, indicating that, as with acetylation, methylation and phosphorylation can target the same molecule.

As well as affecting transcription factors which bind to DNA, methylation can also affect co-activators such as CBP and the related p300 factor, discussed in Chapter 5 (section 5.4.3). Thus, both these factors are modified by methylation on specific arginine residues (for review see Gamble and Freedman, 2002). Most interestingly, such methylation affects the ability of CBP/p300 to bind to the various transcription factors with which they interact. Thus, methylation abolishes the ability of CBP/p300 to bind to the CREB factor but has no effect on its ability to bind to nuclear receptors, such as the steroid receptors. Hence, the competition between different transcription factors for binding to CBP/p300 (see Chapter 6, section 6.5) can be altered by modification of the co-activator, resulting in a different balance between the different factors under different conditions (Fig. 8.28).

As in the case of STAT-1, CBP is modified by phosphorylation, as well as by acetylation. Thus, phosphorylation of CBP on serine 436 enhances its ability to interact with the AP-1 (see Chapter 9, section 9.3.1) and Pit-1 (see Chapter 4, section 4.2.6) transcription factors (for review see Gamble and Freedman, 2002).

Figure 8.28
The ability of CBP/p300
to bind to different
transcription factors is
affected by the
methylation of CBP/
p300 which blocks
binding to the CREB
factor while not affecting
binding to the nuclear
receptors (NR).

The post-translational modification of co-activators can therefore modulate their interaction with different activating molecules, allowing them preferentially to activate different pathways under different conditions. This effect evidently parallels the phosphorylation of transcription factors such as CREB which affects their ability to interact with CBP/p300 and thus produce transcriptional activation (see Chapter 5, section 5.4.3 and section 8.4.2 of this chapter).

8.4.5 UBIQUITINATION

Although phosphorylation, acetylation and methylation all involve the addition of relatively small chemical groups to the transcription factor molecule, it is possible for a much larger entity to be added. Thus, many proteins in the cell, including transcription factors, become modified by the addition of ubiquitin, which is itself a 76 amino acid protein. This small protein is linked to the transcription factor by a covalent bond between the C terminal of ubiquitin and an internal lysine residue of the transcription factor (for review see Freiman and Tjian, 2003).

In many cases, this ubiquitination serves to target the molecule for degradation, since it is recognized by the proteolytic machinery of the cell as marking the protein for destruction. Indeed, in the NFκB/IκB case discussed above (section 8.4.2), phosphorylation of IκB leads in turn to its ubiquitination and hence targets it for destruction, releasing NFκB to activate gene expression (Fig. 8.29) (for review see Maniatis, 1999; Karin and Ben-Neriah, 2000).

An interesting example of such ubiquitin-mediated control of gene expression is provided by the hypoxia inducible factor, HIF-1 (for review see Bruick and McKnight, 2001; Kaelin, 2002). This factor consists of two subunits, HIF-1α and HIF-1β and is activated when cells are exposed to low oxygen. It then activates the expression of genes that are required in this situation. This

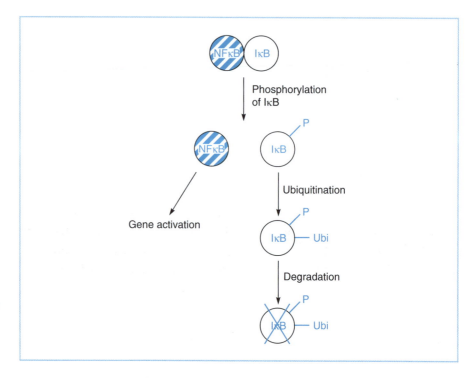

Figure 8.29

Phosphorylation of IκB is followed by its ubiquitination which targets it for destruction.

activation of HIF-1 is controlled at the level of protein degradation. In the presence of oxygen, the HIF-1α subunit is rapidly ubiquitinated and degraded. When oxygen levels fall, HIF-1α is no longer ubiquitinated and can therefore associate with HIF-1β and activate gene transcription (Fig. 8.30).

This obviously leads to the question of how the ubiquitination of HIF-1α is regulated by oxygen. It has been shown that, in the presence of oxygen, HIF-1α is modified by the addition of a hydroxyl (OH) group to a proline amino

Figure 8.30

In the presence of oxygen, the HIF-1α factor is modified by addition of ubiquitin (Ubi) and then degraded. In the absence of oxygen this addition of ubiquitin does not occur and the HIF-1α is stabilized, allowing it to dimerize with HIF-1β and activate transcription.

acid by a proline hydroxylase enzyme. This novel transcription factor modification allows the HIF-1α to be recognized by the von Hippel-Lindau anti-oncogene product (VHL) (see Chapter 9, section 9.4.4) which is part of a multi-protein complex necessary for the addition of ubiquitin. Following a fall in oxygen levels, the proline hydroxylation of HIF-1α does not occur since the activity of the proline hydroxylase enzyme is directly regulated by oxygen. Hence, the VHL product cannot bind and HIF-1α is stabilized (for review see Semenza, 2001; Zhu and Bunn, 2001) (Fig. 8.31).

In this case, therefore, a novel transcription factor modification is recognized by the VHL protein and leads to further modification by ubiquitination. This is evidently analogous to the phosphorylation of IκB discussed above, which is necessary for its subsequent ubiquitination. The structural basis for the role of hydroxyproline in regulating the interaction of HIF-1α and VHL has recently been defined. Thus, the hydroxyproline residue on HIF-1α inserts into a pocket in VHL allowing only hydroxyproline-modified HIF-1α to bind to VHL (Hon *et al.*, 2002; Min *et al.*, 2002) (Fig. 8.32).

Interestingly, the pocket in VHL which binds the hydroxyproline has been shown to be a hot spot for mutations which inactivate VHL and result in cancer. Hence, the anti-oncogenic function of VHL appears to involve its ability to bind to proteins such as HIF-1α via hydroxyproline residues. Indeed, patients with cancer caused by mutation of VHL show expression of HIF-1-activated genes even in the presence of oxygen (see Chapter 9, section 9.4.4 for further discussion).

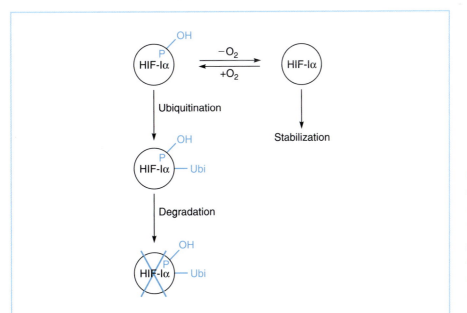

Figure 8.31

Oxygen induces the modification of HIF-1α by addition of a hydroxyl group (OH) on a proline (P) amino acid which results in its ubiquitination and degradation.

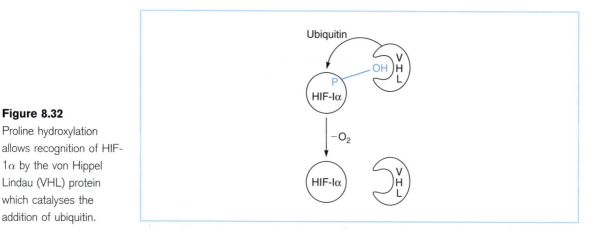

Figure 8.32

Proline hydroxylation allows recognition of HIF-1α by the von Hippel Lindau (VHL) protein which catalyses the addition of ubiquitin.

In the case of HIF-1α therefore, a novel modification involving the hydroxylation of proline residues stimulates ubiquitination and consequent degradation. In addition however, HIF-1α is also modified by a further novel modification involving addition of a hydroxyl group to an asparagine amino acid. Like hydroxylation of proline, this modification is also inhibited by reduced oxygen levels. However, rather than controlling protein stability, the loss of the hydroxyl group on asparagine facilitates the binding of the p300 transcriptional co-activator (for review see Bruick and McKnight, 2002; Kaelin, 2002). This binding of p300 enhances the ability of HIF-1α to activate transcription (see Chapter 5, section 5.4.3 for discussion of CBP/p300). Hence, reduced oxygen levels stabilize the HIF-1α protein by inhibiting hydroxylation of proline and enhance the ability of the stabilized protein to activate transcription by inhibiting hydroxylation of asparagine (Fig. 8.33).

As discussed in Chapter 6 (section 6.4.2), HIF-1α is not the only target for ubiquitination by the VHL complex. Thus, the phosphorylated form of the

Figure 8.33

In the HIF-1α factor removal of oxygen not only blocks the addition of hydroxyl residues to proline (P), preventing VHL binding, but also blocks addition of hydroxyl residues to asparagine (N) promoting the binding of the p300 co-activator molecule. This allows the stabilized protein to stimulate transcription.

large subunit of RNA polymerase II is also ubiquitinated by the VHL complex resulting in its degradation. This specifically blocks the elongation step of transcription, since this phosphorylated form of RNA polymerase II is specifically required for transcriptional elongation (see Chapter 3, section 3.1). As with HIF-1α, the ubiquitination of the large subunit of RNA polymerase II also requires prior proline hydroxylation of the polymerase subunit (Kuznetsova *et al.*, 2003) suggesting that this may be a general mechanism for targeting of proteins by VHL, accounting for its importance in its anti-oncogenic function (see above).

The use of ubiquitination to target proteins such as NFκB, HIF-1α or the large subunit of RNA polymerase II for degradation is not unique to transcription factors but is widely used in the turnover of a variety of different proteins. However, recently a further role of ubiquitin has emerged which is specific to transcription factors. Thus, it has been shown that modification by ubiquitination may be necessary for activation domains to stimulate transcription (see Chapter 5, section 5.2 for a discussion of activation domains). In experiments in yeast, the VP16 acidic activation domain could not activate transcription in a yeast strain which could not add ubiquitin to the VP16 protein. However, if a modified VP16 was prepared in which ubiquitin had already been added to the activation domain, then transcription was activated (Salghetti *et al.*, 2001; Fig. 8.34).

This indicates that modification of the VP16 activation domain by ubiquitination is necessary for it to activate transcription. This effect is not unique to VP16, with the heat shock factor discussed in section 8.3.1 having been shown to be modified by addition of the ubiquitin-related protein SUMO-1. Moreover, this modification stimulates its ability to activate transcription

Figure 8.34

The transcriptional activator VP16 can activate transcription in normal yeast which can modify it by the addition of ubiquitin (Ubi) but not in mutant yeast which cannot carry out this modification. However, if the VP16 is modified by artificial addition of ubiquitin, then it can activate transcription even in the mutant yeast.

(Hong *et al.*, 2001). Hence, modification by addition of ubiquitin or SUMO-1 appears to be widespread among transcription factors (for review see Freiman and Tjian, 2003).

Interestingly, modification of a specific lysine residue in IκB by addition of SUMO-1 has been shown to prevent the addition of ubiquitin and thereby protect IκB from degradation (Desterro *et al.*, 1998) (Fig. 8.35). Hence, different modifications of the same residue may produce opposite effects on transcription factor activity, providing a further mechanism for regulating such activity. As lysine residues are the target for acetylation (section 8.4.3) as well as for addition of ubiquitin or SUMO-1, several different modification enzymes may compete to modify a specific lysine amino acid in a transcription factor with different consequences for its functional activity (for review see Freiman and Tjian, 2003).

Figure 8.35

Modification of a specific lysine residue (L) in IκB by addition of ubiquitin promotes degradation of the protein. In contrast, addition of the ubiquitin-like protein SUMO-1 to the same lysine residue blocks addition of ubiquitin and hence stabilizes IκB.

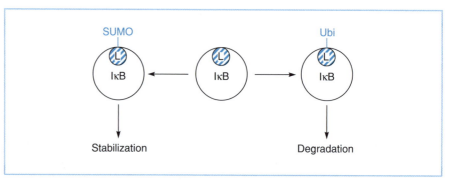

Although the regulation of transcription factor activation domains by ubiquitin may appear unrelated to the role of ubiquitin in protein degradation, this may not be the case. Thus, it has been proposed that modification of an activator by ubiquitination allows it to activate transcription but also targets it for subsequent destruction, after activation has occurred. This would limit the potentially dangerous process of uncontrolled transcriptional activation by ensuring that the activator was rapidly degraded after it had achieved its function of transcriptional activation (for reviews see Tansey, 2001; Conaway *et al.*, 2002) (Fig. 8.36).

The modification of transcription factors by ubiquitin therefore offers a means of regulating both their degradation and their activity. When taken together with modification by phosphorylation, acetylation and methylation discussed above, it is clear that the control of transcription factor activity by post-translational modification is of critical importance, particularly in allowing gene expression to be modulated by specific signalling pathways.

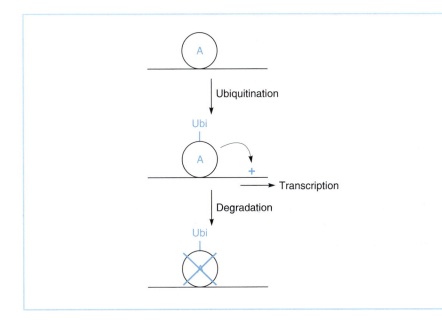

Figure 8.36
Addition of ubiquitin to a transcriptional activator (A) allows it to activate transcription but also targets it for degradation resulting in the transcriptional activation process being self-limiting.

8.5 REGULATION BY PROTEIN DEGRADATION AND PROCESSING

Evidently, a number of the cases discussed in the previous section involve regulating the degradation of a specific factor such as HIF-1α or IκB to ensure that it is stable and can fulfil its function in one situation but is rapidly degraded in another situation and so cannot fulfil its function and a number of other cases of this type have been described (for review see Pahl and Baeuerle, 1996) (see Fig. 8.2d). Hence, regulating the stability of a transcription factor so that it is different in different situations is an important means of regulating transcription factor activity (Fig. 8.37a).

In addition, however, proteolysis can also be used to activate a transcription factor. This can be achieved by cleaving an inactive precursor to produce an active form of the transcription factor (Fig. 8.37b). This form of regulation is also seen in the NFκB family. Thus an NFκB related protein p105 is synthesized as a single molecule in which the NFκB portion is linked to an IκB-like region which inhibits its activity resulting in an inactive precursor protein. Following exposure to an activating stimulus, the IκB-like portion is phosphorylated by the same IκB kinases which phosphorylate IκB. The phosphorylated protein is then cleaved to release active NFκB (Fig. 8.38) (for reviews see Schmitz *et al.*, 2001; Pomerantz and Baltimore, 2002). This mechanism evidently resembles the regulation of NFκB by IκB described above (section 8.4.2), except that in this case the NFκB and IκB-like activities are contained in the same molecule rather than in different molecules.

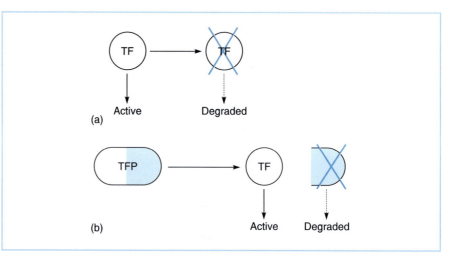

Figure 8.37
Proteolytic cleavage of a transcription factor can be used either (a) to degrade the factor so preventing it from acting or (b) to cleave an inactive precursor molecule to produce an active factor.

This regulatory mechanism is also seen in the case of the SREBP transcription factors which activate gene expression in response to removal of cholesterol (for review see Brown and Goldstein, 1997). In the presence of cholesterol, these factors are anchored in the endoplasmic reticulum by a specific region of the protein. When cells are deprived of cholesterol, this region of the protein is cleaved off, allowing the protein to move to the nucleus and switch on genes whose protein products are required for cholesterol biosynthesis (Fig. 8.39). Interestingly, both of the cases of regulation by

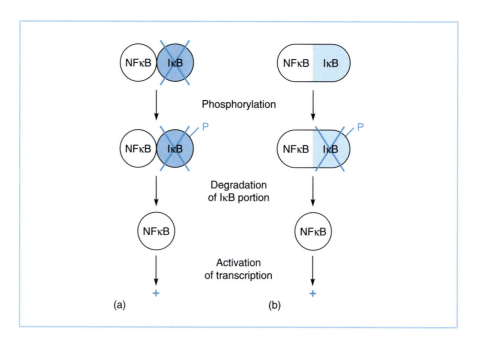

Figure 8.38
In the NFκB family, activation of NFκB can be achieved either (a) by phosphorylation and degradation of an associated IκB protein or (b) by phosphorylation of the IκB-like portion of a large precursor protein (p105) resulting in its proteolytic processing to release active NFκB.

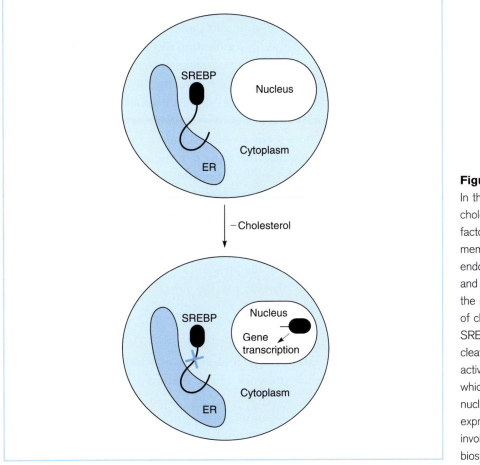

Figure 8.39
In the presence of cholesterol, the SREBP factor is anchored in the membrane of the endoplasmic reticulum and hence cannot enter the nucleus. On removal of cholesterol, the SREBP precursor is cleaved, releasing the active form of the protein which can move to the nucleus and activate the expression of genes involved in cholesterol biosynthesis.

proteolytic cleavage we have described result in a change in localization of the transcription factor, with the NFκB portion of p105 moving from the cytoplasm to the nucleus and the activated SREBP factor moving from the endoplasmic reticulum membrane to the nucleus. This further underlies the importance of changes in transcription factor localization brought about by regulatory processes.

8.6 ROLE OF REGULATED ACTIVITY

In addition to its ability to produce a very rapid activation of gene expression, modification of the activity of a pre-existing protein also allows specific targets for modification to be used in different cases. Thus the various regulatory processes, discussed above, affect the activity of transcription factors at a wide variety of different stages. For example, in the case of phosphorylation (section 8.4.2) we have seen how in different cases a single process can alter the DNA binding ability of a factor, its localization within the cell, its transactivation ability, its ability to associate with another protein, or its degradation.

Clearly therefore post-translational mechanisms for activating pre-existing protein could be used independently to stimulate either the DNA binding or the transcriptional activation activities of a single factor in different situations within a complex regulatory pathway. Indeed, such a combination of mechanisms is actually used to regulate the activity of the yeast GAL4 transcription factor. Thus, as discussed in Chapter 6 (section 6.2.3), the activation of transcription of galactose-inducible genes by GAL4 is mediated by the galactose-induced dissociation of the inhibitory GAL80 protein which exposes the activation domain of DNA bound GAL4. Interestingly, however, this effect only occurs when the cells are grown in the presence of glycerol as the main carbon source. By contrast, however, in the presence of glucose, GAL4 does not bind to DNA and the addition of galactose has no effect (Giniger *et al.*, 1985). Hence by having a system in which glucose modulates the DNA binding of the factor and galactose modulates the activation of bound factor it is possible for glucose to inhibit the stimulatory effect of galactose. This ensures that the enzymes required for galactose metabolism are only induced in the presence of glycerol and not in the presence of the preferred nutrient glucose (Fig. 8.40).

Such a system, in which two different activities of a single factor are independently modulated, could clearly not be achieved by stimulating the *de novo* synthesis of the factor which would simply result in more of it being present. Hence, in addition to its rapidity, the activation of pre-existing factor has the advantage of flexibility in potentially being able to generate different forms of the factor with different activities. It should be noted, however, that this effect

Figure 8.40

Effects of glucose and galactose on GAL4 activity. Note that while galactose stimulates the ability of DNA-bound GAL4 to activate transcription, this effect does not occur in the presence of glucose which results in the release of GAL4 from DNA.

can also be achieved for example, by alternative splicing of the RNA encoding the factor (section 7.3.2) which can, for example, generate forms of the protein with and without the DNA binding domain as in the case of the Era-1 factor, with and without the activation domain as in the case of CREM or Oct-2, or with and without the ligand binding domain as in the case of the thyroid hormone receptor.

8.7 CONCLUSIONS

In this chapter and the previous one, we have discussed how the regulation of gene expression by transcription factors is achieved both by the regulated synthesis or by the regulated activity of these factors. Although there are exceptions, the regulation of synthesis of a particular factor is used primarily in cases of factors which mediate tissue specific or developmentally regulated gene expression where a factor is only required in a small proportion of cell types and is never required in most cell types. In contrast, however, the rapid induction of transcription in response to inducers of gene expression is primarily achieved by the activation of pre-existing inactive forms of transcription factors that are present in most cell types since this process, although more metabolically expensive, provides the required rapidity in response.

Although these two processes have been discussed separately it should not be thought that a given factor can only be regulated either at the level of synthesis or at the level of activity. In fact, in many cases of inducible gene expression which involve activation of pre-existing factors, such activation is supplemented by the slower process of synthesizing new factor in response to the inducing agent. Thus, in the case of the stimulation of genes containing AP-1 sites by phorbol esters discussed above (section 8.4.2), the phorbol ester-induced increase in the DNA binding of pre-existing Jun protein is supplemented by increased synthesis of both Fos and Jun following phorbol ester treatment and such newly synthesized Fos and Jun will clearly eventually become a major part of the increased AP-1 activity observed following phorbol ester treatment (see Chapter 9, section 9.3.1). Similarly the activation of NFκB by dissociation from IκB following treatment with substances such as phorbol esters which activate T cells (see section 8.3.1 and 8.4.2) has been shown to be supplemented by increased synthesis of NFκB and its corresponding mRNA following T-cell activation, while increased synthesis of IκB itself occurs in response to glucocorticoid (section 8.4.2).

Hence in many cases the rapid effects of post-translational processes in activating gene expression are supplemented by *de novo* synthesis of the factor which, although slower, will enhance and maintain the effect. Interestingly, the same factor can be regulated by enhanced synthesis or enhanced activity in different situations. Thus, as described in section 8.4.5, the activity of the HIF-1α factor is enhanced by hypoxia, by means of post-translational modifications which enhance its stability and its association with the p300 co-activator. However, treatment with angiotensin II enhances the synthesis of HIF-1α by enhancing the transcription of the gene encoding it and the translation of its mRNA (Page *et al.*, 2002).

This combination of regulated synthesis and regulated activity is also seen in the case of factors which mediate tissue specific gene expression and which are synthesized in only a few cell types. Thus, in the case of the MyoD factor, which regulates muscle-specific genes, the factor and its corresponding mRNA are synthesized only in cells of the muscle lineage (see Chapter 7, section 7.2.1). The activation of MyoD-dependent genes, which occurs when myoblast cells within this lineage differentiate into myotubes, is not, however, mediated by new synthesis of MyoD which is present at equal levels in both cell types. Rather, it occurs due to the decline in the level of the inhibitory protein Id, resulting in the post-transcriptional activation of pre-existing MyoD and the transcription of MyoD-dependent genes (see Chapter 4, section 4.5.3). Hence, in this case, regulation of synthesis is used to avoid the wasteful production of MyoD in cells of non-muscle lineage while the activation of pre-existing MyoD ensures a rapid response to agents which induce differentiation within cells of the muscle lineage. Thus in a number of cases a combination of both regulated synthesis and regulated activity allows the precise requirements of a particular response to be fulfilled rapidly but with minimum unnecessary wastage of energy.

In summary therefore the different properties of regulated synthesis and regulated activity allow these two processes, both independently and in combination to regulate efficiently the complex processes of inducible, tissue-specific and developmentally regulated gene expression.

REFERENCES

Abu-Amer, Y. and Tondravi, M.M. (1997) NFκB and bone – the breaking point. Nature Medicine 3, 1189–1190.

Ahn, S-G. and Thiele, D. J. (2003) Redox regulation of mammalian heat shock factor 1 is essential for Hsp gene activation and protection from stress. Genes and Development 17, 516–528.

Baeuerle, P.A. (1995) Enter a polypeptide messenger. Nature 373, 661–662.

Baniahmad, A., Tsai, S.Y., O'Malley, B.W. and Tsai, M-J. (1992) Kindred S thyroid hormone receptor is an active and constitutive silencer and a repressor for thyroid hormone and retinoic acid responses. Proceedings of the National Academy of Sciences USA 89, 10633–10637.

Barolo, S. and Posakony, J. W. (2002) Three habits of highly effective signalling pathways: principles of transcriptional control by developmental cell signalling. Genes and Development 16, 1167–1181.

Becker, P.B., Gloss, B., Schmid, W. et al. (1986) In vivo protein-DNA interactions in a glucocorticoid response element require the presence of the hormone. Nature 324, 686–688.

Boyle, W.J., Smeal, T., Defize, L.H.K. et al. (1991) Activation of protein kinase C decreases phosphorylation of c Jun at sites that negatively regulate its DNA binding activity. Cell 64, 573–584.

Brivanlou, A. H. and Darnell, J. E. (2002) Signal transduction and the control of gene expression. Science 295, 813–818.

Brown, M.S. and Goldstein, J.L. (1997) The SREBP pathway: regulation of cholesterol metabolism by proteolysis of a membrane-bound transcription factor. Cell 89, 331–340.

Bruick, R. K. and McKnight, S. L. (2002) Oxygen sensing gets a second wind. Science 295, 807–808.

Brzozowki, A.M., Pike, A.C.W., Dauter, Z. et al. (1997) Molecular basis of agonism and antagonism in the oestrogen receptor. Nature 389, 753–758.

Cantley, L. C. (2001) Translocating Tubby. Science 292, 2019–2021.

Chen, L-f., Fischle, W., Verdin, E. and Greene, W. C. (2001) Duration of nuclear NF-κB action regulated by reversible acetylation. Science 293, 1653–1657.

Conaway, R. C., Brower, C. S. and Conaway, J. W. (2002) Emerging roles of Ubiquitin in transcription regulation. Science 296, 1254–1258.

Costigan, M. and Woolf, C. J. (2002) No DREAM, no pain: closing the spinal gate. Cell 108, 297–300.

Desterro, J. M., Rodriguez, M. S. and Hay, R. T. (1998) SUMO-1 modification of IkappaBalpha inhibits NF-kappaB activation. Molecular Cell 2, 233–239.

Dixit, V. and Mak, T.W. (2002) NF-κB signalling. Many roads lead to Madrid. Cell 111, 615–619.

Dubnicoff, T., Valentine, S.A., Chen, G. et al. (1997) Conversion of Dorsal from an activator to a repressor by the global corepressor Groucho. Genes and Development 11, 2952–2957.

Foo, S. Y. and Nolan, G. P. (1999) NF-κB to the rescue RELs, apoptosis and cellular transformation. Trends in Genetics 15, 229–235.

Freiman, R. N. and Tjian, R. (2003) Regulating the regulators: lysine modifications make their mark. Cell 112, 11–17.

Gallo, G.J., Schuetz, T.J. and Kingston, R.E. (1991) Regulation of heat shock factor in *Schizosaccharomyces pombe* more closely resembles regulation in mammals than in *Saccharomyces cerevisiae*. Molecular and Cellular Biology 11, 281–288.

Gamble, M. J. and Freedman, L. P. (2002) A coactivator code for transcription. Trends in Biochemical Sciences 27, 165–167.

Giniger, E., Varnam, S.M. and Ptashne, M. (1985) Specific DNA binding of GAL4, a positive regulatory protein of yeast. Cell 40, 767–774.

Gu, W. and Roeder, R.G. (1997) Activation of p53 sequence-specific DNA binding by acetylation of the p53 C-terminal domain. Cell 90, 595–606.

He, J. and Furmanski, P. (1995) Sequence specificity and transcriptional activation in the binding of lactoferrin to DNA. Nature 373, 721–724.

Hensold, J.O., Hunt, C.R., Calderwood, S.K. *et al.* (1990) DNA binding of heat shock factor to the heat shock element is insufficient for transcriptional activation in murine erythroleukaemia cells. Molecular and Cellular Biology 10, 1600–1608.

Hill, C.S. and Treisman, R. (1995) Transcriptional regulation by extracellular signals: mechanisms and specificity. Cell 80, 199–211.

Holmberg, C. I., Hietakangas, V., Mikhailov, A. *et al.* (2001) Phosphorylation of serine 230 promotes inducible transcriptional activity of heat shock factor 1. EMBO Journal 20, 3800–3810.

Hon, W-C., Wilson, J. I., Harlos, K. *et al.* (2002) Structural basis for the recognition of hydroxyproline in HIF-1α by pVHL. Nature 417, 975–978.

Hong, Y., Rogers, R., Matunis, M. J. *et al.* (2001) Regulation of heat shock transcription factor 1 by stress-induced SUMO-1 modification. Journal of Biological Chemistry 276, 40263–40267.

Horvath, C.M. (2000) STAT proteins and transcriptional responses to extracellular signals. Trends in Biochemical Sciences 25, 496–502.

Ihle, J.N. (2001) The Stat family in cytokine signalling. Current Opinion in Cell Biology 13, 211–217.

Ip, Y.T. (1995) Converting an activator into a repressor. Current Biology 5, 1–3.

Israël, A. (2000) The IKK complex: an integrator of all signals that activate NF-κB? Trends in Cell Biology 10, 129–133.

Jones, N. (1990) Transcriptional regulation by dimerization: two sides to an incestuous relationship. Cell 61, 9–11.

Kaelin, W. G. (2002) How oxygen makes its presence felt. Genes and Development 16, 1441–1445.

Karin, M. and Ben-Neriah, Y. (2000) Phosphorylation meets ubiquitination: the control of NF-κB activity. Annual Reviews of Immunology 18, 621–663.

Kawasaki, H., Schiltz, L., Chiu, R. *et al.* (2000) ATF-2 has intrinsic histone acetyltransferase activity which is modulated by phosphorylation. Nature 405, 195–200.

Kirov, N.C., Lieberman, P.M. and Rushlow, C. (1996) The transcriptional co-repressor DSP1 inhibits activated transcription by disrupting TFIIA-TBP complex formation. EMBO Journal 15, 7079–7087.

Kuznetsova, A. V., Meller, J., Schnell, P. O. *et al.* (2003) von Hippel-Lindau protein binds hyperphosphorylated large subunit of RNA polymerase II through a proline hydroxylation motif and targets it for ubiquitination. Proceedings of the National Academy of Sciences USA 100, 2706–2711.

Lamb, P. and McKnight, S.L. (1991) Diversity and specificity in transcriptional regulation: the benefits of heterotypic dimerization. Trends in Biochemical Sciences 16, 417–422.

Larson, J.S., Schuetz, T.J. and Kingston, R.E. (1988) Activation *in vitro* of sequence-specific DNA binding by a human regulatory factor. Nature 335, 372–375.

Luo, J., Su, F., Chen, D. *et al.* (2000) Deacetylation of p53 modulates its effect on cell growth and apoptosis. Nature 408, 377–381.

Luscher, B., Christenson, E., Litchfield, D.W. *et al.* (1990) *Myb* DNA binding inhibited by phosphorylation at a site deleted during oncogenic activation. Nature 344, 517–522.

Mandel, G. and Goodman, R. H. (1999) DREAM on without calcium. Nature 398, 29–30.

Maniatis, T. (1999) A ubiquitin ligase complex essential for the NF-κB, Wnt/Wingless, and hedgehog signalling pathways. Genes and Development 13, 505–510.

Marx, J. (1995) How the glucocorticoids suppress immunity. Science 270, 232–233.

May, M. J. and Ghosh, S. (1999) IκB kinases: kinsmen with different crafts. Science 284, 271–273.

McKinsey, T. A., Zhang, C. L. and Olson, E. N. (2002) MEF2: a calcium-dependent regulator of cell division, differentiation and death. Trends in Biochemical Sciences 27, 40–47.

Min, J-H., Yang, H., Ivan, M. *et al.* (2002) Structure of an HIF-1α-pVHL complex: hydroxy-proline recognition in signalling. Science 296, 1886–1889.

Morimoto, R. I. (1998) Regulation of the heat shock transcriptional response: cross talk between a family of heat shock factors, molecular chaperones and negative regulators. Genes and Development 12, 3788–3796.

Mowen, K. A., Tang, J., Zhu, W. *et al.* (2001) Arginine methylation of STAT1 modulates IFNα/β-induced transcription. Cell 104, 731–741.

Nissen, R. M. and Yamamoto, K. R. (2000) The glucocorticoid receptor inhibits NFκB by interfering with serine-2 phosphorylation of the RNA polymerase II carboxy-terminal domain. Genes and Development 14, 2314–2329.

Page, E. L., Robitaille, G. A., Pouysségur, J. and Richard, D. E. (2002) Induction of hypoxia-inducible factor-1α by transcriptional and translational mechanisms. Journal of Biological Chemistry 277, 48403–48409.

Pahl, H.L. and Baeuerle, P.A. (1996) Control of gene expression by proteolysis. Current Opinion in Cell Biology 8, 340–347.

Perkins, N. D. (2000) The Rel/NF-κB family: friend and foe. Trends in Biochemical Sciences 25, 434–440.

Pomerantz, J. L. and Baltimore, D. (2002) Two pathways to NF-κB. Molecular Cell 10, 693–695.

Pratt, W.B. (1997) The role of the hsp90-based chaperone system in signal transduction by nuclear receptors and receptor signalling via MAP kinase. Annual Review of Pharmacology and Toxicology 37, 297–326.

Pratt, W.B. and Toft, D.O. (1997) Steroid receptor interactions with heat shock protein and immunophilin chaperones. Endocrinology Reviews 18, 306–360.

Prives, C. and Manley, J. L. (2001) Why is p53 acetylated? Cell 107, 815–818.

Rabindran, S. K., Haroun, R. I., Clos, J. et al. (1993) Regulation of heat shock factor trimer formation: role of a conserved leucine zipper. Science 259, 230–234.

Renaud, J-P., Rochel, N., Ruff, M. et al. (1996) Crystal structure of the RAR-γ ligand binding domain bound to all trans-retinoic acid. Nature 378, 681–689.

Ronchi, E., Treisman, J., Dostatri, N. et al. (1993) Down-regulation of the Drosophila morphogen bicoid by the torso receptor mediated signal transduction cascade. Cell 74, 347–355.

Salghetti, S. E., Caudy, A. A., Chenoweth, J. G. and Tansey, W. P. (2002) Regulation of transcriptional activation domain function by Ubiquitin. Science 293, 1651–1653.

Schmitz, M. L., Bacher, S. and Kracht, M. (2001) IκB-independent control of NF-κB activity by modulatory phosphorylations. Trends in Biochemical Sciences 26, 186–190.

Semenza, G. L. (2001) HIF-1, O_2, and the 3 PHDs: how animal cells signal hypoxia to the nucleus. Cell 107, 1–3.

Sorger, P.K., Lewis, M.J. and Pelham, H.R.B. (1987) Heat shock factor is regulated differently in yeast and HeLa cell. Nature 329, 81–84.

Stewart, S. and Crabtree, G. R. (2000) Regulation of the regulators. Nature 408, 46–47.

Tansey, W. P. (2001) Transcriptional activation: risky business. Genes and Development 15, 1045–1050.

Thanos, D. and Maniatis, T. (1995) NFκB: a lesson in family values. Cell 80, 529–532.

Thiele, D.J. (1992) Metal regulated transcription in eukaryotes. Nucleic Acids Research 20, 1183–1191.

Ting, A. Y. and Endy, D. (2002) Decoding NF-κB signalling. Science 298, 1189–1190.

Treisman, R. (1996) Regulation of transcription by MAP kinase cascades. Current Opinion in Cell Biology, 8, 205–215.

Tsukiyama, T., Becker, P.B. and Wu, C. (1994). ATP-dependant nucleosome disruption at a heat-shock promoter mediated by DNA binding of GAGA transcription factor. Nature 367, 525–532.

Vandromme, M., Gauthier-Rouviere, C., Lamb, N. and Fernadez, A. (1996) Regulation of transcription factor localisation: fine tuning of gene expression. Trends in Biochemical Sciences 21, 59–64.

Weatherman, R. V., Fletterick, R. J. and Scanlan, T. S. (1999) Nuclear-receptor ligands and ligand-binding domains. Annual Reviews of Biochemistry 68, 559–581.

Wilmann, T. and Beato, M. (1986) Steroid-free glucocorticoid receptor binds specifically to mouse mammary tumour DNA. Nature 324, 688–691.

Wurtz, J-M., Bourget, W., Renaud, J-P. et al. (1996) A canonical structure for the ligand binding domain of nuclear receptors. Nature Structural Biology 3, 87–94.

Zhu, H. and Bunn, H. F. (2001) How do cells sense oxygen? Science 292, 449–451.

Zimarino, V. and Wu, C. (1987) Induction of sequence-specific binding of Drosophila heat-shock activator protein without protein synthesis. Nature 327, 727–730.

Zou, J., Guo, Y., Guettouche, T. et al. (1998) Repression of heat shock transcription factor HSF1 activation by HSP90 (HSP90 complex) that forms a stress-sensitive complex with HSF1. Cell 94, 471–480.

CHAPTER 9

TRANSCRIPTION FACTORS AND HUMAN DISEASE

9.1 DISEASES CAUSED BY TRANSCRIPTION FACTOR MUTATIONS

In previous chapters we have discussed a number of examples of the involvement of transcription factors in normal cellular regulatory processes, for example constitutive, inducible, cell type-specific or developmentally regulated transcription. It is not surprising that aspects of this complex process can go wrong and that the resulting defects in transcription factors can result in disease (for reviews see Engelkamp and van Heyningen 1996; Latchman, 1996).

For example, mutations in several classes of transcription factor that control gene expression during development have been shown to result in human developmental disorders. Thus, mutations in the gene encoding the POU family transcription factor Pit-1 (see Chapter 4, section 4.2.6) result in a failure of pituitary gland development leading to congenital dwarfism, while mutations in the genes encoding the Pax family transcription factors Pax3 and Pax6 (see Chapter 4, section 4.2.7) result in eye defects. Similarly, mutations in genes encoding homeobox proteins (see Chapter 4, section 4.2) result in a variety of congenital abnormalities (for review see Boncinelli, 1997).

Interestingly, the mutations in Pit-1 and Pax6 discussed above are both dominant with one single copy of the mutant gene being sufficient to produce the disease, even in the presence of a functional copy. However, this dominance arises for different reasons (for review see Latchman, 1996). In the case of Pit-1, the mutant Pit-1 can bind to its DNA binding site but cannot activate gene expression. It therefore not only fails to stimulate transcription of its target genes but can also act as a dominant negative factor inhibiting gene activation by preventing the wild type protein from binding to DNA (Fig. 9.1a). This mechanism is similar to one mode of action of transcriptional repressors which act by preventing an activator from binding to its DNA target site (see Chapter 6, section 6.2.1). In contrast, the dominant nature of the Pax6 mutation does not reflect any dominant negative action of the

Figure 9.1

Different mechanisms by which mutations in genes encoding transcription factors can be dominant, producing disease in the presence of a functional copy of the gene. (a) In the case of Pit-1, the mutation produces a dominant negative form of the factor (square) which binds to the appropriate binding site and not only fails to activate transcription but also prevents binding and activation by the functional protein (circle). (b) In the case of Pax-6, one functional gene cannot produce enough functional protein (circle) to activate its target genes (dotted circle).

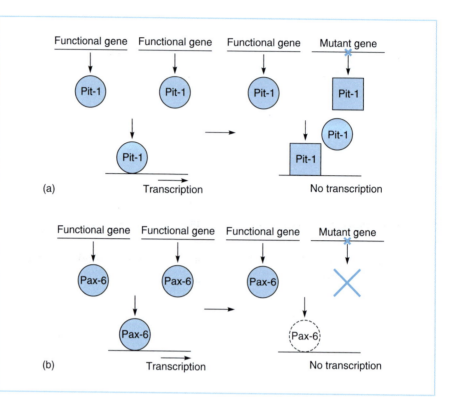

mutant protein since such mutations often involve complete deletion of the gene. Rather it reflects a phenomenon known as haploid insufficiency in which the amount of protein produced by a single functional copy of the gene is not enough to allow it to activate its target genes effectively (Fig. 9.1b).

As well as resulting from mutations in the genes encoding DNA binding factors, developmental disorders can also result from mutations in genes encoding other types of transcription factors such as components of the basal transcriptional complex, co-activators or factors which alter chromatin structure. Thus, for example, mutation in the gene encoding the SNF2 factor which is part of the SWI/SNF chromatin remodelling complex (see Chapter 1, section 1.2.3) results in a lack of α-globin gene expression and a variety of other symptoms such as mental retardation. This indicates that this factor is necessary for opening the chromatin structure of the α-globin genes and a number of other genes so preventing their transcription when it is absent (Gibbons *et al.*, 1995) (Fig. 9.2).

Similarly, mutations in specific subunits of the basal transcription factor TFIIH (see Chapter 3, section 3.5) result in the skin disease xeroderma pigmentosum. Interestingly, the mutant TFIIH proteins found in these patients show defects in their ability to respond to the transcriptional activator FBP

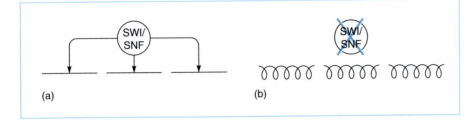

(a)

(b)

Figure 9.2
(a) Functional SWI/SNF alters the chromatin structure of its target genes to a more open configuration (solid line).
(b) If SWI/SNF is inactive these genes remain in a closed chromatin structure (wavy line) and are thus not transcribed.

and the transcriptional repressor FIR (Fig. 9.3) (Liu *et al.*, 2001). Hence, in this case disease results from an inability of a component of the basal transcriptional complex to respond appropriately to activating or repressing signals.

As well as affecting DNA binding factors, chromatin remodelling factors and components of the basal transcriptional complex, mutation can also affect co-activator molecules which transmit the signal between DNA binding factors and the basal complex. Thus, mutation in the gene encoding the CBP factor, which acts as a co-activator for a variety of other transcription factors (see Chapter 5, section 5.4.3), results in the severe developmental disorder known as Rubinstein-Taybi syndrome which is characterized by mental retardation and physical abnormalities (for review see D'Arcangelo and Curran, 1995) indicating that CBP is an important co-activator for developmentally regulated as well as inducible gene expression. Interestingly, no individuals with mutations inactivating both copies of the CBP gene have ever been identified

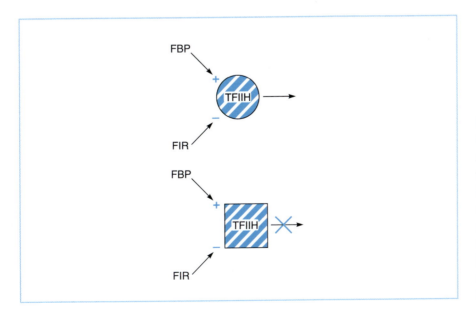

Figure 9.3

In the human disease xeroderma pigmentosum, the mutant TFIIH (square) has lost the ability of the wild type protein (circle) to respond appropriately to the activating factor FBP and the inhibitory factor FIR.

and it is likely that a lack of functional CBP is incompatible with life. Individuals with Rubinstein-Taybi syndrome have a single functional CBP gene and a single mutant gene indicating that the mutation is dominant. As with Pax6, however, this dominance apparently reflects a haploid insufficiency in which a single copy of the CBP gene cannot produce enough functional protein. This is not surprising since, as discussed in Chapter 6 (section 6.5), the amount of CBP in the cell is limited and different transcription factors compete for it.

As well as being the cause of Rubinstein-Taybi syndrome, CBP is also involved in the neurodegenerative disease, Huntington's chorea. However, in this case, the CBP protein is entirely normal and the disease is caused by mutations in a protein known as Huntingtin. Although Huntingtin is not a transcription factor, it can bind to CBP and sequester it into protein aggregations (Nucifora *et al.*, 2001). Since the amounts of CBP in the cell are limiting, this prevents it binding to transcriptional activators and hence causes disease.

Hence, inactivation of CBP can occur by mutation or by its binding of another protein and consequent inactivation (Fig. 9.4). Interestingly, it has recently been shown that as well as targeting CBP, the mutant Huntingtin can also disrupt the interaction between the DNA binding factor Sp1 and the transcriptional co-activator $TAF_{II}130$ (for review see Freiman and Tjian, 2002) indicating that it can target DNA binding factors as well as co-activators.

Hence, developmental disorders can arise from mutations in genes encoding DNA binding activator proteins such as Pit-1 and Pax-6, components of the basal transcriptional complex such as TFIIH, co-activators such as CBP or components of chromatin modulating complexes such as SNF2 (Fig. 9.5).

As well as such developmental defects, mutations in the genes encoding the nuclear receptor transcription factor family (see Chapter 4, section 4.4) can produce a failure to respond to the hormone which normally binds to the receptor and regulates transcription. Such mutations have been reported, for example in the receptors for glucocorticoid, thyroid hormone and vitamin D

Figure 9.4

CBP can be inactivated by mutation producing Rubinstein-Taybi syndrome or by binding to mutant Huntingtin protein in Huntington's disease.

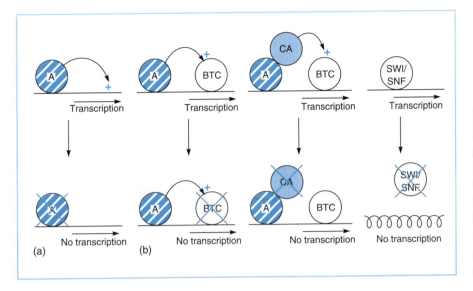

Figure 9.5

Mutations can occur in genes encoding (a) DNA binding activators (A), (b) components of the basal transcriptional complex (BTC), (c) co-activators (CA) or (d) factors which alter chromatin structure (SWI/SNF).

(for review see Latchman, 1996). Similarly, mutations in the gene encoding the peroxisome proliferator-activated receptor γ (PPARγ), another member of the nuclear receptor family, have been identified in patients who show resistance to insulin, diabetes and high blood pressure (for review see Schwartz and Kahn, 1999; Kersten *et al.*, 2000).

9.2 CANCER

Despite the existence of transcription factor mutations producing developmental defects or non-responsiveness to hormone, a special place in the human diseases which can involve alterations in transcription factors is occupied by cancer. Thus, because this disease results from growth in an inappropriate place or at an inappropriate time, it can be caused not only by deficiencies in particular genes but also by the enhanced expression or activation of specific cellular genes involved in growth regulatory processes which are normally only expressed at low levels or very transiently.

Interestingly, many cancer-causing genes of this type, known as oncogenes (for general reviews see Bourne and Varmus, 1992; Broach and Levine, 1997; Hunter, 1997), were originally identified within cancer-causing retroviruses which had picked them up from the cellular genome. Within the virus, the oncogene has become activated either by over-expression or by mutation and is therefore responsible for the ability of the virus to transform cells to a cancerous phenotype. In contrast, the homologous gene within the cellular genome is clearly not always cancer-causing since all cells are not cancerous. It can be activated, however, into a cancer-causing form either by over-expression or by mutation and hence these genes can play an important role in the generation of human cancer (Fig. 9.6). The form of the oncogene isolated from the retrovirus and from the normal cellular genome is distinguished by the prefixes v and c respectively, as in v-onc and c-onc.

Despite this potential to cause cancer, the c-onc genes are highly conserved in evolution, being found not only in the species from which the original virus was isolated but in a wide range of other eukaryotes. This indicates that the products of these oncogenes play a critical role in the regulation of normal cellular growth processes, their malregulation or mutation resulting therefore in abnormal growth and cancer. In agreement with this idea oncogenes identified in this way include genes encoding many different types of protein involved in growth control such as growth factors, growth factor receptors and G proteins. They also include, however, several genes encoding cellular transcription factors which normally regulate specific sets of target genes. Similarly, a number of other genes encoding transcription factors have been identified at the break points of the chromosomal translocations characteristic of human leukaemias with their activation being involved in the resulting cancer. Section 9.3 of this chapter therefore discusses several cases of this type and the insights they have provided into the processes regulating gene expression in normal cells and their malregulation in cancer (for reviews see Rabbits, 1994; Latchman, 1996).

Following the discovery of cellular oncogenes, it subsequently became clear that another class of genes existed whose protein products appeared to restrain cellular growth. The deletion or mutational inactivation of these so-

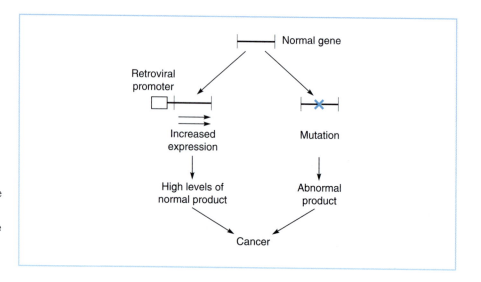

Figure 9.6

A cellular proto-oncogene can be converted into a cancer-causing oncogene by increased expression or by mutation.

called anti-oncogenes therefore results in the abnormal unregulated growth characteristic of cancer cells (for reviews see Knudson, 1993; Weinberg, 1993). As some of these anti-oncogenes also encode transcription factors, they are discussed in section 9.4 of this chapter.

9.3 CELLULAR ONCOGENES AND CANCER

9.3.1 FOS, JUN AND AP1

The AP1 binding site is a DNA sequence that renders genes that contain it inducible by treatment with phorbol esters such as TPA. The activity binding to this site is referred to as AP1 (activator protein 1). It is clear, however, that preparations of AP1 purified by affinity chromatography on an AP1 binding site contain several different proteins (for reviews see Kerppola and Curran, 1995; Karin *et al.*, 1997; Shaulian and Karin, 2002).

A possible clue as to the identity of one of these AP1 binding proteins was provided by the finding that the yeast protein GCN4, which induces transcription of several yeast genes involved in amino acid biosynthesis, does so by binding to a site very similar to the AP1 site (Fig. 9.7). In turn, the DNA binding region of GCN4 shows strong homology at the amino acid level to v-*jun*, the oncogene of avian sarcoma virus ASV17 (Fig. 9.8). This suggested therefore that the protein encoded by the cellular homologue of this gene, c-*jun*, which was known to be a nuclearly located DNA binding protein, might be one of the proteins which bind to the AP1 site.

```
          DNA-binding site
          GCN4   5′  T  G   A  C/G T  C  A  T   3′
          AP1    5′  T  G   A   G  T  C  A  G   3′
```

Figure 9.7

Relationship of the DNA-binding sites for the yeast transcription factor GCN4 and the mammalian transcription factor AP1.

```
  Jun   206 P L F P I D M E S Q E R I K A E R K R M R N R I A A S K S R K
  GCN4  216 P L S P I V P E S S D P     A A L K R A R N T E A A R R S R A

            R K L E R I A R L E E K V K T L K A Q N S E L A S T A N M L R
            R K L Q R M K Q L E D K V             E E L L S K N Y H L E

            E Q V A Q L K Q K V M N H V N S G C Q L M L T Q Q L G T F 296
            N E V A R L K K L V G E R 281
```

Figure 9.8

Comparison of the carboxyl-terminal amino-acid sequences of the chicken Jun protein and the yeast transcription factor GCN4. Boxes indicate identical residues.

In agreement with this, antibodies against the Jun protein react with purified AP1-binding proteins, while Jun protein expressed in bacteria can bind to AP1 binding sites. Hence the Jun protein is capable of binding to the AP1 binding site and constitutes one component of purified AP1 preparations which also contain other Jun-related proteins such as Jun B (Fig. 9.9). Moreover, co-transfection of a vector expressing the Jun protein with a target promoter resulted in increased transcription if the target gene contained AP1 binding sites but not if it lacked them, indicating that Jun was capable of stimulating transcription via the AP1 site (Fig. 9.10). Hence the Jun oncogene product is a sequence specific transcription factor capable of stimulating transcription of genes containing its binding site.

In addition to Jun and Jun-related proteins, purified AP1 preparations also contain the product of another oncogene c-*fos*, as well as several Fos-related

Figure 9.9

Passage of total cellular proteins through a column containing an AP1 site results in the purification of several cellular proteins including Jun, Jun B, Fos and Fos-related antigens (Fras) which are capable of binding to the AP1 site either alone or in combination.

Figure 9.10

Artificial expression of the Jun protein in an expression vector results in the activation of a target promoter containing several AP1 binding sites but has no effect on a similar promoter lacking these sites, indicating that Jun can specifically activate gene expression via AP1 binding sites.

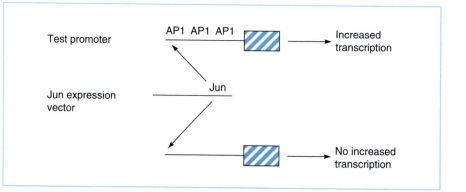

proteins known as the Fras (Fos-related antigens; see Fig. 9.9). Unlike Jun, however, Fos cannot bind to the AP1 site alone but can do so only in the presence of another protein p39, which is identical to Jun (see Chapter 4, section 4.5). Hence, in addition to its ability to bind to AP1 sites alone, Jun can also mediate binding to this site by the Fos protein. Such DNA binding by Fos and Jun is dependent on the formation of a dimeric molecule. Although Jun can form a DNA binding homodimer, Fos cannot do so. Hence DNA binding by Fos is dependent upon the formation of a heterodimer between Fos and Jun which binds to the AP1 site with approximately thirtyfold greater affinity than the Jun homodimer (Fig. 9.11).

It is clear therefore that both Fos and Jun which were originally isolated in oncogenic retroviruses are also cellular transcription factors which play an important role in inducing specific cellular genes following phorbol ester treatment (for review see Ransone and Verma, 1990). Increased levels of Fos and Jun occur in cells following treatment with phorbol esters indicating that these substances act, at least in part, by increasing the levels of Fos and Jun which, in turn, bind to the AP1 sites in phorbol ester-responsive genes and activate their expression.

Similar increases in the levels of Fos and Jun as well as Jun-B and the Fos-related protein Fra-1 are also observed when quiescent cells are stimulated to grow by treatment with growth factors or serum, indicating that these substances act, at least in part, by increasing the levels of Fos and Jun which in turn will switch on genes whose products are necessary for growth itself (Fig. 9.12). In agreement with this idea, cells derived from mice in which the c-*jun* gene has been inactivated grow very slowly in culture and the mice themselves die early in embryonic development (Johnson *et al.*, 1993). Hence Fos and Jun play a critical role in normal cells, as transcription factors inducing phorbol ester or growth dependent genes.

Normally levels of Fos and Jun increase only transiently following growth factor treatment resulting in a period of brief controlled growth. Clearly

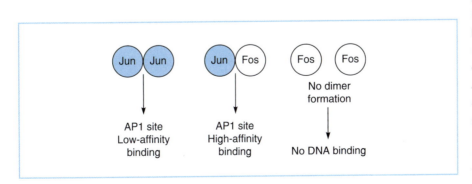

Figure 9.11

Heterodimer formation between Fos and Jun results in a complex capable of binding to an AP1 site with approximately thirtyfold greater affinity than a Jun homodimer while a Fos homodimer cannot bind to the AP1 site.

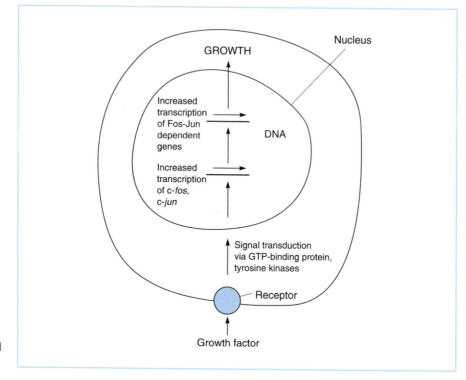

Figure 9.12

Growth factor stimulation of cells results in increased transcription of the c-*fos* and c-*jun* genes which in turn stimulates transcription of genes which are activated by the Fos-Jun complex.

continuous elevation of these proteins, such as would occur when cells become infected with a retrovirus expressing one of them, would result in cells which exhibited continuous uncontrolled growth and were not subject to normal growth regulatory signals. Since such uncontrolled growth is one of the characteristics of cancer cells it is relatively easy to link the role of Fos and Jun in inducing genes required for growth with their ability to cause cancer. Normally, however, the transformation of a cell to a transformed cancerous phenotype requires more than simply its conversion to a continuously grow-ing immortal cell (for review see Land *et al.*, 1983). Since repeated treatments with phorbol esters can promote tumour formation in immortalized cells, the prolonged induction of phorbol ester responsive genes by elevated levels of Fos and Jun may therefore result in the conversion of already continuously growing cells into the tumorigenic phenotype characteristic of cancer cells (Fig. 9.13).

Hence the ability of Fos and Jun to cause cancer represents an aspect of their ability to induce transcription of specific cellular genes. In agreement with this idea, mutations in Fos which abolish its ability to dimerize with Jun and hence prevent it from binding to AP1 sites also abolish its ability to transform cells to a cancerous phenotype. It should be noted, however, that

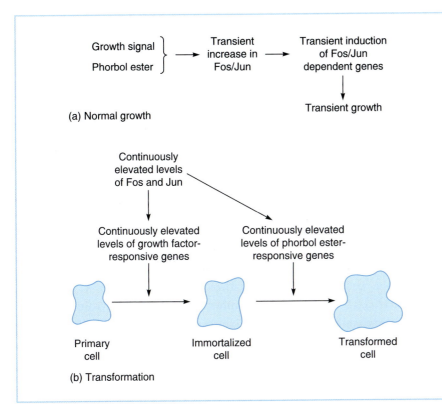

Figure 9.13

Effects of Fos and Jun on cellular growth. In normal cells (a) a brief exposure to a growth signal or phorbol ester will lead to a brief period of growth via the transient induction of Fos and Jun and hence of Fos/Jun-dependent genes. In contrast the continuous elevation of Fos and Jun produced for example by infection with a retrovirus expressing Fos or Jun results in continuous unlimited growth and cellular transformation (b).

in addition to their over-expression within a retrovirus, there is also some evidence that mutational changes render the viral proteins more potent transcriptional activators than the equivalent cellular proteins. Thus the v-Jun protein appears to activate transcription more efficiently than c-Jun due to a deletion in a region which is involved in targeting the c-Jun protein for degradation (Treier *et al.*, 1994) and which also mediates its interaction with a negatively acting cellular factor (Baichwal and Tjian, 1990).

Interestingly, in addition to its central role in the growth response, the Fos, Jun, AP1 system also appears to represent a target for other oncogenes. Thus, for example, the *ets* oncogene which, like *fos* and *jun* encodes a cellular transcription factor, acts via a DNA binding site known as PEA3 which is located adjacent to the AP1 site in a number of TPA-responsive genes such as collagenase and stromelysin. Moreover, the Ets protein cooperates with Fos and Jun to produce high level activation of these promoters (Wasylyk *et al.*, 1990).

In addition to interacting positively with other factors, the Fos/Jun complex can also inhibit the action of other transcription factors. Thus, as described in Chapter 6 (section 6.5), the Fos/Jun complex requires the CBP co-activator in order to activate transcription. It therefore competes with the activated glucocorticoid receptor for CBP hence preventing the receptor from activating transcription. Similarly both Fos and Jun can inhibit the activation of muscle specific promoters by the MyoD transcription factor (see Chapter 7, section 7.2.1) thereby preventing cells from differentiating into non-dividing muscle cells and allowing cellular proliferation to continue (Li *et al.*, 1992).

Hence the Fos and Jun oncogene products play a critical role in the regulation of specific cellular genes in normal cells, interacting with the products of other transcription factors to produce the controlled activity of their target genes necessary for normal controlled growth.

9.3.2 v-erbA AND THE THYROID HORMONE RECEPTOR

The v-*erbA* oncogene is one of two oncogenes carried by avian erythroblastosis virus (AEV). The cellular equivalent of this oncogene c-*erbA*, has been shown to encode the cellular receptor for thyroid hormone (Sap *et al.*, 1986; Weinberger *et al.*, 1986) which is a member of the steroid/thyroid hormone receptor super family discussed in Chapter 4 (section 4.4). Following the binding of thyroid hormone, the receptor/hormone complex binds to its appropriate recognition site in the DNA of thyroid hormone responsive genes and activates their transcription (Fig. 9.14).

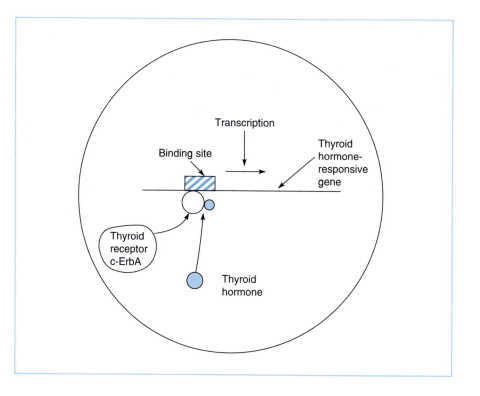

Figure 9.14

The c-*erbA* gene encodes the thyroid hormone receptor and activates transcription in response to thyroid hormone.

Hence the protein encoded by the c-*erbA* gene represents a *bona fide* cellular transcription factor involved in the activation of thyroid hormone responsive genes. Unlike the case of the *fos* and *jun* gene products, which regulate genes involved in growth, it is not immediately obvious how the form of thyroid hormone receptor encoded by the viral v-*erbA* gene can transform cells to a cancerous phenotype.

The solution to this problem is provided by a comparison of the cellular ErbA protein, which is a functional thyroid hormone receptor, and the viral ErbA protein encoded by AEV. Thus, in addition to being fused to the retroviral gag protein at its N terminus, the viral ErbA protein contains several mutations in the regions of the receptor responsible for binding to DNA and for binding thyroid hormone as well as a small deletion in the hormone binding domain (Fig. 9.15).

Interestingly, although these changes do not abolish the ability of the viral ErbA protein to bind to DNA, they do prevent it from binding thyroid hormone and thereby becoming converted to a form which can activate transcription (Sap *et al.*, 1986, 1989). However, these changes do not affect the inhibitory domain which, as discussed in Chapter 6 (section 6.3.2), allows the thyroid hormone receptor to repress transcription. Hence, the viral v-ErbA protein can inhibit the induction of thyroid hormone responsive genes when cells are treated with thyroid hormone by binding to the thyroid hormone response elements in their promoters and dominantly repressing their transcription, as well as preventing binding of the activating complex of thyroid hormone and the cellular ErbA protein (Fig. 9.16). In agreement with this critical role for repression in producing transformation by v-ErbA, a mutation in v-ErbA which abolishes its ability to repress transcription by preventing it binding its co-repressor (see Chapter 6, section 6.3.2) also abolishes its ability to transform cells (Perlmann and Vennstrom, 1995).

Hence the viral ErbA protein acts as a dominant repressor of thyroid hormone responsive genes being both incapable of activating transcription itself and able to prevent activation by intact receptor. This mechanism of

Figure 9.15

Relationship of the cellular ErbA protein and the viral protein. The black dots indicate single amino acid differences between the two proteins while the arrow indicates the region where nine amino acids are deleted in the viral protein.

Figure 9.16
Inhibitory effect of the viral ErbA protein on gene activation by the cellular protein, in response to thyroid hormone. The viral protein both inhibits binding by the activated c-ErbA protein and also dominantly represses transcription by means of its inhibitory domain. Note the similarity to the action of the alpha-2 form of the c-ErbA protein, illustrated in Figure 6.14.

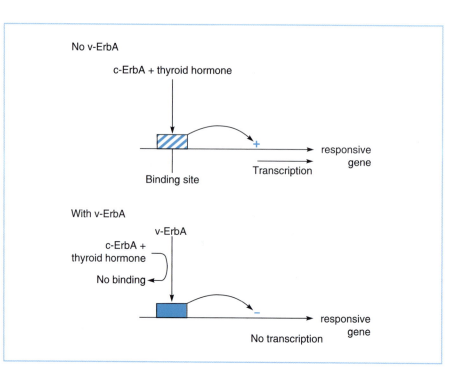

action is clearly similar to the repression of thyroid hormone responsive genes by the naturally occurring alternatively spliced form of the thyroid hormone receptor which, as discussed in Chapter 6 (section 6.3.2), lacks the hormone binding domain and therefore cannot bind hormone. Thus the same mechanism of gene repression by a non-hormone binding receptor is used naturally in the cell and by an oncogenic virus.

One of the targets for repression by the viral ErbA protein is the erythrocyte anion transporter gene (Zenke *et al.*, 1988), which is one of the genes normally induced when avian erythroblasts differentiate into erythrocytes. This differentiation process has been known for some time to be inhibited by the ErbA protein and it is now clear that it achieves this effect by blocking the induction of the genes needed for differentiation. In turn such inhibition will allow continued proliferation of these cells rendering them susceptible to transformation into a tumour cell type by the product of the other AEV oncogene v-*erbB* which encodes a truncated form of the epidermal growth factor receptor (Downward *et al.*, 1984) and therefore renders cell growth independent of external growth factors (Fig. 9.17).

The two cases of Fos/Jun and ErbA therefore represent contrasting examples of the involvement of transcription factors in oncogenesis both in terms of the mechanism of transformation and the manner in which the cellular form of the oncogene becomes an active transforming gene. Thus, in the case

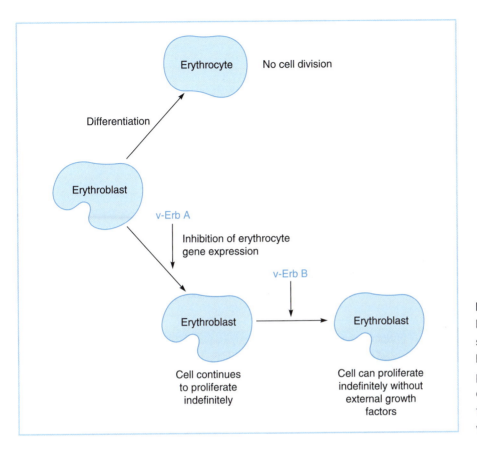

Figure 9.17

Inhibition of erythrocyte-specific gene expression by the v-ErbA protein prevents erythrocyte differentiation and allows transformation by the v-ErbB protein.

of Fos and Jun, transformation is achieved by the continuous activation of genes necessary for growth in normal cell types. Moreover, it occurs, at least in part, via the natural activity of the cellular oncogene in inducing these genes being enhanced by their over-expression such that it occurs at an inappropriate time or place (Fig. 9.18a). In contrast in the ErbA case transformation is achieved by inhibiting the expression of genes whose products are required for the differentiation of a particular cell type therefore allowing growth to continue. Moreover, this occurs via the activity of a mutated form of the transcription factor which, rather than carrying out its normal function more efficiently, actually interferes with the normal role of the thyroid hormone receptor in inducing thyroid hormone responsive genes required for differentiation (Fig. 9.18b).

9.3.3 THE myc ONCOGENE

Interestingly, for a considerable period, the techniques of molecular biology failed in the case of the c-*myc* oncogene, which was one of the earliest cellular

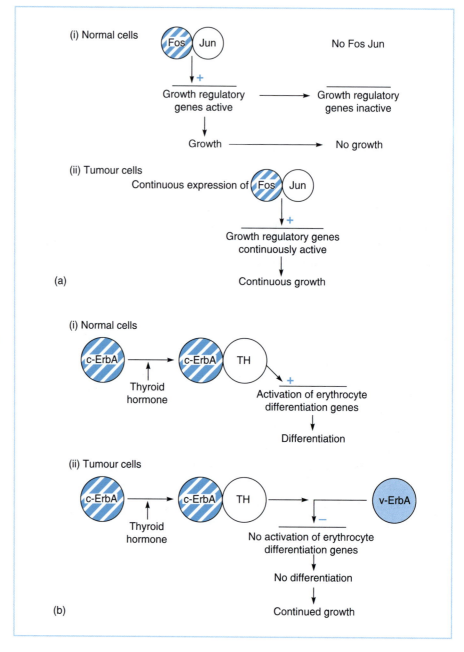

Figure 9.18

Transformation mechanisms of Fos/Jun (panel a) and ErbA (panel b). Note that Fos/Jun-induced transformation occurs because the proteins induce the continual activation of growth regulatory genes which are normally expressed only transiently while v-ErbA-induced transformation occurs because the protein interferes with the action of its cellular homologue and hence inhibits the induction of genes involved in erythrocyte differentiation.

oncogenes to be identified, with its expression being dramatically increased in a wide variety of transformed cells (for review see Grandori *et al.*, 2000; Eisenman, 2001). Thus the Myc protein has a number of properties suggesting that it is a transcription factor, notably nuclear localization, the possession of several motifs characteristic of transcription factors such as the helix-loop-helix and leucine zipper elements (see Chapter 4, section 4.5) and the ability to activate target promoters in co-transfection assays. Despite exhaustive efforts, however, no DNA sequence to which the Myc protein binds could be defined, rendering its mechanism of action uncertain.

The solution to this problem was provided by the work of Blackwood and Eisenman (1991) who identified a novel protein, Max, which can form hetero-dimers with the Myc protein via the helix-loop-helix motif present in both proteins. Myc/Max heterodimers can bind to DNA and regulate transcription, whereas Myc/Myc homodimers cannot do so (for reviews see Grandori *et al.*, 2000; Baudino and Cleveland, 2001) (Fig. 9.19). This effect evidently parallels the requirement of the Fos protein for dimerization with Jun in order to bind with high affinity to AP1 sites (see section 9.3.1).

The Max protein therefore plays a critical role in allowing the DNA binding of Myc and the structure of a Myc/Max heterodimer bound to DNA has recently been defined (Nair and Burley, 2003). Moreover, the ability to inter-act with Max, bind to DNA and modulate gene expression is critical for the ability of the Myc protein to transform since mutations in Myc which abolish its ability to heterodimerize with Max also abolish its transforming ability. Hence, as was previously speculated, the Myc protein is a transcription factor whose over-expression causes transformation, presumably via the activation of genes whose protein products are required for cellular growth (for reviews see Zornig and Evan, 1996; Grandori and Eisenman, 1997; Levens, 2002).

Interestingly, the Max protein does not appear to represent a passive part-ner which merely serves to deliver Myc to the DNA of target genes. Rather it plays a key role in regulating the activity of target genes containing the appro-priate binding site. Thus, it has been shown that, whereas Myc/Max hetero-dimers can activate transcription, Max/Max homodimers can bind to the same site and weakly repress transcription. Moreover, Max can also hetero-dimerize with another member of the helix-loop-helix family, known as Mad, to form a strong repressor of transcription (for review see Bernards, 1995) (Fig. 9.20).

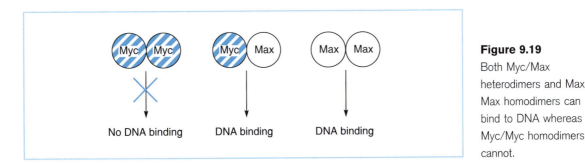

Figure 9.19
Both Myc/Max heterodimers and Max/Max homodimers can bind to DNA whereas Myc/Myc homodimers cannot.

Figure 9.20

Functional effects of Max/Max homodimers and of Myc/Max or Mad/Max heterodimers. Note that Max/Max homodimers repress transcription only weakly by passively blocking activator binding, whereas Mad/Max heterodimers actively repress transcription and therefore have a much stronger effect.

Activation   Weak repression   Strong repression

The Max/Max homodimer appears to act as a weak repressor simply by preventing the Myc/Max activator from binding to its appropriate binding sites and thereby preventing it from activating transcription. In contrast, the Mad/Max heterodimer appears to act as an active repressor which is capable of reducing transcription below that which would be observed in the absence of any activator binding (see Chapter 6, for discussion of the mechanisms of transcriptional repression). Thus, it has been shown that the Mad protein can bind the same complex of N-CoR, mSIN-3 and mRPD3, which mediates active repression by nuclear receptors such as the thyroid hormone receptor in the absence of hormone (see Chapter 6, section 6.3.2) (for review see Wolffe, 1997). As this complex includes the mRPD3 protein, which has histone deacetylase activity, it is possible that the Mad/Max heterodimer may repress transcription, at least in part, by recruiting a complex which deacetylates histones thereby organizing a more tightly packed chromatin structure (Fig. 9.21).

In the case of the nuclear receptors, the switch from the repressed state of target genes to their activation is mediated by the addition of hormone. In contrast, however, in the case of the Myc family it is mediated by signals which produce a rise in Myc expression and a corresponding fall in the expression of Mad. Thus Myc is expressed at very low levels in resting cells and its expression is induced when cells begin to grow, whereas Max is expressed at similar high levels in both resting and proliferating cells and Mad is expressed at high levels only in resting cells and not in proliferating cells. Hence in resting cells

Figure 9.21

The Max/Mad heterodimer can recruit a co-repressor complex (Co-R) with histone deacetylase activity which can produce a more tightly packed chromatin structure (compare with Fig. 6.26).

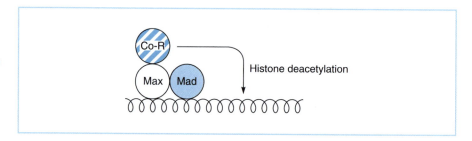

Histone deacetylation

Mad and not Myc will be expressed and the expression of Myc dependent genes will be repressed by Mad/Max homodimers. In contrast expression will be activated by Myc/Max heterodimers as the cells receive signals to proliferate resulting in increased Myc expression and decreased Mad expression (Fig. 9.22). Clearly the over-expression of the Myc gene, which is observed in many cancer cells, would result in a similar production of activating Myc/Max heterodimers leading to gene activation. Hence, as in the case of the Fos/Jun system, transformation by the Myc oncogene appears to depend primarily on its over-expression resulting in the activation of genes required for cellular growth.

Interestingly, it has been shown that Myc can also interact with another transcription factor, Miz-1 (Myc interacting zinc finger protein-1), which is a zinc finger protein (see Chapter 4, section 4.3 for discussion of this type of protein). Unlike the situation with Max, however, in the absence of Myc, Miz-1 acts as an activator of genes promoting growth arrest. In the presence of Myc, however, this activity of Miz-1 is inhibited resulting in the repression of these genes (Peukert *et al.*, 1997). Hence the rise in Myc levels in transformed cells stimulates the activity of growth promoting genes via Myc/Max-mediated gene activation and represses growth inhibitory genes via a repression of Miz-1 activity.

As well as regulating growth by altering the transcription of specific protein-coding genes by RNA polymerase II, another means by which Myc can alter growth has recently been demonstrated. Thus, it has been shown that Myc interacts with the TFIIIB transcription factor which is essential for transcription by RNA polymerase III (see Chapter 3, section 3.4) and stimulates the transcription of the genes encoding tRNA and 5S ribosomal RNA (Gomez-Roman *et al.*, 2003). Since these RNAs are essential for protein synth-

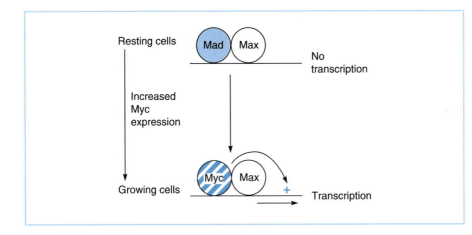

Figure 9.22

In resting cells, Myc-dependent genes will be repressed by a Mad/Max heterodimer. As cells begin to grow, the expression of Myc increases resulting in the formation of Myc/Max heterodimers which activate transcription.

esis by the ribosome and hence for cellular growth, this provides a further means by which transcriptional regulation by Myc can regulate cellular growth.

9.3.4 OTHER ONCOGENIC TRANSCRIPTION FACTORS

In view of the likely need for multiple different transcription factors to regulate genes involved in cellular growth processes, it is not surprising that several other genes encoding transcription factors have also been identified as oncogenes as well as playing a key role in gene expression in specific cell types. Thus, for example, the *myb* oncogene and the *maf* oncogene, both of which were originally isolated from avian retroviruses, play key roles in gene regulation in monocytes and erythroid cells respectively (for reviews see Graf, 1992; Blank and Andrews, 1997; Motohashi *et al.*, 1997). Similarly, the rel oncogene of the avian retrovirus Rev-T is a member of the NFκB family of transcription factors discussed in Chapter 8 (for review see Baeurale and Baltimore, 1996; Foo and Nolan, 1999) while the Bcl-3 oncogene is a member of the IκB family which interacts with the NFκB proteins (Bours *et al.*, 1993).

Interestingly, the Bcl-3 factor illustrates another facet of the mechanisms by which transcription factor genes become oncogenic. This factor was not identified as a retroviral oncogene but on the basis that it was located at the break point of chromosomal rearrangements which resulted in its translocation to a position adjacent to the immunoglobulin gene in some B cell chronic leukaemias. A number of other transcription factors have also been shown to be capable of causing cancer when translocated in this way. This can occur because their expression is increased due to their being translocated to a highly expressed locus such as the immunoglobulin gene loci in B cells or the T-cell receptor gene loci in T cells (Fig. 9.23a). Alternatively it can occur because the translocation results in the production of a novel form of the transcription factor due to its truncation or its linkage to another gene (encoding either another transcription factor or another class of protein) following the translocation (Fig. 9.23b).

Factors translocated in these ways include both factors which were originally identified in oncogenic retroviruses and others which had not previously been shown to have oncogenic potential (for reviews see Rabbits, 1994; Latchman, 1996; Look, 1997). Thus for example, expression of the c-*myc* oncogene (section 9.3.3) is dramatically increased by its translocation into the immunoglobulin heavy chain locus which occurs in the human B-cell malignancy known as Burkitt's lymphoma (for review see Spencer and Groudine, 1991) while the gene encoding the Ets transcription factor, discussed above (section 9.3.1), is fused to the gene for the platelet derived

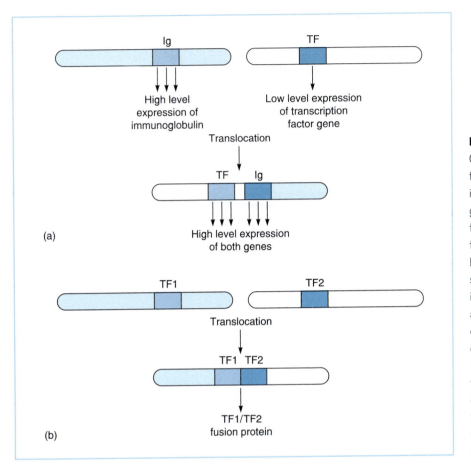

(a)

(b)

Figure 9.23

Chromosomal translocations can result in cancer when (a) the gene encoding a transcription factor is translocated next to a highly transcribed locus such as the immunoglobulin gene (Ig) and is therefore expressed at a high level or (b) the translocation results in the fusion of the genes encoding two different transcription factors resulting in a fusion protein with oncogenic properties.

growth factor receptor to create a novel oncogenic fusion protein in patients with chronic myelomonocytic leukaemia (for review see Sawyers and Denny, 1994).

Similarly, expression of the homeobox gene Hox11 (see Chapter 4 section 4.2.5) is activated in cases of acute childhood T-cell leukaemia while the CBP co-activator (see Chapter 5, section 5.4.3) is fused to the MLL gene in acute myeloid leukaemia (Sobulo *et al.*, 1997). Interestingly, the PBX factors, which are the mammalian homologues of the *Drosophila* extradenticle factor discussed in Chapter 4 (section 4.2.4), were originally identified on the basis of the fact that the gene encoding PBX1 was found fused to the E2A gene (which encodes the E12 and E47 proteins discussed in Chapter 4, section 4.5.2) in a human leukaemia (for review see Mann and Chan, 1996).

These findings provide further evidence that transcription factor genes are not only rendered oncogenic by transfer into a retrovirus but are also involved in the causation of human cancers playing a key role, for example, in the oncogenic effects of the chromosome translocations which are characteristic of specific cancers.

9.4 ANTI-ONCOGENES AND CANCER

9.4.1 NATURE OF ANTI-ONCOGENES

As noted in section 9.2, a number of genes exist whose normal function is to encode proteins that function in an opposite manner to those of oncogenes, acting to restrain cellular growth. The deletion or mutational inactivation of these anti-oncogenes (also known as tumour suppressor genes) therefore results in cancer (for reviews see Knudson, 1993; Fearon, 1997; Hunter, 1997) (Fig. 9.24). This effect evidently parallels the production of cancer by the over-expression or mutational activation of cellular proto-oncogenes (compare Figs. 9.6 and 9.24).

A number of anti-oncogenes of this type have been defined and several encode transcription factors. The two best characterized of these act by different mechanisms. Thus, p53 acts by binding to the DNA of its target genes and regulating their expression, whereas the retinoblastoma gene product (Rb-1) acts primarily via protein–protein interactions with other DNA binding transcription factors. The p53 and Rb-1 proteins are therefore discussed in sections 9.4.2 and 9.4.3 as examples of these two mechanisms of action. Other anti-oncogenes encoding transcription factors are discussed in section 9.4.4.

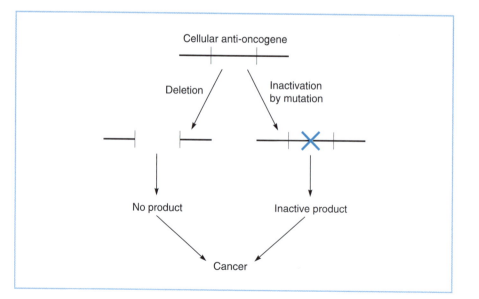

Figure 9.24

Cancer can result from the deletion of specific anti-oncogenes or their inactivation by mutation.

9.4.2 p53

The gene encoding the 53 kilo-dalton protein known as p53 is mutated in a very wide variety of human tumours, especially carcinomas (for review see Ko and Prives, 1996; Levine, 1997; Vogelstein *et al.*, 2000; Haupt *et al.*, 2002; Sharpless and DePinho, 2002). In normal cells expression of this protein is induced by agents which cause DNA damage and its over-expression results in growth arrest of cells containing such damage or their death by the process of programmed cell death (apoptosis). Hence p53 has been called the 'guardian of the genome' (Lane, 1992), which allows cells to proliferate only if they have intact undamaged DNA. This would prevent the development of tumours containing cells with mutations in their DNA and the inactivation of the p53 gene by mutation would therefore result in an enhanced rate of tumour formation. In agreement with this idea, mice in which the p53 gene has been inactivated do not show any gross abnormalities in normal development but do exhibit a very high rate of tumour formation, leading to early death (for review see Berns, 1994).

The molecular analysis of the p53 gene product showed that it contains a DNA binding domain and a region capable of activating transcription. The majority of the mutations in p53 which occur in human tumours are located in the DNA binding domain (Friend, 1994; Anderson and Tegtmeyer, 1995). These mutations result in a failure of the mutant p53 protein to bind to DNA, indicating that this ability is crucial for the ability of the normal p53 protein to regulate cellular growth and suppress cancer.

The p53 protein therefore functions, at least in part, by activating the expression of genes whose protein products act to inhibit cellular growth (Fig. 9.25a). The absence of functional p53 either due to gene deletion (Fig. 9.25b) or to its inactivation by mutation (Fig. 9.25c) results in a failure to express these genes leading to uncontrolled growth.

In addition, functional p53 can also be prevented from activating gene transcription by interaction with the MDM2 oncoprotein (Fig. 9.25d). Thus MDM2 masks the activation domain of p53 preventing it activating transcription (Fig. 9.26a). Moreover, MDM2 when bound to p53 also actively inhibits transcription by interacting with the basal transcriptional complex to reduce its activity (Thut *et al.*, 1997) (Fig. 9.26b).

The major inhibitory effect of the interaction of MDM2 with p53, however, is that it results in the rapid degradation of p53. Thus, MDM2 causes the addition of ubiquitin residues to p53, thereby promoting its degradation (Haupt *et al.*, 1997; for review see Lane and Hall, 1997) (Fig. 9.26c). Hence, several of the different inhibitory mechanisms, discussed in Chapter 6, are involved in the inhibitory effect of MDM2 on p53 (Fig. 9.26) (for review see

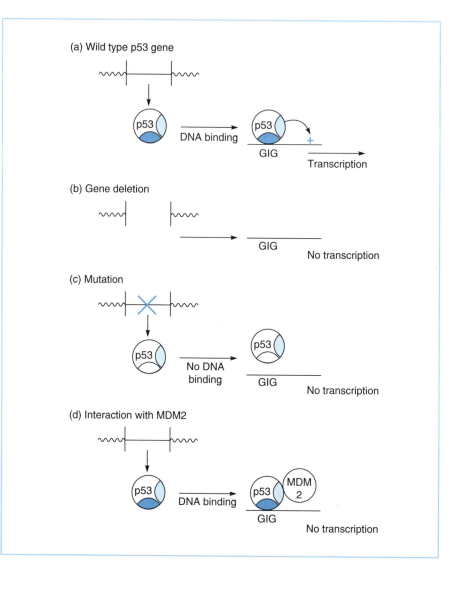

Figure 9.25

The ability of wild type p53 to activate genes encoding growth inhibiting proteins (GIG) (panel a) can be abolished by deletion of the p53 gene (panel b), mutations in the DNA binding domain (solid) which prevent it binding to DNA (panel c) or by the interaction of functional p53 with the MDM2 protein which prevents it from activating transcription (panel d).

Oren, 1999). Interestingly, the addition of ubiquitin to p53, targeting it for degradation is paralleled by the addition of the ubiquitin-related protein SUMO-1 to MDM2. This modification of MDM2 paradoxically enhances its ability to add ubiquitin to p53 and thereby induce p53 degradation (Buschmann *et al.*, 2000) (see Chapter 8, section 8.4.5 for discussion of the regulation of transcription factors by the addition of ubiquitin or SUMO-1).

The inhibitory effect of MDM2 on p53 brought about by these multiple mechanisms is of particular importance in many human soft tissue sarcomas where the p53 gene is intact and encodes wild type p53 but the protein is functionally inactivated due to the high levels of MDM2 resulting from ampli-

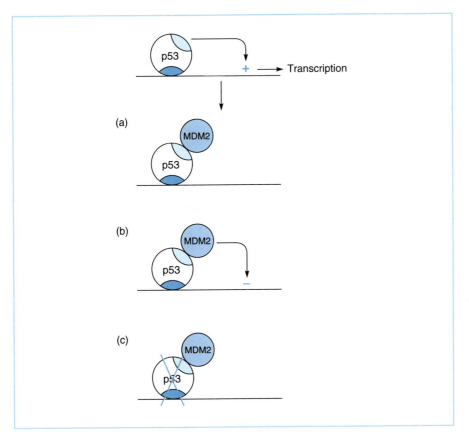

Figure 9.26

Multiple mechanisms by which MDM2 inhibits p53 involving (a) masking of its activation domain (pale blue), (b) direct inhibition of transcription by MDM2 itself and (c) targeting p53 for degradation by proteolytic enzymes.

fication of the *mdm2* gene encoding it. Indeed, the major function of MDM2, even in normal cells, may be to inhibit the action of p53 by interacting with it. Thus mice in which the gene encoding MDM2 is inactivated are non-viable but can be rendered viable by the additional inactivation of the p53 gene (de Oca Luna *et al.*, 1995).

Interestingly, both partners in the p53/MDM2 interaction are subject to modification by phosphorylation and these modifications affect their interaction with one another (for reviews see Prives, 1998; Mayo and Donner, 2002). Thus, following exposure to DNA damage/stress, p53 is phosphorylated. This enhances its retention in the nucleus and inhibits its interaction with MDM2, so allowing it to activate transcription of its target genes. Conversely, stimuli that inhibit apoptosis, lead to phosphorylation of MDM2. This promotes its movement from the cytoplasm to the nucleus and hence allows it to inhibit p53 and its pro-apoptotic effect (Fig. 9.27) (for review see Gottifredi and Prives, 2001).

As well as affecting binding to MDM2, phosphorylation also enhances the binding of p53 to the Pin1 protein (Zheng *et al.*, 2002). Pin1 is a member of

Figure 9.27

The interaction of p53 and MDM2, which inactivates p53, is promoted by phosphorylation of MDM2 but inhibited by phosphorylation of p53.

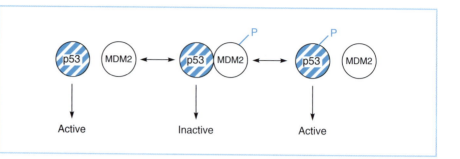

the class of proteins known as peptidyl prolyl isomerases which have the ability to change the structure of the peptide bond between proline residues and adjacent amino acids in a process known as cis-trans isomerization. In the case of p53, interaction with Pin1 and the consequent isomerization of peptide bonds within the p53 protein, stimulates the DNA binding and trans-activation ability of p53. This therefore provides a second mechanism for phosphorylation to stimulate the activity of p53 (Fig. 9.28) (for further details of the effect of phosphorylation on transcription factors see Chapter 8, section 8.4.2).

It should be noted that this effect of a prolyl isomerase enzyme on a transcription factor is not unique to p53. Thus, DNA binding activity of the c-*myb* proto-oncogene protein (see section 9.3.4) has been shown to be nega-tively regulated by its interaction with the peptidyl prolyl isomerase, Cyp40 (for review see Hunter, 1998).

Hence, signals such as DNA damage/stress can activate p53 by inducing its phosphorylation. As noted in Chapter 8 (section 8.4.3) p53 is also subject to acetylation, which stimulates its activity (for review see Prives and Manley, 2001). Recently, it has been demonstrated that histone deacetylase enzymes,

Figure 9.28

Phosphorylation of p53 blocks its interaction with MDM2 so stabilizing the protein and also enhances its interaction with the peptidyl prolyl isomerase Pin1, which stimulates the activity of p53.

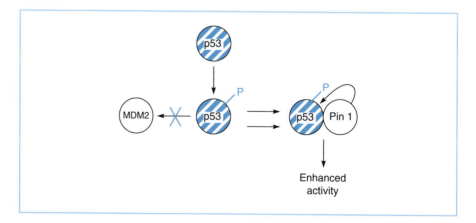

such as Sir2, can specifically deacetylate p53, thereby reducing its ability to activate transcription (Luo *et al.*, 2001; Vaziri *et al.*, 2001). Indeed, it appears that MDM2 exists in a complex with a histone deacetylase enzyme and that deacetylation of p53 actually enhances its degradation by MDM2 (Ito *et al.*, 2002; Li *et al.*, 2002). Hence, this inhibitory complex can deacetylate p53, reducing its activity and targeting it for degradation by MDM2 (Fig. 9.29).

The activity of p53 is thus regulated, in part, by the balance between its acetylation by molecules such as the p300 co-activator (as discussed in Chapter 8, section 8.4.3) and its deacetylation by molecules such as Sir2. In turn, this represents part of the multiple modifications used to alter the activity of the p53/MDM2 system which include many of the modifications that can affect transcription factors, such as phosphorylation, acetylation and modification by addition of ubiquitin or SUMO-1 (see Chapter 8, section 8.4 for a discussion of the modulation of transcription factor activity by post-translational modifications).

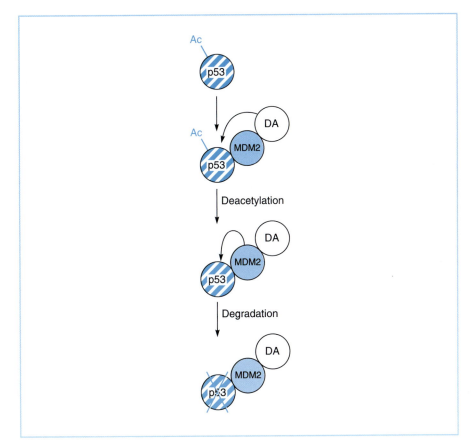

Figure 9.29

Binding to p53 of a complex of MDM2 and a histone deacetylase enzyme (DA) results in deacetylation of p53 which promotes its subsequent degradation by MDM2.

The interaction of p53 with the MDM2 oncogenic protein is paralleled by its interaction with the transforming proteins of several DNA viruses. Indeed p53 was originally discovered as a protein that interacted with the large T oncoprotein of the DNA tumour virus SV40. The functional inactivation of p53 produced by this interaction appears to play a critical role in the ability of these DNA viruses to transform cells paralleling the similar action of MDM2. These interactions suggest that functional antagonism between oncogene and anti-oncogene products is likely to be critical for the control of cellular growth with changes in this balance which activate oncogenes or inactivate anti-oncogenes resulting in cancer.

These considerations evidently focus attention on the genes that are activated by p53. One such gene is that encoding a 21 kilo-dalton protein (p21) which acts as an inhibitor of cyclin-dependent kinases (for review of p53-dependent genes see Ko and Prives, 1996; Vogelstein *et al.*, 2000). As the cyclin-dependent kinases are enzymes that stimulate cells to enter cell division, the finding that p53 stimulates the expression of an inhibitor of these enzymes is entirely consistent with its role in restraining growth, since the inhibition of the cyclin-dependent kinases will prevent cells replicating their DNA and undergoing cell division (Fig. 9.30).

Interestingly, p53 also stimulates expression of the *mdm2* gene whose protein product interferes with the activity of p53 as described above. This effect is likely to be part of a negative feedback loop in which p53, having fulfilled its

Figure 9.30

p53 activates the gene for the 21kD inhibitor of cyclic dependent kinases (cdk). This inhibitor then prevents the cyclin dependent kinases from stimulating DNA synthesis and consequent cell division.

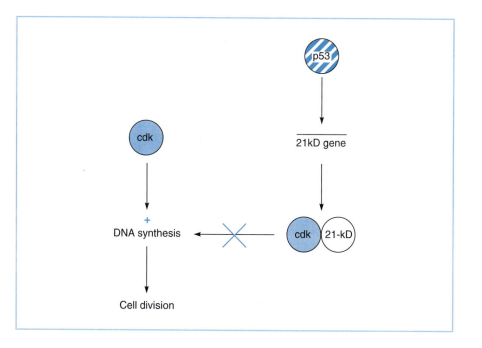

function, activates *mdm2* expression resulting in p53 inactivation (Fig. 9.31) (for review see Oren, 1999). This would allow, for example, cells which had repaired the damage to their DNA to inactivate p53 and resume cell division. Similarly, p53 also stimulates the expression of the bax gene whose protein product stimulates programmed cell death or apoptosis allowing p53 to promote the death of cells whose damaged DNA is irreparable. Further studies have also identified several genes involved in the generation of toxic reactive oxygen species whose expression is induced by p53 during this process, indicating that p53 may also promote apoptosis by inducing the production of these species (Wyllie, 1997).

As with AP1 (section 9.3.1) transcriptional activation by p53 requires the CBP co-activator or the closely related p300 protein (Avantaggiati *et al.*, 1997). Hence, as with AP1 and the steroid receptors (see Chapter 6, section 6.5), AP1 and p53 can compete for CBP/p300 resulting in antagonism between the oncogenic activity of AP1 and the anti-oncogenic activity of p53.

Hence the p53 gene product plays a key role in regulating cellular growth by binding to DNA and activating the expression of specific genes (for review see Almog and Rotter, 1997). Its inactivation by mutation or by interaction with oncogene products is likely to play a critical role in most human cancers. Interestingly, two novel p53-related proteins which encode transcription factors known as p73 and p63, have been described (for reviews see Lohrum and Vousden, 2000; Morrison and Kinoshita, 2000; Yang *et al.*, 2002). It is currently unclear whether either p63 or p73 play a role as anti-oncogenes whose inactivation results in human cancers. However, inactivation of p63 or p73 in knock out mice results in gross developmental abnormalities whereas this is

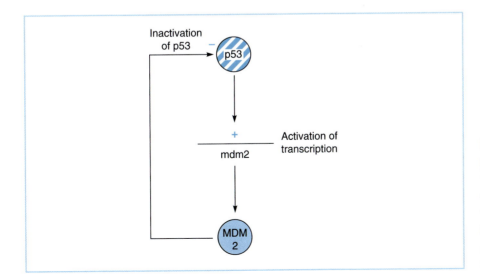

Figure 9.31

p53 activates the gene for the MDM2 protein which acts in a negative feedback loop to inactivate p53 by the mechanisms illustrated in Figure 9.26.

not the case for p53 knock out mice (see above). Similarly, inactivation of p63 is the cause of EEC syndrome (ectrodactyly, ectodermal dysplasia and cleft lip) in humans, in which patients have limb defects and facial clefts (Celli *et al.*, 1999). These findings further emphasize the importance of p53 and the proteins related to it in the regulation of normal embryonic development and cellular proliferation/survival and in the development of cancer.

9.4.3 THE RETINOBLASTOMA PROTEIN

The retinoblastoma gene (Rb-1) was the first anti-oncogene to be defined and is so named because its inactivation in humans results in the formation of eye tumours known as retinoblastomas (for reviews see Lipinski and Jacks, 1999; Harbour and Dean, 2000a). Like p53 the Rb-1 gene product is a transcription factor which exerts its anti-oncogenic effect by modulating the expression of specific target genes. In contrast to p53, however, it exerts this effect via protein–protein interactions with other transcription factors rather than by direct DNA binding.

One of the major targets for Rb-1 is the transcription factor E2F which plays a critical role in stimulating the expression of genes encoding growth promoting proteins such as Myc (section 9.3.3), DNA polymerase α and thymidine kinase (for reviews see Harbour and Dean, 2000b; Müller and Helin, 2000; Morris and Dyson, 2001) and the structure of Rb-1 bound to E2F has recently been defined (for review see Münger, 2003). The association of Rb-1 and E2F does not inhibit the DNA binding of E2F but prevents it from stimulating the transcription of these growth promoting genes and hence inducing growth arrest (Fig. 9.32a).

It appears that Rb-1 exerts its inhibiting effect on transcription in two distinct ways. First, it acts as an indirect repressor by blocking the ability of DNA-bound E2F to activate transcription. This is achieved by Rb-1 binding resulting in the masking of several key residues in the activation domain of E2F, thereby preventing transcriptional activation (Lee *et al.*, 2002) (see Chapter 6, section 6.2.3 for a discussion of this quenching mechanism of transcriptional repression).

Secondly, the Rb/E2F complex acts directly to inhibit transcription, by organizing a tightly packed chromatin structure incompatible with transcription (Ross *et al.*, 2001). This involves the ability of Rb-1 to recruit histone deacetylases and methyltransferases which, as discussed in Chapter 1 (section 1.2.3), promote a more tightly packed chromatin structure (for review see Harbour and Dean, 2000b; Ringrose and Paro, 2001) (Fig. 9.33). It appears that this second effect, involving chromatin structure, maybe of greater impor-

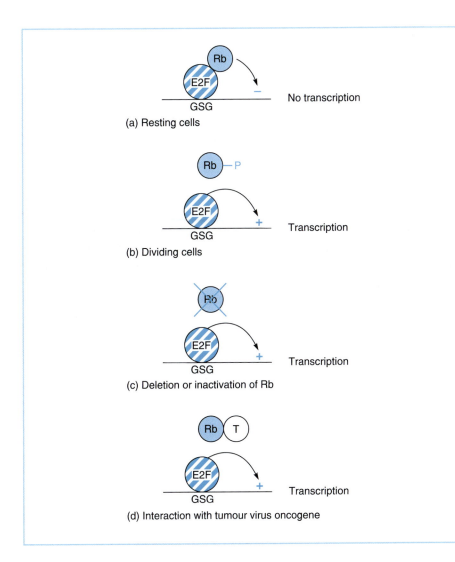

(a) Resting cells — No transcription

(b) Dividing cells — Transcription

(c) Deletion or inactivation of Rb — Transcription

(d) Interaction with tumour virus oncogene — Transcription

Figure 9.32
In resting cells, the Rb-1 protein binds to E2F and prevents it activating the transcription of genes encoding growth stimulating proteins (GSG) as well as directly inhibiting transcription of these genes (panel a). In normal dividing cells, the Rb-1 protein is phosphorylated at the G1/S transition in the cell cycle which prevents it from interacting with E2F and hence allows E2F to activate transcription (panel b). This release of E2F can also occur in tumour cells where the Rb-1 gene is deleted or inactivated by mutation (panel c) or following the interaction of Rb-1 with tumour virus oncogenes (T) (panel d).

tance since it has been shown to be essential for the growth-arresting effect of Rb-1 (Zhang *et al.*, 1999).

Hence Rb-1 exerts its anti-oncogenic effect by inhibiting the transcription of growth promoting genes using both indirect and direct inhibiting mechanisms (see Chapter 6) rather than, as with p53, promoting the transcription of growth inhibiting genes. In normal dividing cells, this interaction of Rb-1 and E2F is inhibited as cells move from G to S phase in the cell cycle. This effect is dependent on the phosphorylation of Rb-1 which prevents it interacting with E2F (see Fig. 9.32b). Hence the controlled growth of normal cells can be regulated by the regulated phosphorylation of Rb-1 which in turn regulates its ability to interact with E2F and modulate its activity.

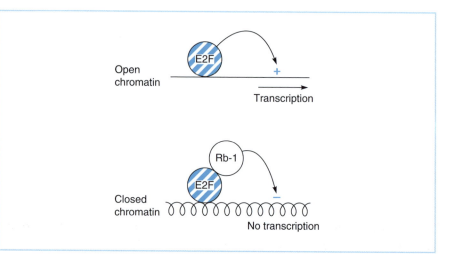

Figure 9.33

Binding of Rb-1 to E2F represses transcription both indirectly by inhibiting activation by E2F and directly by organizing a closed chromatin structure incompatible with transcription.

Interestingly, the phosphorylation of Rb-1 in the cell cycle is carried out by the cyclin-dependent kinases (for review see Sherr, 1994). This provides a link between p53 and the regulation of Rb-1 activity since, as noted above (section 9.4.2) p53 activates the gene encoding the p21 protein which inhibits cyclin-dependent kinases and would thus prevent the phosphorylation of Rb-1 and cell cycle progression (Fig. 9.34). To add to the complexity still further, it appears that the activity of both p53 itself and E2F is also altered following phosphorylation by cyclin-dependent kinases, indicating that a complex net-

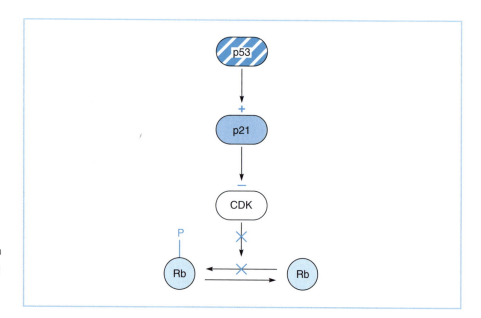

Figure 9.34

By activating the p21 gene whose protein product inhibits cyclin dependent kinases (CDK), p53 produces a fall in CDK activity which results in more Rb being in the growth inhibitory unphosphorylated form.

work of interacting transcription factors, kinases and their inhibitors regulates cellular growth (for review see Dynlacht, 1997).

Interestingly, Rb-1 can be modified by acetylation as well as by phosphorylation (Chan *et al.*, 2001). Such acetylation reduces the ability of the cyclin-dependent kinases to phosphorylate Rb-1. Hence, as with p53 (see section 9.4.2) Rb-1 is modified by multiple post-translational modifications which interact with one another (Fig. 9.35).

Clearly abolishing the activity of Rb-1, either by deletion of its gene or by mutation, will result in the unregulated activity of E2F leading to the uncontrolled growth, which is characteristic of cancer cells (Fig. 9.32c). Interestingly, the inactivation of Rb-1 can also be achieved by the transforming proteins of DNA tumour viruses, such as SV40 or adenovirus. These proteins bind to the Rb-1 protein resulting in the dissociation of the Rb-1/E2F complex releasing free E2F which can activate gene expression (see Fig. 9.32d).

Although E2F is a major target of Rb-1, there are also other factors with which Rb-1 interacts. Thus Rb-1 has been shown to inactivate the UBF factor which plays a critical role in transcription of the ribosomal RNA genes by RNA polymerase I (see Chapter 3, section 3.3). Due to the need for these ribosomal RNAs for the effective functioning of the ribosomes, the inactivation of UBF by Rb-1 will lead to a decrease in the levels of total protein synthesis which would in turn lead to the arrest of cell growth. In agreement with this idea, the inactivation of UBF by Rb-1 appears to play a critical role in

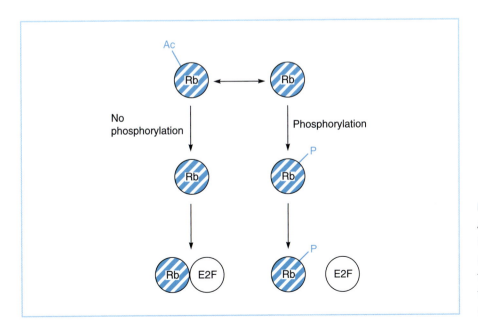

Figure 9.35

Acetylation of Rb-1 inhibits its phosphorylation and thereby promotes the formation of the Rb/E2F complex.

the growth arrest and associated differentiation of U937 monocytic cells (for review see Dynlacht, 1995).

Obviously, the 5S ribosomal RNA and the transfer RNAs which are produced by RNA polymerase III are also necessary for ribosomal function and protein synthesis. Indeed, it has been shown that Rb can also inhibit RNA polymerase III transcription by interacting directly with the polymerase III transcription factor TFIIIB (see Chapter 3, section 3.4) and inhibiting its activity (Larmine *et al.*, 1997). This is evidently the opposite effect to that produced by interaction of the Myc protein with TFIIIB, which stimulates transcription of the genes encoding tRNA and 5S RNA (see section 9.3.3).

Hence Rb-1 can directly inhibit transcription of genes involved in cellular growth both by inhibiting the transcription of E2F-dependent genes by RNA polymerase II and the transcription of all the genes transcribed by RNA polymerases I and III. It therefore has a remarkable ability to modulate transcription by all three RNA polymerases (for review see White, 1997; Brown *et al.*, 2000) (Fig. 9.36) and is likely to play a critical role not only in preventing cancer but also in normal cells by promoting the growth arrest which is necessary for terminal differentiation (see Fig. 9.32).

In agreement with this idea, mice in which the Rb-1 gene has been inactivated die before birth and show gross defects in cellular differentiation (Lee *et al.*, 1992; Wu *et al.*, 2003; for review see Dyson, 2003). This indicates that Rb-1 plays a key role in normal development as well as acting as an anti-oncogene and contrasts with the viability of mice in which the p53 gene has been inactivated (see section 9.4.2). Interestingly, many of the developmental defects observed in mice lacking Rb-1 can be rescued by also inactivating the gene encoding Id2, which is an inhibitory transcription factor having a helix-loop-helix motif but lacking a DNA-binding domain (see Chapter 4, section 4.5 and Chapter 6, section 6.2.2). This indicates that during normal development, Id2

Figure 9.36

Rb can inhibit all transcription by RNA polymerases I and III by inhibiting the activity of UBF and TFIIIB as well as inhibiting the activity of E2F and hence inhibiting the ability of RNA polymerase II to transcribe genes whose protein products stimulate growth.

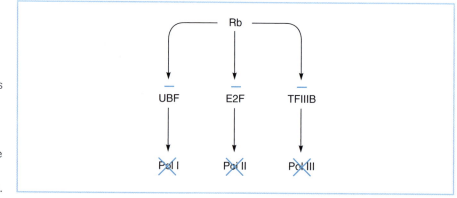

and Rb-1 antagonize one another so that the effects of inactivating both are less severe than inactivating Rb-1 alone (Lasorella *et al.*, 2000). In agreement with this, Id2 has been shown to interact directly with the non-phosphorylated form of Rb-1 via a protein–protein interaction and inactivate it.

Hence, the correct balance between the antagonistic factors Id2 and Rb-1 is essential for normal development. It has been shown that in cells over-expressing the Myc oncogene protein (see section 9.3.3), the expression of Id2 is transcriptionally activated by Myc. The excess Id2 then inactivates Rb-1, thereby promoting tumour formation (Fig. 9.37).

Hence, the Rb protein plays a key role in regulating cellular growth and differentiation by interacting with transcription factors involved in transcription by RNA polymerases I, II and III. Its inactivation either by mutation or by specific oncogenes therefore results in uncontrolled proliferation and cancer. When taken together with the similar role of p53 in growth regulation and as a target for oncogenes, this suggests that anti-oncogenes are likely to play a key role in regulating cellular growth which is likely to be controlled by the balance between the antagonistic effects of oncogene and anti-oncogene products.

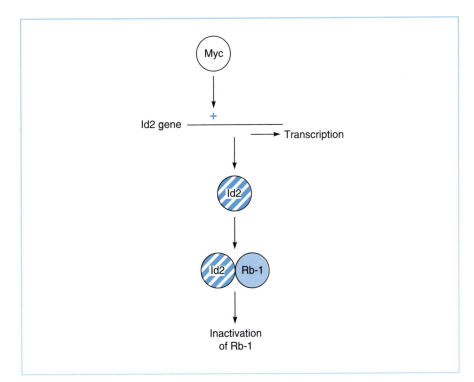

Figure 9.37
The Myc oncogene product can transcriptionally activate the gene encoding the Id2 transcription factor. Id2 then binds to Rb-1 and inhibits its tumour suppressor function.

9.4.4 OTHER ANTI-ONCOGENIC TRANSCRIPTION FACTORS

Normally, anti-oncogenes are identified on the basis of their inactivation in specific human cancers and their functional role subsequently characterized. For some time only three anti-oncogene products were known to be transcription factors namely the p53 and Rb-1 proteins discussed in previous sections and the Wilms' tumour gene product (for review see Hastie, 2001).

More recently, however, other anti-oncogene products have also been implicated in transcriptional control. Thus, while the BRCA-1 and BRCA-2 anti-oncogenes, which are mutated in many cases of familial breast cancer, appear to function primarily in controlling the repair of damaged DNA, there is also evidence that they may influence transcription. For example, both BRCA-1 and BRCA-2 contain regions which can act as activation domains and stimulate transcription (for review see Marx, 1997) (for discussion of such domains see Chapter 5, section 5.2). Moreover, BRCA-1 appears to be a component of the RNA polymerase holoenzyme which also contains RNA polymerase II and basal transcription factors (see Chapter 3, section 3.5.2) again suggesting that this factor is involved in transcriptional control (Scully *et al.*, 1997).

In contrast to these features suggesting that BRCA-1 can influence transcription rates within the nucleus, the adenomatous polyposis coli (APC) anti-oncogene, which is mutated in most human colon tumours (for review see Moon and Miller, 1997), appears to influence transcription indirectly. Thus APC acts by interacting with a protein known as β-catenin which is involved both in cell adhesion and also acts as a transcription factor (for review see Peifer, 1997). This interaction between APC and β-catenin results in the export of β-catenin to the cytoplasm and its rapid degradation (Fig. 9.38a) (Rosin-Abersfeld *et al.*, 2000).

In normal cells, specific secreted proteins known as WNT proteins (or wingless proteins after the first member of the family which was discovered in *Drosophila*) activate a kinase enzyme, glycogen synthase kinase, and this kinase phosphorylates and thereby stabilizes β-catenin preventing it from being degraded (for review see Hunter, 1997; Nusse, 1997; Polakis, 2000; Taipale and Beachy, 2001). The β-catenin then moves to the nucleus and interacts with the LEF-1 transcription factor discussed in Chapter 1 (section 1.3.6) and stimulates its ability to activate transcription (Fig. 9.38b). One of the genes activated by the LEF-1/β-catenin complex is that encoding the Pitx2 transcription factor which, in turn, activates the cyclin D2 gene, thereby stimulating cellular proliferation (Kioussi *et al.*, 2002).

In a normal situation, therefore, this ability of β-catenin to interact with LEF-1 and stimulate its activity, is tightly regulated by the presence or absence

Figure 9.38
(a) Interaction of the anti-oncogenic protein APC and the oncogenic protein β-catenin resulting in degradation of β-catenin. (b) Following activation of glycogen synthase kinase (GSK) by WNT proteins, β-catenin is stabilized. It then moves to the nucleus and interacts with the LEF-1 transcription factor promoting its ability to stimulate transcription. (c) In cancer, the APC factor is inactivated resulting in the constitutive activation of β-catenin.

of WNT proteins so ensuring appropriate control of cellular growth. Any change which causes this pathway to become constitutively active results in cancer. For example, if the APC gene is mutated so that APC cannot inactivate β-catenin, cancer will result from the constitutive activation of β-catenin (see Fig. 9.38c). Hence APC acts as an anti-oncogene whose inactivation by mutation causes cancer.

As well as illustrating how an anti-oncogene can act indirectly to influence transcription, this example also illustrates how oncogene products interact with one another. Clearly, mutations in the β-catenin gene which enhance β-catenin stability or mutations in the WNT genes which result in their over-expression will also cause cancer and hence the genes encoding β-catenin or the WNT proteins are oncogenes whose products act in the same pathway as the APC anti-oncogene product.

In addition to being stabilized by phosphorylation (see above), β-catenin is also acetylated by the CBP co-activator (see Chapter 5, section 5.4.3). This acetylation apparently reduces the ability of β-catenin to activate one of its target genes, the c-*myc* proto-oncogene (see section 9.3.3) providing an example of CBP acting to inhibit transcription rather than its normal role as a co-activator (Wolf *et al.*, 2002). Interestingly, the lysine residue in β-catenin, which is a target for acetylation, is often found mutated to a non-acetylated form in human cancers and, as expected, from its non-acetylatability, this mutant protein is a strong activator of c-*myc* oncogene expression (Fig. 9.39).

Hence, β-catenin offers an example of a proto-oncogene whose activity is regulated both by phosphorylation and acetylation, paralleling the similar

Figure 9.39

The ability of β-catenin to activate the c-*myc* gene is reduced by its acetylation but enhanced by its mutation to a non-acetylatable form (square).

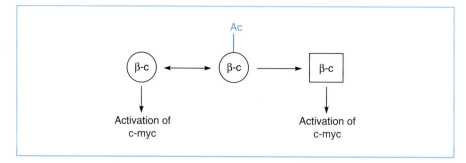

regulation of the anti-oncogene proteins p53 (section 9.4.2) and Rb-1 (section 9.4.3) by multiple post-translational modifications.

Like the majority of transcription factors, the anti-oncogenic proteins discussed so far act by directly or indirectly altering the rate at which transcription is initiated by RNA polymerase. This is apparently not the case, however, for the von Hippel-Lindau anti-oncogene protein which is mutated in multiple forms of cancer. Thus, as discussed in Chapter 6 (section 6.4.2), this factor acts to inhibit transcriptional elongation by promoting the degradation of the large subunit of RNA polymerase II.

Interestingly, the mutant forms of the von Hippel-Lindau protein found in human tumours do not inhibit transcriptional elongation indicating that the anti-oncogenic action of the protein is mediated, at least in part, by its effect on transcriptional elongation. This is likely to be because several oncogenes such as c-*fos* and c-*myc* are regulated at the stage of transcriptional elongation with many of the RNA transcripts which are initiated, not being elongated to produce a full length functional mRNA. It is possible therefore that in the absence of the von Hippel-Lindau protein, too much full length mRNA is produced resulting in over-production of the corresponding oncogenic proteins.

However, as discussed in Chapter 8 (section 8.4.5), the von Hippel-Lindau protein also acts to promote the degradation of the hypoxia-inducible factor HIF-1α thereby ensuring that it activates its target genes only in response to lowered oxygen levels. The mutations of the von Hippel-Lindau protein which occur in cancer also block its interaction with HIF-1α and, in these tumour cells, HIF-dependent genes are expressed at high levels even in the presence of oxygen. As one of the roles of HIF-1 is to activate genes involved in blood vessel formation in response to falling oxygen levels in tissues, the inappropriate activation of HIF-1 and its target genes in tumours, may enhance the blood supply to the tumour, allowing it to grow more rapidly.

Hence, the von Hippel-Lindau protein may target multiple pathways promoting the degradation of proteins which are important in regulating normal growth, including transcriptional activation by HIF-1α and transcriptional elongation. Interestingly, the gene encoding another transcriptional elongation factor, ELL, is found at the break point of chromosomal translocations in several leukaemias (see section 9.3.4) indicating that it can be oncogenic under certain circumstances (for review see Conaway and Conaway, 1999). The existence of both oncogenic and anti-oncogenic transcription factors which modulate transcriptional elongation indicates that this is an important target for processes which regulate normal cellular growth and hence for malregulation in cancer (for review see Li and Green, 1996).

More generally, the examples given in this section add considerable variety to the three 'classical' anti-oncogenes encoding transcription factors (p53, Rb-1 and the Wilm's tumour gene) and indicate the key role of such gene products in different forms of transcriptional regulation in normal cells and in cancer.

9.5 CONCLUSIONS

The ability to affect cellular transcriptional regulatory processes is crucial to the ability of many different viruses to transform cells. Thus, for example, the large T oncogenes of the small DNA tumour viruses SV40 and polyoma and the Ela protein of adenovirus can all affect cellular gene expression and this ability is essential for the transforming ability of these viruses (for review see Moran, 1993).

In this chapter we have seen that several RNA viruses also have this ability, containing transcription factors which can act as oncogenes either by promoting the expression of genes required for growth or by inhibiting the expression of genes required for the production of non-proliferating differentiated cells.

Although the oncogenes of both DNA and some RNA tumour viruses can therefore affect transcription, their origins are completely different. Thus while the oncogenes of the DNA viruses do not have equivalents in cellular DNA and appear to have evolved within the viral genome, the oncogenes of retroviruses have, as we have seen, been picked up from the cellular genome. The fact that despite their diverse origins, both types of oncogenes can affect transcription, indicates therefore that the modulation of transcription represents an effective mechanism for the transformation of cells.

In addition, however, the origin of retroviral oncogenes from the cellular genome allows several other features of transcription to be studied. Thus, for example, the conversion of a normal cellular transcription factor into a cancer-causing viral oncogene allows insights to be obtained into the processes whereby oncogenes become activated.

In general such oncogenes, whether they encode growth factors, growth factor receptors or other types of protein, can be activated within a virus either by over-expression driven by a strong retroviral promoter or by mutation. The transcription factors we have discussed in this chapter illustrate both these processes. Thus the Fos, Jun and Myc oncogenes, for example, become cancer-causing both by continuous expression of proteins which are normally made only transiently, leading to constitutive stimulation of genes required for growth as well as in some cases by mutations in the viral forms of the protein which render them more potent transcriptional activators. Similarly the ErbA oncogene is activated by deletion of a part of the protein coding region leading to a protein with different or enhanced properties. Although such effects of mutation or over-expression have initially been defined in tumorigenic retroviruses, it is clear that such changes can also occur within the cellular genome, over-expression of the c-*myc* oncogene, for example, being characteristic of many different human tumours (for review see Spencer and Groudine, 1991), while several other transcription factor genes are activated by the translocations characteristic of particular human leukaemias (see section 9.3.4).

In addition, since cellular oncogenes clearly also play an important role in the regulation of normal cellular growth and differentiation, their identification via tumorigenic retroviruses has, paradoxically, greatly aided the study of normal cellular growth regulatory processes. Thus, for example the prior isolation of the c-*fos* and c-*jun* genes greatly aided the characterization of the AP1 binding activity and of its role in stimulating genes involved in cellular growth.

A similar boost to our understanding of growth regulation in normal cells has also emerged from studies of the anti-oncogene proteins. Thus studies on the Rb-1 gene, which was originally identified on the basis of its inactivation in retinoblastomas, have led to an understanding of its key role in regulating the balance between cellular growth and differentiation. Similarly, work on p53, which was originally identified as a protein interacting with the product of the SV40 large T oncogene, has led to the identification of its key role as the so-called 'guardian of the genome'.

The interaction of p53 and SV40 large T indicates another aspect of anti-oncogenes, namely their antagonistic interaction with oncogene products. Thus both p53 and Rb-1 have been shown to bind cellular and viral oncogene proteins with the activity of the anti-oncogene product being inhibited by this interaction. Such interactions are not confined to the oncogenes and anti-oncogenes which encode factors directly regulating transcription. Thus, as discussed in section 9.4.4, the APC anti-oncogene protein does not act directly as a transcription factor. Rather it interacts with the β-catenin oncogene product to promote its degradation. Hence cancer can result from mutations in the β-catenin oncogene which enhance the stability of its protein product or directly enhance its ability to stimulate transcription or from mutations in the APC protein which inactivate it and prevent it interfering with the function of β-catenin. Similarly, it has been shown that the Jun oncoprotein can promote tumour formation by antagonizing the pro-apoptotic effect of p53 (Eferl *et al.*, 2003).

Hence the interaction between oncogene and anti-oncogene products is likely to play a key role in regulating cellular growth and survival. The uncontrolled growth characteristic of cancer cells therefore results from changes in this balance, due either to over-expression or mutational activation of oncogenes or to deletion or mutational inactivation of anti-oncogenes.

It is clear therefore that as with other oncogenes and anti-oncogenes, the study of the oncogenes and anti-oncogenes that encode transcription factors can provide considerable information on both the processes regulating normal growth and differentiation and on how these processes are altered in cancer. When taken together with the involvement of transcription factor mutations in disorders of development or hormone responses discussed in section 9.1, they illustrate the key role played by transcription factors and the manner in which alterations in their activity can result in disease.

REFERENCES

Almog, N. and Rotter, V. (1997) Involvement of p53 in differentiation and development. Biochemica et Biophysica Acta 1333, F1–F27.

Anderson, M.E. and Tegtmeyer, P. (1995) Giant leap for p53, small step for drug design. Bioessays 17, 3–7.

Avantaggiati, M.L., Ogryzko, V., Gardner, K. *et al.* (1997) Recruitment of p300/CBP in p53 dependent signalling pathway. Cell 89, 1175–1184.

Baeurale, P.A. and Baltimore, D. (1996) NFκB: Ten years after. Cell 87, 13–20.

Baichwal, U.R. and Tjian, R. (1990) Control of c-Jun activity by interaction of a cell-specific inhibitor with regulatory domain delta: Differences between v- and c-Jun. Cell 63, 815–825.

Baudino, T. A. and Cleveland, J. L. (2001) The Max network gone mad. Molecular and Cellular Biology 21, 691–702.

Bernards, R. (1995) Flipping the Myc switch. Current Biology 5, 859–861.

Berns, A. (1994) Is p53 the only real tumour suppressor gene? Current Biology 4, 137–139.

Blackwood, E.M. and Eisenman, R.N. (1991) Max: a helix-loop-helix zipper protein that forms a sequence-specific-DNA binding complex with myc. Science 251, 1211–1217.

Blank, V. and Andrews, N.C. (1997) The Maf transcription factors: regulators of differentiation. Trends in Biochemical Sciences 22, 437–441.

Boncinelli, E. (1997) Homeobox genes and disease. Current Opinion in Genetics and Development 7, 331–337.

Bourne, H.R. and Varmus, H.E. (eds) (1992) Oncogenes and cell proliferation. Current Opinion in Genetics and Development 2, 1–57.

Bours, V., Franzoso, G., Azarenko, V. et al. (1993) The oncoprotein Bcl-3 directly transactivates through κB motifs via association with DNA-binding p50B homodimers. Cell 72, 729–739.

Broach, J.R. and Levine, A.J. (1997) Oncogenes and cell proliferation. Current Opinion in Genetics and Development 7, 1–6.

Brown, T. R. P., Scott, P. H., Stein, T. et al. (2000) RNA Polymerase III transcription: its control by tumour suppressors and its deregulation by transforming agents. Gene Expression 9, 15–28.

Buschmann, T., Fuchs, S. Y., Lee, C-G. et al. (2000) SUMO-1 modification of Mdm2 prevents its self-ubiquitination and increases Mdm2 ability to ubiquitinate p53. Cell 101, 753–762.

Celli, J., Duijf, P., Hamel, B. C. J. et al. (1999) Heterozygous germline mutations in the p53 homologue p63 are the cause of EEC syndrome. Cell 99, 143–153.

Chan, H. M., Krstic-Demonacos, M., Smith, L. et al. (2001) Acetylation control of the retinoblastoma tumour-suppressor protein. Nature Cell Biology 3, 667–674.

Conaway, J. W. and Conaway, R. C. (1999) Transcription elongation and human disease. Annual Reviews of Biochemistry 68, 301–319.

D'Arcangelo, G. and Curran, T. (1995) Smart transcription factors. Nature 389, 149–152.

de OcaLuna, R.M., Wagner, D.S. and Lozano, G. (1995) Rescue of early embryonic lethality in mdm2-deficent mice by deletion of p53. Nature 378, 203–206.

Downward, J., Yarden, Y., Mayes, E. et al. (1984) Close similarity of epidermal growth factor receptor and v-erb-B oncogene protein sequences. Nature 307, 521–527.

Dynlacht, B.D. (1995) PolI gets repressed. Nature 374, 114.

Dynlacht, B.D. (1997) Regulation of transcription by proteins that control the cell cycle. Nature. 389, 149–152.

Dyson, N. (2003) A twist in a mouse tale. Nature 421, 903–904.

Eferl, R., Ricci, R., Kenner, L. et al. (2003) Liver tumour development: c-Jun antagonises the proapoptotic activity of p53. Cell 112, 181–192.

Eisenman, R. N. (2001) Deconstructing Myc. Genes and Development 15, 2023–2030.

Engelkamp, D. and van Heyningen, V. (1996) Transcription factors in disease. Current Opinion in Genetics and Development 6, 334–342.

Fearon, E.R. (1997) Human cancer syndromes: clues to the origin and nature of cancers. Science 278, 1043–1050.

Foo, S. Y. and Nolan, G. P. (1999) NF-kB to the rescue RELs, apoptosis and cellular transformation. Trends in Genetics 15, 229–235.

Freiman, R. N. and Tjian, R. (2002) A glutamine-rich trail leads to transcription factors. Science 296, 2149–2150.

Friend, S. (1994) p53: a glimpse at the puppet behind the shadow play. Science 265, 334–335.

Gibbons, R.J., Picketss, D.S., Villard, L. and Higgs, D.R. (1995) Mutations in a putative global regulator cause X-linked mental retardation with α-Thalassaemia (ATR-X) syndrome. Cell 80, 837–845.

Gomez-Roman, N., Grandori, C., Eisenman, R. N. and White, R. J. (2003) Direct activation of RNA polymerase III transcription by c-myc. Nature 421, 290–294.

Gottifredi, V. and Prives, C. (2001) Getting p53 out of the nucleus. Science 292, 1851–1852.

Graf, T. (1992) Myb, a transcriptional activator linking proliferation and differentiation in haematopoietic cells. Current Opinion in Genetics and Development 2, 249–255.

Grandori, C. and Eisenman, R.N. (1997) Myc target genes. Trends in Biochemical Sciences 22, 177–181.

Grandori, C., Cowley, S. M., James, L. P. and Eisenman, R. N. (2000) The Myc/Max/Mad network and the transcriptional control of cell behaviour. Annual Review of Cell & Developmental Biology 16, 653–699.

Harbour, J. W. and Dean, D. C. (2000a) Rb function in cell-cycle regulation and apoptosis. Nature Cell Biology 2, E65–E67.

Harbour, J. W. and Dean, D. C. (2000b) The Rb/E2F pathway: expanding roles and emerging paradigms. Genes and Development 14, 2393–2409.

Hastie, N. D. (2001) Life, sex, and WT1 isoforms – three amino acids can make all the difference. Cell 106, 391–394.

Haupt, Y., Maya, R., Kazaz, A. and Oren, M. (1997) Mdm2 promotes rapid degradation of p53. Nature 387, 296–299.

Haupt, Y., Robles, A. I., Prives, C. and Rotter, V. (2002) Deconstruction of p53 functions and regulation. Oncogene 21, 8223–8231.

Hunter, T. (1997) Oncoprotein networks. Cell 88, 333–346.

Hunter, T. (1998) Prolyl isomerases and nuclear function. Cell 92, 141–143.

Ito, A., Kawaguchi, Y., Lai, C-H. *et al.* (2002) MDM2-HDAC1-mediated deacetylation of p53 is required for its degradation. EMBO Journal 21, 6236–6245.

Johnson, R.S., Van Lingen, B., Papaivannou, V.E. and Speigelmann, B.M. (1993) A null mutation at the c-*jun* locus causes embryonic lethality and retarded cell growth in culture. Genes and Development 7, 1309–1317.

Karin, M., Liu, Z-G. and Zandi, E. (1997) AP-1 function and regulation. Current Opinion in Cell Biology 9, 240–246.

Kerppola, T. and Curran, T. (1995) Zen and the art of Fos and Jun. Nature 373, 199–200.

Kersten, S., Desvergne, B. and Wahli, W. (2000) Roles of PPARs in health and disease. Nature 405, 421–424.

Kioussi, C., Briata, P., Baek, S. H. *et al.* (2002) Identification of a Wnt/Dvl/b-Catenin → Pitx2 pathway mediating cell-type-specific proliferation during development. Cell 111, 673–685.

Knudson, A.G. (1993) Anti-oncogenes and human cancer. Proceedings of the National Academy of Sciences, USA 90, 10914–10921.

Ko, L.J. and Prives, C. (1996) p53: puzzle and paradigm. Genes and Development 10, 1054–1072.

Land, H., Parada, L.F. and Weinberg, R. (1983) Cellular oncogenes and multistep carcinogenesis. Science 222, 771–778.

Lane, D.P. (1992) p53, guardian of the genome. Nature 358, 15–16.

Lane, D.P. and Hall, P.A. (1997) MDM2 – arbitor of p53's destruction. Trends in Biochemical Sciences 22, 373–374.

Larmine, C.G.C., Cairns, C.A., Mital, R. *et al.* (1997) Mechanistic analysis of RNA polymerase III regulation by the retinoblastoma protein. EMBO Journal 16, 2061–2071.

Lasorella, A., Noseda, M., Beyna, M. and Lavarone, A. (2000) Id2 is a retinoblastoma protein target and mediates signalling by Myc oncoproteins. Nature 407, 592–598.

Latchman, D.S. (1996) Transcription factor mutations and human diseases. New England Journal of Medicine 334, 28–33.

Lee, C., Chang, J. H., Lee, H.S. and Cho, Y. (2002) Structural basis for the recognition of the E2F transactivation domain by the retinoblastoma tumour suppressor. Genes and Development 16, 3199–3212.

Lee, E.Y-H.P., Chang, C.Y. Hu, N. *et al.* (1992) Mice deficient for Rb are non viable and show defects in neurogenesis and haematopoesis. Nature, 359, 289–294.

Levens, D. (2002) Disentangling the MYC web. Proceedings of the National Academy of Sciences USA 99, 5757–5759.

Levine, A.J. (1997) p53, the cellular gatekeeper for growth and division. Cell 88, 323–331.

Li, L., Chambard, J.C., Karin, M. and Olson, E.M. (1992) Fos and Jun repress transcriptional activation by myogenin and MyoD: the amino terminus of Jun can mediate repression. Genes and Development 6, 676–689.

Li, M., Luo, J., Brooks, C. L. and Gu, W. (2002) Acetylation of p53 inhibits its ubiquitination by Mdm2. Journal of Biological Chemistry 52, 50607–50611.

Li, X-Y. and Green, M.R. (1996) Transcriptional elongation and cancer. Current Biology 6, 943–944.

Lipinski, M. M. and Jacks, T. (1999) The retinoblastoma gene family in differentiation and development. Oncogene 18, 7873–7882.

Liu, J., Akoulitchev, S., Weber, A. et al. (2001) Defective interplay of activators and repressors with TFIIH in Xeroderma Pigmentosum. Cell 104, 353–363.

Lohrum, M. A. E. and Vousden, K. H. (2000) Regulation and function of the p53-related proteins: same family, different rules. Trends in Cell Biology 10, 197–202.

Look, A.T. (1997) Oncogenic transcription factors in the human acute leukaemias. Science 278, 1059–1064.

Luo, J., Nikolaev, A. Y., Imai, S-I. et al. (2001) Negative control of p53 by Sir2a promotes cell survival under stress. Cell 107, 137–148.

Mann, R.S. and Chan, S.K. (1996) Extra specificity from extradenticle: the partnership between Hox and PBX/EXD homeodomain proteins. Trends in Genetics 12, 258–262.

Marx, J. (1997) Possible function found for breast cancer genes. Science 276, 531–532.

Mayo, L. D. and Donner, D. B. (2002) The PTEN, Mdm2, p53 tumour suppressor-oncoprotein network. Trends in Biochemical Sciences 27, 462–467.

Moon, R.T. and Miller, J.R. (1997) The APC tumour suppressor in development and cancer. Trends in Genetics 13, 256–258.

Moran, E. (1993) DNA tumour virus transforming proteins and the cell cycle. Current Opinion in Genetics and Development, 3, 63–70.

Morris, E. J. and Dyson, N. J. (2001) Retinoblastoma protein partners. Advances in Cancer Research 82, 1–54.

Morrison, R. S. and Kinoshita, Y. (2000) p73 – guilt by association? Science 289, 257–258.

Motohashi, H., Shavit, J.A., Igarashi, K. et al. (1997) The world according to MAF. Nucleic Acids Research 25, 2953–2959.

Müller, H. and Helin, K. (2000) The E2F transcription factors: key regulators of cell proliferation. Biochimica et Biophysica Acta 1470, M1–M12.

Münger, K. (2003) Clefts, grooves and (small) pockets: the structure of the retinoblastoma tumour suppressor in complex with its cellular target E2F unveiled. Proceedings of the National Academy of Sciences USA 100, 2165–2167.

Nair, S. K. and Burley, S. K. (2003) X-ray structures of Myc-Max and Mad-Max recognising DNA: molecular bases of regulation by proto-oncogenic transcription factors. Cell 112, 193–205.

Nucifora, F. C. Jr, Sasaki, M., Peters, M. F. *et al.* (2001) Interference by Huntingtin and Atrophin-1 with CBP-mediated transcription leading to cellular toxicity. Science 291, 2423–2428.

Nusse, R. (1997) A versatile transcriptional effector of wingless signalling. Cell 89, 321–323.

Oren, M. (1999) Regulation of the p53 tumour suppressor protein. Journal of Biological Chemistry 274, 36031–36034.

Peifer, M. (1997) β-caterin as oncogene: the smoking gun. Science 275, 1752–1753.

Perlmann, T. and Vennstrom, B. (1995) Nuclear receptors: the sound of silence. Nature 377, 387–388.

Peukert, K., Staller, P., Schneider, A. *et al.* (1997) An alternative pathway for gene regulation by Myc. EMBO Journal 16, 5672–5686.

Polakis, P. (2000) Wnt signalling and cancer. Genes and Development 14, 1837–1851.

Prives, C. (1998) Signalling to p53: breaking the MDM2-p53 circuit. Cell 95, 5–8.

Prives, C. and Manley, J. L. (2001) Why is p53 acetylated? Cell 107, 815–818.

Rabbits, T.H. (1994) Chromosomal translocations in human cancer. Nature 372, 143–149.

Ransone, L.J. and Verma, I.M. (1990) Nuclear proto-oncogenes. Fos and Jun. Annual Review of Cell Biology 6, 531–557.

Ringrose, L. and Paro, R. (2001) Cycling silence. Nature 412, 493–494.

Rosin-Arbesfeld, R., Townsley, F. and Bienz, M. (2000) The APC tumour suppressor has a nuclear export function. Nature 406, 1009–1012.

Ross, J. F., Näär, A., Cam, H. *et al.* (2001) Active repression and E2F inhibition by pRB are biochemically distinguishable. Genes and Development 15, 392–397.

Sap, J., Munoz, A., Damm, K. *et al.* (1986) The c-*erb*-A protein is a high-affinity receptor for thyroid hormone. Nature 324, 635–640.

Sap, J., Munoz, A., Schmitt, A. *et al.* (1989) Repression of transcription at a thyroid hormone response element by the v-*erbA* oncogene product. Nature 340, 242–244.

Sawyers, C.L. and Denny, C.T. (1994) Chronic myelomonocytic leukemia. Tel-a-kinase what it's all about. Cell 77, 171–173.

Schwartz, M. W. and Kahn, S. E. (1999) Insulin resistance and obesity. Nature 402, 860–861.

Scully, R., Anderson, S.F., Chao, D.M. *et al.* (1997) BRCA1 is a component of the RNA polymerase II holoenzyme. Proceedings of the National Academy of Sciences USA 94, 5605–5610.

Sharpless, N. E. and DePinho, R. A. (2002) p53: Good Cop/Bad Cop. Cell 110, 9–12.

Shaulian, E. and Karin, M. (2002) AP-1 as a regulator of cell life and death. Nature Cell Biology 4, E131–E136.

Sherr, C.J. (1994) G1 phase progression: cycling on cue. Cell 79, 551–555.

Sobulo, O.M., Borrow, J., Tomek, R. *et al.* (1997) MLL is fused to CBP a histone acetyl-transferase, in therapy-related acute myeloid leukaemia with a t (11; 16) (q23; p13.3). Proceedings of the National Academy of Sciences, USA. 94, 8732–8737.

Spencer, C.A. and Groudine, M. (1991) Control of c-*myc* regulation in normal and neo-plastic cells. Advances in Cancer Research 56, 1–48.

Taipale, J. and Beachy, P. A. (2001) The Hedgehog and Wnt signalling pathways in cancer. Nature 411, 349–354.

Thut, C.J., Goodrich, J.A. and Tjian, R. (1997) Repression of p53 mediated transcription by MDM2: a dual mechanism. Genes and Development 11, 1974–1986.

Treier, M., Staszewski, L.M. and Bohmann, D. (1994) Ubiquitin – dependent c-Jun degradation *in vivo* is mediated by the δ domain. Cell 78, 787–798.

Vaziri, H., Dessain, S. K., Eaton, E. N. *et al.* (2001) $hSIR2^{SIRT1}$ Functions as an NAD-dependent p53 deacetylase. Cell 107, 149–159.

Vogelstein, B., Lane, D. and Levine, A. J. (2000) Surfing the p53 network. Nature 408, 307–310.

Wasylyk, B., Wasylyk, C., Flores, P. *et al.* (1990) The c-*ets* proto-oncogenes encode transcription factors that cooperate with c-Fos and c-Jun for transcriptional activation. Nature 346, 191–193.

Weinberg, R. A. (1993) Tumour suppressor genes. Neuron 11, 191–196.

Weinberger, C., Thompson, C.C., Ong, E.S. *et al.* (1986) The c-*erb*-A gene encodes a thyroid hormone receptor. Nature 324, 641–646.

White, R.J. (1997) Regulation of RNA polymerases I and II by the retinoblastoma protein: a mechanism for growth control. Trends in Biochemical Sciences 22, 77–80.

Wolf, D., Rodova, M., Miska, E. A. *et al.* (2002) Acetylation of β-Catenin by CREB-binding protein (CBP). Journal of Biological Chemistry 277, 25562–25567.

Wolffe, A. (1997) Sinful repression. Nature 387, 16–17.

Wu, L., de Bruin, A., Saavedra, H. I. *et al.* (2003) Extra-embryonic function of Rb is essential for embryonic development and viability. Nature 421, 942–947.

Wyllie, A. (1997) Clues in the p53 murder mystery. Nature 389, 237–238.

Yang, A., Kaghad, M., Caput, D. and Mckeon, F. (2002) On the shoulders of giants: p63, p73 and the rise of p53. Trends in Genetics 18, 90–95.

Zenke, M., Kahn, D., Disela, C. *et al.* (1988) v-*erb A* specifically suppresses transcription of the avian erythrocyte anion transporter (Band 3) gene. Cell 52, 107–119.

Zhang, H. S., Postigo, A. A. and Dean, D. C. (1999) Active transcriptional repression by the Rb-E2F complex mediates G1 arrest triggered by p16^{INK4a}, TGFβ and contact inhibition. Cell 97, 53–61.

Zheng, H., You, H., Zhou, X. Z. *et al.* (2002) The prolyl isomerase Pin1 is a regulator of p53 in genotoxic response. Nature 419, 849–853.

Zornig, M. and Evan, G.I. (1996) On target with Myc. Current Biology 6, 1553–1556.

CHAPTER 10

CONCLUSIONS AND FUTURE PROSPECTS

At the time the first edition of this book was published (1991) enormous progress had been made in understanding the nature and role of transcription factors. Thus the roles of specific factors in processes such as constitutive, inducible, tissue specific and developmentally regulated gene expression had been defined, as had their involvement in diseases such as cancer. Moreover, by studying these factors in detail, it proved possible to analyse how they fulfill their function in these processes by binding to specific sites in the DNA of regulated genes and activating or repressing transcription, as well as the regulatory processes which result in their doing so only at the appropriate time and place. Moreover, the regions of individual factors which mediate these effects and the critical amino acids within them which are of importance had been identified in a number of cases.

In the intervening years, up to the current publication of the fourth edition, much further progress has been made in these areas. In addition, the ability to prepare 'knock out' mice, in which the gene encoding an individual factor has been inactivated, has allowed the *in vivo* functional role of many factors to be directly assessed while numerous studies have elucidated the structure of specific factors either in isolation or bound to DNA, as illustrated in the colour plate section. It has become increasingly clear, however, that the activity of a particular factor cannot be considered in isolation. Thus, very often the activity of a factor can be stimulated either positively or negatively by its interaction with another factor. For example, the Fos protein needs to interact with the Jun protein to form a DNA binding complex (see Chapter 4, section 4.5 and Chapter 9, section 9.3.1). Conversely, the DNA binding ability of the glucocorticoid receptor is inhibited by its association with hsp90 (see Chapter 8, section 8.2.2) while that of the MyoD factor is inhibited by its association with Id (Chapter 4, section 4.5.3 and Chapter 6, section 6.2.2).

Additionally, however, it has become clear that as well as stimulating or inhibiting factor activity such protein–protein interactions can also alter the specificity of a factor. Thus, differences in the ability to interact with other proteins can affect the DNA binding specificity of particular factors and hence the target genes to which they bind. This can result in factors with identical DNA binding specificities having entirely different functional effects as in the case of Ubx and Antp (see Chapter 4, section 4.2.4). Alternatively, it may completely change the factor from activator to repressor as in the case of the dorsal/DSP1 interaction (see Chapter 8, section 8.3.3). Hence, by altering the specificity of particular factors, interactions of this type are likely to play a crucial role in the complex regulatory networks which allow a relatively small number of transcription factors to control highly complex processes such as development.

As well as such regulatory interactions between different factors it has become increasingly clear in recent years that many activating transcription factors need to interact with other factors, such as the mediator complex and/ or co-activators in order to stimulate transcription. The most important of such co-activators, CBP, was originally characterized as being required for transcriptional activation in response to cyclic AMP treatment, mediated via the CREB transcription factor (see Chapter 5, section 5.4.3). It is also involved, however, in transcriptional activation mediated via a number of other transcription factors activated by different signalling pathways. In turn, because of the limiting amounts of CBP in the cell, the different transcription factors and signalling pathways compete for CBP resulting in mutual antagonism between, for example, the signalling pathways mediated by AP1 and the glucocorticoid receptor (see Chapter 6, section 6.5).

Thus, the critical dependence of many activating factors on a specific co-activator can result in a functional link between two different factors which do not themselves interact but which compete for the same co-activator. Moreover, the activity of a transcription factor can be regulated by controlling its ability to interact with its co-activator. For example, in the absence of thyroid hormone, the thyroid hormone receptor has an inhibitory effect on transcription because it binds co-repressor molecules which act to inhibit transcription. Following exposure to thyroid hormone however, the receptor undergoes a conformational change which allows it to bind co-activator molecules and hence activate transcription (see Chapter 6, section 6.3.2). Similarly, only the phosphorylated form of CREB can interact with CBP and therefore activate transcription, whereas the non-phosphorylated form does not interact with CBP and is thus inactive (see Chapter 5, section 5.4.3).

This regulation of CREB by phosphorylation is only one example of a plethora of post-translational modifications which have been shown in recent years to modulate transcription factor activity. In addition to phosphorylation, these modifications include methylation, acetylation and ubiquitination. Moreover, they have been shown to modify transcription factor activity via altering processes as diverse as DNA binding, cellular localization and stability as well as via modulating their interaction with other factors (see Chapter 8, section 8.4).

Moreover, individual factors can be modulated by more than one post-translational modification as seen in the cases of the oncogenic transcription factor β-catenin (see Chapter 9, section 9.4.4) and the anti-oncogenic transcription factors, p53 (see Chapter 9, section 9.4.2) and Rb-1 (see Chapter 9, section 9.4.3). Similarly, these modifications can also target co-activators as well as DNA binding transcription factors. Thus, as discussed in Chapter 8 (section 8.4.4), CBP can be modified by methylation and phosphorylation and this differentially affects its binding to different activating transcription factors. The competition between different activators for binding to CBP can therefore be regulated by modifying CBP itself, as well as by modifying the transcription factors themselves.

Hence, the interaction between activators and co-activators plays a critical role in the activation of transcription and its regulation. Although co-activators are likely to act in some cases by interacting with the basal transcriptional complex (see Chapter 5, section 5.3), the finding that many co-activators have histone acetyltransferase activity (see Chapter 5, section 5.5.1) indicates that they may stimulate transcription via altering chromatin structure. Hence such factors could act by acetylating histones thereby altering the chromatin structure to a more open structure able to support active transcription (see Chapter 1, section 1.2.3). Similarly, activators and co-activators may also recruit chromatin remodelling complexes such as SWI/SNF which use ATP-dependent processes to open up the chromatin (see Chapter 1, section 1.2.2 and Chapter 5, section 5.5.1). Conversely, co-repressors which have histone deacetylase activity may act by producing a more closed chromatin structure incompatible with transcription (see Chapter 6, section 6.4.1).

The activation of a target promoter is likely therefore to require the recruitment of activating molecules, histone acetyltransferases and chromatin remodelling complexes as well as of the basal transcriptional complex. As discussed in Chapter 5 (section 5.6) this is a highly ordered process and at each promoter, a series of events will occur with the recruitment of each factor facilitating the next stage in the transcription activation process.

Ultimately, therefore, the understanding of transcription factor function will require a knowledge of the nature and effect of interactions between different transcription factors, their co-activators and co-repressors which is as good as that now available for individual factors. Moreover, it will be necessary to establish how such changes modulate the activity of the basal transcriptional complex and alter chromatin structure. Clearly much work remains to be done before this is achieved. The rapid progress since the first edition of this work was published suggests, however, that an eventual understanding in molecular terms of the manner in which transcription factors control highly complex processes such as *Drosophila* and even mammalian development can ultimately be achieved.

INDEX

a1 protein, α2 interaction, repressive
  effects, 89–90
Acanthamoeba, transcription, 58–60, 62
ACE1, 255
  copper-induced activation, 247, 249
Acetylation, 272–4
  histones, 5–7
  phosphorylation interaction, 274
  protein, 273
  transcription factors, 272
'Acid blobs', 138
Acidic activation domains, 137–9, 140–1
Activation domains, 135–6, 174
  acidic, 137–9, 140–1
  basal transcriptional complex
    interaction, 142–3
  chromatin remodelling and, 167–8
  co-activator interactions, 149–61
  domain-swop experiments, 44–5, 135
  factor activity, 146–9
  factor binding, 143–6
  function of different, 140–1
  glutamine-rich, 137, 139–40, 140–1
  ligand binding domain and, 253, 255
  nature of, 136–41
  proline-rich, 137, 140–1
Activation of transcription, 135–81, 342
  chromatin remodelling and, 162–70,
    173
  by stimulation of elongation, 170–4
  targets for transcriptional activators,
    161–2
  see also Activation domains
Activity of transcription factors,
    regulated, 245–91
  evidence for, 245–7

protein degradation/processing,
    281–3
  protein-ligand binding, 247–55
  protein modification, 265–81
  protein-protein interactions, 255–65
  role of, 283–5
ACTR, steroid receptor interactions, 167
Adenine (A) residues, 31–2, 84
Adenomatous polyposis coli (APC) gene,
    328
Adenovirus, 325, 331
  E1A protein, 109, 110
Adipocyte differentiation, 190–1
ADR1 gene, 103
ADR1 protein, 104
AEBP1, 190–1
Affinity chromatography, DNA, 34–5, 36
Albumin gene, 231
Alcohol dehydrogenase (Adh) gene, 87
α2 protein
  a1 interaction, repressive effects,
    89–90, 264
  α1 interaction, repressive effects, 194
  MCM1 interaction, 90
  repression of transcription, 194
α-Amanitin, 55
Ames dwarf, 99
Amphioxus, 222
Angiotensin II, 286
Aniridia, 101
Antennapedia (Antp) gene, 79, 80, 81
  Bithorax gene cluster, 92
  expression patterns, 221–2
  mutations, 78
Antennapedia (Antp) protein
  activation domain, 139

Antennapedia (Antp) protein (*cont.*)
  DNA sequence recognition, 85–6
  homeodomain, 81–2
  in regulation of development, 89
Anti-oncogenes, 204
  cancer and, 314–31
α-l-Antitrypsin, 231
AP1, 270, 271, 285, 299–301, 332
  binding site, 299
  CBP interaction, 274
  co-activators, 160
  competition for CBP/p300, 321
  p53 interaction, 321
  Sp1 interaction, 17–18
AP2, 12, 124, 140
APC gene, 329
APC protein, 332–3
Apoptosis, 315, 321
Arabidopsis Athb proteins, 122–3
Archaebacteria, TBP homologue, 72
Assembly factors, 60–1, 62, 72
ATF-2, 274
ATGAATAA/T motif, 12
ATGCAAAT, *see* Octamer motif
ATP hydrolysis, 4, 167
Avian erythroblastosis virus (AEV), 304, 306
Avian retrovirus Rev-T, 312
Avian sarcoma virus ASV17, 299
5-Azacytidine, 187, 188, 212, 213, 214, 215

B lymphocytes
  differentiation, 246
  immunoglobulin genes, *see*
    Immunoglobulin genes
  Oct-2 mRNA, 235, 237
  TAFs, 156
Bacteriophage proteins, recognition
    helix, 83, 86
Basal transcriptional complex (BTC)
  activation domain interactions, 142–3
  inhibitory factors, 191–2, 196, 198–200
  mutations, 296, 297
  stepwise assembly, 63–6, 143
  steroid receptor interactions, 168
  target promoters, 73

Basic DNA-binding domain, 117–24, 126, 127
  CREM, 233, 234
  helix-loop-helix motif and, 121–2
  MyoD, 121, 121–2, 216
  see also DNA binding
bax gene, 321
bcl-3 oncogene, 312
Bicoid (bcd)
  concentration gradient, 227–9
  DNA binding specificity, 84
  as morphogen, 228
  phosphorylation, 270
  regulation, 228
Bithorax, expression patterns, 221–2
Bithorax-Antennapedia complex, 92
Blood pressure, high, 297
Blood vessel growth, 106
Bone development, 246
Brahma mutation, 3, 164
BRCA-1 anti-oncogene, 328
BRCA-2 anti-oncogene, 328
Breast cancer, 328
Brn-3, 46
Brn-3a, 99
Brn-3b, 99
Brn-4, 99
Burkitt's lymphoma, 312

C/EBP
  DNA-binding domain, 117, 119, 120, 125–6
  gene cloning, 39–40
  regulation of synthesis, 231, 239–40
CAAT box sequence, 126
Caenorhabditis elegans, homeobox genes, 90–1
Calcium, 271
  binding to DREAM, 249, 267
Cancer, 297–331
  anti-oncogenes and, 314–31
  chromosomal translocations, 312, 313, 331
  oncogenes and, 299–314, 331–2
  role of Fos/Jun, 301–3, 306–7, 308
  role of viral ErbA protein, 306–7, 308
Cap site, 8

β-Catenin, 328–30, 332–3
Caudal, 228
CBP (CREB-binding protein), 273, 342, 343
   competition for, 206, 303–4, 321
   CREB interaction, 157–8, 159–60, 160–1, 267
   inhibiting transcription, 329
   methylation, 274, 275
   mutations, 295–6
   p53 requirements, 321
   phosphorylation, 274
   steroid receptor interactions, 167
CCAAT box, 8, 9
CCAAT box transcription factor, see CTF/NF1
CCAAT/enhancer binding protein, see C/EBP
CD4, 16
CDK9 kinase enzyme, 172
Cell cycle, Rb-1 phosphorylation, 323–4
Cell division, inhibition by MyoD, 216
Cell type-specific expression of transcription factor, 230–1, 238
Cell type-specific gene expression, 211
   regulation, 245, 285
CF1a, 99
Chaperone protein, 260
ChIP (chromatin immunoprecipitation) assay, 48–50, 143
Cholesterol, 282, 283
Chromatin, 2–7
   remodelling
      in activation of transcription, 162–70, 173, 343
      factors, 3–5
      histone acetyltransferases in, 68
      MyoD-mediated, 215–16
      in repression of transcription, 200–3
      during transcription, 66, 68
   structure, 2–3
      histone acetylation/modification and, 5–7
      modulation of, 200–3, 324, 343
      mutations in factors altering, 294, 295, 297
Chromatin immunoprecipitation (ChIP) assay, 48–50, 143

Chromosomal translocations, 312, 313, 331
cis-trans isomerization, 318
CLIM co-activator molecule, 197–8
Cloning, genes, 38–41
   protein production from cloned genes, 42
   use of cloned genes, 41–50
Co-activators, 68, 149–61, 253, 342–3
   co-repressor interchange, 197–8
   competition for, 206
   histone acetyltransferase activity, 160–1, 176, 215
   methylation, 274
   mutations, 295, 297
Co-repressors, 343
   co-activator interchange, 197–8
   histone deacetylase activity, 202
   inhibitory domains and, 196–7
Collagenase gene, 205–6, 303
Colon tumours, 328
Complementary DNA (cDNA)
   clones
      applications, 41–50
      isolation, 38–41
   libraries, 38, 39, 40
Copper, 247
Corticosterone, 250
CRE (cyclic AMP response element), 157–8, 186, 233, 235, 236
Creatine kinase (CK), 188
Creatine kinase gene, 215
CREB (cyclic AMP response element binding protein), 233, 234, 235, 236, 342
   CBP interaction, 157–8, 159–60, 160–1, 267
   Jun interaction, 264
   phosphorylation, 158–9, 266, 267, 342
CREB-binding protein, see CBP
CREM, 285
   activating/inhibiting forms, 233–5, 236
   phosphorylation, 267
   promoters, 234–5
CTF/NF1
   activation domains, 137, 140
   binding sites, 166
   DNA-binding domain, 124, 125–6

CTF/NF1 (*cont.*)
  gene cloning, 38
  purification, 36–8
CYC1, 142
Cyclic AMP-inducible genes, 266
Cyclic AMP-inducible transcription, 1
Cyclic AMP response element (CRE),
    157–8, 186, 233, 235, 236, 264
Cyclic AMP response element binding
    protein, *see* CREB
Cyclin-dependent kinases, 324–5
  p21 inhibitor, 216, 320
Cyclin genes, 155
Cycloheximide, 245, 258
Cyp40, 318

D2 gene, 328
DBP, 231
Dbx1, 225, 226
Dbx2, 225, 226
Deafness, 99, 101
Decapentaplegic (dpp) gene, 89
Deformed gene, 92
Degradation, protein
  regulation, 281–3
  repression by, 190–1
  ubiquitination and, 275–6, 277, 279
Demethylation, 212–13
Deoxyribonuclease I, *see* DNase I
Developmental disorders, human, 293–7
Developmentally regulated gene
    expression, 211, 285
Diabetes, 297
Diethylpyrocarbonate, 31
Dimerization
  disulphide bond-mediated, 119
  helix-loop-helix-mediated, 121–2
  leucine zipper-mediated, 117, 118, 120
  in regulation of activity, 251, 252
  role in DNA binding, 122–4
Dimethyl sulphate (DMS)
  in vivo footprinting assay, 33
  methylation interference assay, 30–2
  protection footprinting, 29–30
2, 4-Dinitrophenol (DNP), 258
Disease, human, 293–340
Disulphide bridges, 119, 120
Diurnal variations, 231

DNA
  affinity chromatography, 34–5
  bending, factors causing, 18–19
  competitor, 25–7, 28
  complementary, *see* Complementary
    DNA
  looping, 2–3, 4–5, 17, 18
  methylation, 212
  packaging, 2
  protein interactions, methods of study,
    23–34
  repair, 65
DNA binding, 39
  homeobox proteins, 86, 88–90
  leucine zipper/basic domain, 117–24
  mapping methods, 43
  motifs, 124–5, 125–7
  not activating transcription, 43
  repression mechanisms, 183–7
  specificity, uncharacterized factor,
    45–7
  by transcription factors, 127
  zinc finger motif, 104–16
  see also Basic DNA-binding domain
DNA helicase, 65, 68
DNA mobility shift assay, 23–7, 30, 256
  in methylation interference assay, 30
  octamer motif, 27
DNA polymerase α, 322
DNA sequence elements, 7–19, 126
  basic transcription process, 8–9
  enhancers, 13–14
  factor interaction, 17–19
  gene promoters, 7–9
  insulators, 16–17
  locus control regions, 14–16
  negatively acting, 16–17
  regulated transcription, 9–12
  silencers, 16
DNase I
  footprinting assay, 27–30, 249, 250
  hypersensitive sites, 28, 164, 166
Domain mapping, 41–5
Domain-swop experiments, 44–5, 135
Dorsal protein, 264
Dorsal switch protein (DSP1), 264, 265
Dr1, 63, 198–200
Dr2, 63

DRE (DREAM response element), 249
DREAM, 248, 249, 255, 267
  calcium binding, 249, 267
Drosophila
  activation domains, 139
  brahma mutation, 3, 164
  fork head factor, 125
  heat-shock element, 10–11, 163
  heat-shock factor, 258–9, 261, 262
  homeobox genes, 78–9
  homeobox proteins, 79
  mutations, 78
  PCNA gene, 72
  POU protein, 99
  regulation of activity, 264
  repression of transcription, 191, 192,
      196, 200
  RNA polymerase, 56
  TLFs, 72
  transcription factors in development,
      78–9
  trithorax gene, 164
  zinc finger proteins, 79, 103–4, 105
  see also specific genes and proteins
DSP1 (dorsal switch protein), 264, 265
Dwarfism, 99, 293
Dynorphin, 248, 249, 267

E1A protein, 331
E2A gene, 313
E2F, 322, 323, 325, 326
  inhibition by Rb-1, 189, 193, 270
E4BP4, 196
E12, 121–2
E47, 121
EEC (ectrodactyly, ectodermal dysplasia
      and cleft lip) syndrome, 322
Egr 1, 103
8S complex, 250, 251
ELL, 331
Elongation, transcriptional, 170–4
  factors promoting, 170–4, 204, 330
  inhibition of, 204–5, 330
  von Hippel-Lindau (VHL) protein and,
      204, 330
emc gene, 187
engrailed gene (eng), 79, 92
Engrailed protein (eng), 184, 185, 187

Enhanceosome, 14, 19
Enhanceosome complex, 176
Enhancers, 13–14
  IFN-β gene, 175–6
era-1 gene, 232, 285
c-erbA oncogene, 304
c-ErbA protein, 305, 306
  see also Thyroid hormone receptor
v-erbA oncogene, 194, 195, 255, 304–7,
      332
v-ErbA protein, role in cancer, 305,
      306–7, 308
v-erbB oncogene, 306
Erythrocyte anion transporter gene, 306
Erythrocyte differentiation, 306, 307
Erythroid precursors, 14
Escherichia coli, RNA polymerase, 56
ets gene (Ets), 125, 303, 312
Ets protein, 303
eve (even-skipped), 192, 195, 196
  expression pattern, 228, 229
Evi 1, 103
Extradenticle (exd) protein, 89, 264, 313
Eye defects, 293
  retinoblastomas, 322
Eye development, 101

FBP, 294
'Finger swop' experiments, 111–13
FIR, 295
5S ribosomal RNA genes, 55, 60–2
Foggy protein, 204–5
Footprinting, 27–30
  dimethyl sulphate (DMS) protection,
      29–30
  DNaseI, 27–30
  hydroxyl radical, 28–9
    in vivo, 32–4
  phenanthroline-copper, 28–9
Fork head factor, 125
Fos, 332
  DNA binding, 117, 119, 123
  factors stimulating synthesis, 285
  glucocorticoid-mediated inhibition,
      205
  mutations, 302
  oncogenes targeting, 303
  role in cancer, 301–3, 306–7, 308

Fos/Jun complex, 263, 341
    competition for CBP, 206, 303–4
    DNA binding, 123, 124
    inhibitory effects, 303
    leucine zipper regions, 118
Fos-related antigens (Fras), 300, 301
c-*fos*, 300, 302, 330, 332
    promoter, repression by MyoD, 216,
        217
4S receptor protein, 251
Fras (Fos-related antigens), 300, 301
Fushi-tarazu (Ftz) gene, 79, 80
Fushi-tarazu (Ftz) protein, 85–6, 219
    DNA binding, 87, 88
    DNA sequence recognition, 85–6
    interactions with other factors, 184,
        185
Fusion proteins, oncogenic, 313

Gag protein, 305
GAGA, 163, 164–6, 169–70, 202
GAL4, 109, 110
    activation domain, 136, 137, 138–9,
        140
    defective responses to, 147
    regulation of activity, 189, 245, 284
    squelching phenomenon, 142
    TFIID interaction, 148–9
    transcriptional elongation and, 173
Gal11, 145
GAL80, 189–90, 245, 284
Galactose, 189, 284
Gap genes, 78, 228, 229
GCN4, 299
    activation domains, 135, 136, 137–8
    DNA binding, 43
    DNA-binding domains, 117, 118, 119,
        120
    domain-swop experiment, 44–5, 135
    Jun homology, 299
    regulation of synthesis, 238–9, 241
Gel retardation assay, *see* DNA mobility
        shift assay
Gene promoters, 7–9
Genes
    cloned
        protein production from, 42
        use of, 41–50

cloning, 38–41
hormone responsive, 111
regulation of expression
    chromatin structure and, 2–3
    role of transcription, 1
solenoid structure, 2
target, identification for transcription
    factors, 47–50
transcription, *see* Transcription
see also individual genes
Giant gene, 228, 229
GIG (growth-inhibiting proteins), 316
Glass gene, 103
Globin genes, 14–15, 294
Glucocorticoid hormones, 160, 206,
    269
Glucocorticoid receptor, 194, 253
    activation, 251
    activation domains, 135, 136, 137
    activation of gene expression, 194
    chromatin remodelling, 167, 169
    competition for CBP, 160, 206,
        303–4
    DNA binding, 110–14, 250
    domain-swop experiments, 135
    hsp90 interaction, 250–1, 252
    inhibition of gene expression, 185,
        194, 206, 207
    ligand binding, 253–4
    mutations, 296–7
    structure, 109
Glucocorticoid response element (GRE),
    8, 11, 108, 185–6
    for negative regulation (nGRE), 185–6,
        186
Glucose, 284
Glutamine, 84
Glutamine-rich activation domains, 137,
    139–40, 140–1
Glycerol, 284
Glycogen synthase kinase (GSK), 328,
    329
Glycoprotein hormone alpha subunit
    gene, 186
Groucho, 264
Growth factors, 298, 302
Growth-inhibiting proteins (GIG), 316
Growth-stimulating proteins, 322–3

Guanine (G) residues, 29–30, 30–1, 32, 33, 84

Hac1p, 237, 238
HAP1, 140
HAP2, 140
Haploid insufficiency, 294, 296
Heat-inducible transcription, 267
Heat-shock element (HSE), 8, 10–12, 163, 258, 259
    enhancer function, 14
Heat-shock factor (HSF), 164
    activation domain, 261
    DNA binding, 260, 261
    heat-induced activation, 245, 257–8
    phosphorylation, 259, 260, 261, 262, 267
    response to stress, 260
    transcriptional elongation and, 172
Heat-shock protein 70 gene, *see hsp70* gene
Heavy metals, 12
HeLa cells, 245, 246
Helix-loop-helix motif, 121–2
    and dimerization, 122–4
    muscle-specific factors, 216
    in repression of transcription, 187, 188
Helix-turn-helix motif
    in *Drosophila* homeobox proteins, 80, 81–3
    POU-specific domain, 93–5
    winged, 125, 126
Herpes simplex virus
    immediate-early (IE) gene promoters, 95
    VP16 (Vmw65) protein, *see* VP16
HIF-1 (hypoxia inducible factor), 275
HIF-1α, 275–8, 286, 330, 331
    ubiquitination, 279
HIF-1β, 275–6
High mobility group (HMG) proteins, 125
Histone acetyltransferases, 5, 161, 343
    in chromatin remodelling, 68
    co-activators, 160–1, 176, 215
    steroid receptor interactions, 167
Histone code, 7, 176
Histone deacetylases, 271, 318–19, 322

co-repressors, 202–3
Histones
    acetylation, 5–7, 167, 176, 202–3
    H2B, 93
    methylation, 5, 7
    modifications, 5–7
HL60 cells, 170, 171
HMG box, 125
HMG (high mobility group) proteins, 125
HMGI(Y) DNA binding transcription factor, 19, 175
HNF-3, 125
Homeobox (homeodomain), 79–81, 126
    helix-turn-helix motif, *see* Helix-turn-helix motif
    Pax proteins, 100–1
    POU proteins, 93
    transcription factor synthesis regulation, 219–29
Homeobox genes, 79
    expression patterns, 220–1, 224–5, 226
    homologues in other organisms, 92
    Hox11, 313
    mutations, 293
    see also Hoxb genes; *and other specific genes*
Homeobox proteins, 79, 184
    concentration gradients, 227–9
    in *Drosophila* development, 192
    interactions between, 88–90
Homeodomain, 78–101
Homeotic genes, 78, 78–9, 79
Hormone response elements, 108, 193
Hox11, 313
Hoxb genes
    alternative splicing, 232
    expression patterns, 220–1
    mouse, 220–3
    organization, 92
    retinoic acid-mediated activation, 222–3
HoxD genes, 224
hRFX1, 125
HSE, *see* Heat-shock element
HSF, *see* Heat-shock factor
hsp70 gene
    enhancer, 14

hsp70 gene (*cont.*)
  heat-induced expression, 9–12
  promoter, 8, 9, 63
  transcriptional elongation, 172
  upstream promoter elements, 8, 9,
      9–10
hsp90, 250–1, 253
  glucocorticoid receptor interaction,
      251, 252
Human disease, 293–340
Human immunodeficiency virus (HIV)
  DNaseI footprinting assay, 27–8, 29
  methylation interference assay, 30–2
  promoter, 246
  Tat, 172, 173
Hunchback gene, 103, 228, 229
Huntingtin, 296
Huntington's chorea, 296
Hydroxyl radical footprinting, 28–9
Hydroxyproline, 277
4-Hydroxytamoxifen, 253
Hypertension, 297
Hypoxia inducible factor, *see* HIF

ICERs (inducible cyclic AMP early
      repressors), 234, 235, 236
Id, 124, 187, 188, 219, 286
Id2, 326–7, 327
IFN, *see* Interferon
IκB
  acetylation, 273
  kinases, 269
  in NFκB activation, 256, 257, 268–70
  phosphorylation, 268–70
  protein degradation and processing,
      281, 282
Immunoglobulin genes, 1
  B cell-specific expression, 245
  cancer-inducing translocations, 312,
      313
  enhancer elements, 14
  κ gene, 245, 246
  promoter elements, 12
Immunoprecipitation, 46, 47–8, 49
In vivo footprinting assay, 32–4
Inducible gene expression, 245, 285
Inhibitory domains, 106, 192–8
Initiation site, transcription, 9

Initiator element, 9
Insulators, 16–17, 202
Insulin resistance, 297
Interferon-α, 266
Interferon-β (IFN-β), 175, 176, 264–5
Interferon-γ, 266
Interferon-responsive genes, 266
Interleukin-2 α receptors, 246
Irx3, 225, 226
Isoleucine, 118
Isomerization, cis-trans, 318

Jun, 299–304, 332
  activation domains, 140
  as AP1 component, 299–300
  DNA binding, 117, 119, 123, 124
  factors stimulating synthesis, 270, 271,
      285
  glucocorticoid-mediated inhibition,
      205
  oncogenes targeting, 303
  role in cancer, 301–3, 306–7, 308
  see also Fos/Jun complex
c-*jun* gene, 299, 301, 302, 332
c-Jun protein, 303
v-*jun* gene, 299
v-Jun protein, 303
Jun B protein, 300

Kidney, nuclear run-on assay, 231
Kinase enzymes, 266, 268, 271
Knirps gene, 78, 79
Knock-out mice, 341
Krox 20 protein, 105–6
Kruppel gene, 79, 103, 228, 229
Kruppel protein, 103, 103–4, 196

LAC9, 109, 110
Lactoferrin, 247
Lactotrope cells, 97
Lambda cro protein, 82
LEF-1, 18–19, 328, 329
Leucine residues, 102, 111
Leucine zipper motif, 117–21, 157–8
  CREM, 233, 234
  and dimerization, 122–4
  DNA binding, 117–24
  heat-shock factor, 259, 260

Leukaemia, 313
Lex A protein, 44
Lhx3, 219–20
Lhx4, 219–20
Libraries, cDNA, 38, 39, 40
Ligand binding domain, activation domain and, 253, 255
Ligand–protein binding, 247–55
LIM homeodomain transcription factors, 197–8, 219
Lipopolysaccharide, 245, 256
Liver, nuclear run-on assay, 231
Liver activator protein (LAP), 239
Liver inhibitor protein (LIP), 239
Liver-specific gene expression, 231
Locus control regions (LCRs), 14–16, 224
Lysine, 84–5, 176
Lysine amino acid, demethylation, 5–7
Lysozyme gene, chicken, 198

Mad, 309, 310–11
maf oncogene, 312
Mammals
    heat-shock element, 10–11, 163
    heat-shock factor, 259, 261, 262
    RNA polymerase, 56
    zinc finger proteins, 110
Mating types, yeast, 89–90
Matrix attachment regions (MATs), 2–3, 15
Max, 263, 309–11
MCM1, α2 interaction, 90, 264
MDM2
    inhibition of p53, 189, 190, 192–3, 315, 316, 317, 319, 321
    phosphorylation, 317, 318
mdm2 gene, 317, 320–1
mec-3 gene, 90
Mediator complex, 66, 149–51
MEF2, 271, 272
MEF2A, 122
MEL (murine erythroleukaemia) cells, 259
Mental retardation, 294, 295
Metal response elements (MREs), 8, 12
Metallothionein IIA gene
    activation of transcription, 247, 249

interactions between bound factors, 17–18
    promoter, 8, 9
    regulatory elements, 11–12
    upstream promoter elements, 8, 9
Metals, heavy, 12
Methylation, 274–5
    histones, 5, 7
    transcription factors, 274
Methylation interference assay, 30–2
Methyltransferases, 322
Mice, *see* Mouse
Miz-1, 311
MK1, 103
MK2, 103
MLL gene, 313
d-MN, 219–20
v-MN, 219–20
MORE sequence, 96, 97
Mot 1, 199–200
Mouse
    dwarfism, 99
    homeobox genes, 91
    Hoxb genes, 92, 220–3
    Pax6 and, 101
    splotch mutant, 101
    TBP/TLF independent transcription, 72
    zinc finger proteins, 103
Mouse mammary tumour virus long terminal repeat promoter (MMTV-LTR), 166–7, 250
mRNA splicing, alternative, *see* RNA splicing, alternative
mRNAs, encoding MyoA, MyoD, MyoH, 218
mRPD3, 203
MSX1, 217
Murine erythroleukaemia (MEL) cells, 259
Muscle-specific gene expression, 121–2, 212–19
Mutations, diseases caused by, 293–7
myb oncogene, 312
c-*myb*, 312
myc oncogene, 307–12
myc-CF1, 189
c-*myc* oncogene, 16, 307–8, 312, 329–30

c-*myc* oncogene (*cont.*)
    repression of transcription, 189
    transcriptional elongation, 170
myc-PRF, 189
Myc protein, 308–12, 322, 326, 332
    DNA binding, 117, 263
    Id2 activation, 327
MyoA, 213–15, 218
Myoblasts, 187
MyoD, 187, 188, 212–19
    activation of own expression, 218
    basic DNA-binding domain, 121,
        121–2, 216
    control of gene expression by, 215
    Fos/Jun actions, 304
    Pax3 and, 101, 217
    regulation, 286
Myogenesis, 217
MyoH, 213–15, 218
Myosin, 215

N-CoR, 197, 203
N-terminal inhibitory domain, 194–6, 235
'Negative noodles', 138
Nervous system development, 99–100,
        101
Neuronal cells, Oct-2 expression, 100,
        235
Neutrophils, mammalian, 247
NF1 (nuclear factor 1), *see* CTF/NF1
NFκB, 125, 175, 285, 312
    acetylation, 273
    activation, 245–6, 282, 283
    IκB activation, 256, 257, 268–70
    inactive form, 246
    phosphorylation-mediated regulation,
        268–70
    protein degradation and processing,
        281, 282, 283
    protein-protein interactions regulating,
        255–7, 265, 268
    ubiquitination, 279
NGF-1A, 103
Nkx homeobox genes, 225, 226
Northern blotting, 42, 218
Nuclear localization, 283, 284
Nuclear matrix, 2
Nuclear receptor gene superfamily, 108

co-activators, 160
    DNA binding, 109–16
    ligand binding, 248
    mutations, 296–7
    see also Steroid receptors; Thyroid
        hormone receptor
Nuclear receptor transcription factor
        family, identification of target
        genes, 66
Nuclear run-on assays, 231
Nucleosome remodelling factor (NURF),
        164–5
Nucleosomes, 2, 162–3
    displacement, 4
    remodelling, 4
    sliding, 4
NuRD, 202
NURF (nucleosome remodelling factor),
        164–5

OBF-1, 96
OCA-B, 136
Oct-1, 93
    activation domains, 140
    DNA binding, 43, 95–6, 97
    DNA mobility shift assay, 25, 26–7
    gene cloning, 40
    POU domains, 93
    VP16 (Vmw65) interaction, 98–9, 136,
        137
Oct-2, 93, 285
    activation domains, 140
    in developing brain, 100
    DNA binding, 43
    DNA mobility shift assay, 25, 27
    gene cloning, 40
    inhibitory domain, 196
    POU domains, 93
    regulated synthesis, 235, 237
    VP16 (Vmw65) interaction, 98–9
Octamer motif (ATGCAAAT), 12
    DNA mobility shift assay, 27
Octapeptide, 100
Oestrogen receptor, 253
    activation, 251, 254
    DNA binding, 110–15, 253
    ligand binding, 253–4
    structure, 109

transcriptional activation by, 153, 154
Oestrogen response element, 108
Oligonucleotide probes, gene cloning, 38–40
Oligonucleotides, 46, 47
    degenerate, 41
Oncogenes, 297
    cancer and, 299–314, 331–2
    cellular (*c-onc*), 298
    interactions, 329
    viral (*v-onc*), 298
    see also individual oncogenes

P box, 158
p21, 320, 324
p21 inhibitor, 216
p39, 301
p53, 160, 314, 315–22, 332–3
    acetylation, 273, 318
    competition for CBP/p300, 321
    deacetylation, 319
    DNA-binding domain, 125
    gene deletions, 315, 316
    inhibition by MDM2, 189, 190, 192–3, 315, 316, 317, 319, 321
    mutations, 315
    phosphorylation, 317, 318
    Rb-1 regulation, 324–5
p53-related proteins, novel, 321
p62^{TCF}, 267
p63, 321
p73, 321
p105, 281, 283
p300, 159–60, 278, 286
    methylation, 274, 275
    MyoD interaction, 215
    p53 acetylation, 273, 319
Paired domain, 101–2, 126
Paired protein (prd), 184
Pancreas, 101
Pax3, 101
    mutations, 101, 293
    MyoD activation, 101, 217
Pax6, 101, 225, 226
    mutations, 293, 293–4, 296
Pax8, alternative splicing, 232–3
Pax factors, 100
Pax proteins, 100–1

DNA binding, 125
PBX proteins, 313
PCAF, steroid receptor interactions, 167
PEA3, 303
Peptidyl prolyl isomerases, 318
Peroxisome proliferator-activated receptor γ (PPARγ), 297
Phage 434 repressor, 82
Phenanthroline-copper footprinting, 28–9
Phenylalanine residues, 102, 111
PHO4, 169, 170
PHO5 gene, 169
Phorbol esters, 206, 246
    activation of Fos/Jun synthesis, 285, 301
    AP1 binding site, 299
    NFκB activation, 256–7
    tumour promotion, 302
Phospholipase C, 262, 263
Phosphorylation, 266–72, 284
    acetylation interaction, 274
    CBP, 274
    CREB, 158–9, 266, 267, 342
    CREM, 267
    heat-shock factor (HSF), 259, 260, 261, 262, 267
    histones, 5
    IκB, 268–70
    MDM2, 317, 318
    p53, 317, 318
    Rb-1, 270, 323–5
    regulating transcription rate, 56
    retinoic acid receptor, 65–6
    role of TFIIH, 65–6, 68
PHYL, 191
PI(4,5)P$_2$, 262, 263, 269
Pin1, 317–18
Piperidine, 30, 31, 33
Pit-1, 12, 93
    CBP interaction, 274
    DNA binding, 96–7, 98
    mutations, 99, 293, 294, 296
    POU domains, 93
Pituitary gland, 12, 97, 99, 293
Pitx2, 220, 328
Platelet-derived growth factor receptor, 312–13

Polycomb, 200–2, 274
Polymerase chain reaction (PCR), 33, 46
Polyoma, large T oncogene, 331
POMC (pro-opiomelanocortin) gene,
    185, 186
PORE sequence, 96, 97
POU domain, 93, 126
POU proteins, 93–100
    development of specific cell types, 99
    DNA-binding site, 46–7, 125
    mutations, 99, 293
    novel, 99
    regulation of developmental gene
        expression, 99
POU-specific domain, 93, 125
POU transcription factors, 41
PPARγ (peroxisome proliferator-
    activated receptor γ), 297
Pro-opiomelanocortin (POMC) gene,
    185, 186
Probes, gene cloning, 38–40
Progesterone, 253
Progesterone receptor, 109
Programmed cell death (apoptosis), 315,
    321
Prokaryotes, 7
Prolactin gene, 185
Proline hydroxylation, 278, 279
Proline-rich activation domains, 137,
    140–1
Prolyl isomerase enzymes, 318
Promoters, gene, 7–9
    enhancers and, 13–14
    regulatory elements, 9–12
    RNA polymerase binding, 62
    selective activation, 155
    upstream promoter elements, 8, 9,
        9–10, 15
Prophet of Pit-1, 99
Protein kinase A, 266, 267
Protein kinase C, 270
Protein kinases, TFIIH activity, 65, 68
Proteins
    degradation, 281–3
    DNA interactions, methods of study,
        23–34
    inhibitory, dissociation of, 248, 268
    ligand binding, 247–55

modification, 248, 268, 272
post-translational modification, 265–81
processing, 281–3
-protein interactions, 145, 160, 198,
    255–65, 268, 341–2
protein-protein interactions, 69–70, 70,
    71–2
purification, 34–8
regulation of synthesis, 238
synthesis from cloned genes, 42
Proteolysis, see Degradation, protein
Proto-oncogenes, cellular, 298, 314
PPRI, 109, 110
PRTF, see MCM1
PU-1, 125
Purification, of proteins, 34–8

Quenching phenomenon, 189–90

Raloxifene, 254
Rat, zinc finger proteins, 103
Rb-1, see Retinoblastoma protein
Rb-1 gene, 332
    inactivation, 323, 326
Recognition helix, 83–4, 85, 111
    bacteriophage proteins, 83, 86
rel oncogene, 312
Repression of transcription, 183–200
    by activator degradation, 184, 190–1
    direct, 183, 184, 191–200
    by forming non-DNA binding complex,
        187–9
    indirect, 183–91
    by masking of DNA binding site, 183–7
    mechanisms, 184, 191–2
    by quenching activator function,
        189–90
Retinoblastoma anti-oncoprotein, 62
Retinoblastoma protein (Rb-1), 314,
    322–7, 332
    acetylation, 325
    E2F interaction, 189, 193, 270
    inhibiting effect, 322
    phosphorylation, 270, 323–5
    repressive activity, 189
    targets, 322–6
Retinoblastomas, 322
Retinoic acid, 253

concentration gradient, 222, 223
  Hoxb gene activation, 222–3
  as a morphogen, 222
9-cis Retinoic acid, 108, 115
Retinoic acid receptor, 115, 223
  DNA binding, 115
  ligand binding, 253
  phosphorylation, 65–6, 267
Retinoic acid response element, 108, 223
Retinoid X-receptor (RXR), 115–16
Retroviruses, cancer-causing, 297, 312,
    331
Rev-T, 312
Ribosomal RNA genes, 55
  5S, transcriptional complex, 60–2
  regulation by Rb-1, 325
  RNA polymerase I transcriptional
    complex, 58–60
RLIM co-repressor molecule, 197–8
RNA polymerase I, 55–6, 58–60
  Dr1 and, 199
  Rb-1 actions, 326
  role of TBP, 71, 71–2, 199
  stable transcriptional complex, 58–60,
    62
RNA polymerase II, 9, 55–6, 63–7
  activator interactions, 147–9
  basal transcriptional complex, see Basal
    transcriptional complex
  C-terminal domain (CTD),
    phosphorylation, 65, 151, 172–3
  Dr1 and, 199–200
  heptapeptide repeat, 56, 147
  holoenzyme complex, co-activators,
    160
  non-TBP/TLF transcription, 72
  phosphorylation, 56, 65–6
  role of TBP, 68–70, 71–2, 199
  stable transcriptional complex, 63–7
  ubiquitination, 279
RNA polymerase III, 55–6
  Rb-1 actions, 326
  role of TBP, 70, 71–2, 199
  stable transcriptional complex, 60–2
  transcription process, 61–2
  transcription units, 60–2
RNA polymerase holoenzyme complex,
    66–7

activator actions, 145, 150–1, 162
  C-terminal domain, phosphorylation,
    204
RNA polymerases, 55–7
  DNA-binding site, 62
  role of TBP, 71–2
RNA splicing
  alternative
    CREM, 233–5
    thyroid hormone receptor (ErbA),
      236
    transcription factors, 232–8
  polymerase and, 57
Rubinstein–Taybi syndrome, 295–6

Saccharomyces cerevisiae, heat-shock factor
    (HSF), 261
SATBI protein, 4–5
Schizosaccharomyces pombe, heat-shock
    factor (HSF), 261
Serum response element (SRE), MyoD
    action, 216, 217
Serum response factor (SRF), 267
  gene cloning, 38
  purification, 38
Shh, 225–7
Silencer elements, 16, 198
Simian virus 40, see SV40
SIN3, 202, 203
SINA, 191
Sir2, 319
Skn-l, 119
SLl, 59–60, 70
Snail gene, 103
SNF2 mutations, 294, 296
SnRNA, 93
Somatostatin, 1
Sonic hedgehog (Shh), 225–7
Southern blotting, 42, 91
Sp1, 296
  activation domain, 137, 139, 152
  AP1 interaction, 17–18
  in DNA mobility shift assay, 27
  gene cloning, 38, 39
  purification, 34–5, 36
  repression by Sp3, 184
  zinc finger motif, 102–3, 103
Sp1 box, 9

Sp3, 184, 185, 187
Splicing, alternative, *see* RNA splicing,
    alternative
Splotch mutant, 101
Squelching, 142–3, 149–50, 153
SRC-1
    nuclear receptor interactions, 160
    steroid receptor interactions, 167
SREBP, 282–3
SRF, *see* Serum response factor
Stable transcriptional complex, 57–67
    characteristics, 57–8
    RNA polymerase I, 58–60
    RNA polymerase II, 63–7
    RNA polymerase III, 60–2
STATs, 266, 274
Steroid hormones, steroid receptor
    activation, 166–9
Steroid receptors
    activation, 166–9, 253, 254
    activation of gene expression, 206–7
    DNA binding, 107–9, 113–14, 166
    ligand binding, 248–50
    see also Glucocorticoid receptor;
        Oestrogen receptor
Steroid-thyroid hormone receptor gene
    family, *see* Nuclear receptor gene
    superfamily
Stromelysin gene, 303
Sugl, 160
SUMO-1, 279–80, 316, 319
SV40, 325
    enhancer, Oct-1 binding, 95
    large T oncogenes/oncoprotein, 320,
        331, 332
SW15 gene, 103
SWI/SNF complex
    chromatin remodelling, 3–4, 7, 164,
        170, 171, 176
    mutations affecting function, 294, 295,
        297
    nuclear receptor interactions, 167
    RNA polymerase holoenzyme and,
        66
Synthesis of transcription factors,
    regulated, 211–29, 286
    mechanisms regulating, 229–40
    RNA splicing, 232–8

role, 240–1
transcription, 230–2
translation, 238–40

T-cell receptor α gene enhancer, 19
T lymphocytes, 245, 246
    activation of gene expression, 246
    CD4 expression, 16
    LEF-1 factor expression, 18–19
TAFs (TBP-associated factors), 68, 151–7,
    161–2
    $TAF_{II}18$, 156
    $TAF_{II}30$, 156
    $TAF_{II}31$ ($TAF_{II}40$), 138–9, 152, 160,
        162
    $TAF_{II}55$, 152
    $TAF_{II}110$, 152
    $TAF_{II}130$ ($TAF_{II}105$), 156, 296
    $TAF_{II}145$, 155
    $TAF_{II}250$, 154, 161
Tailless gene, 79
TATA-binding protein, *see* TBP
TATA box, 8–9
    activation of transcription and, 155
    in RNA polymerase II promoters, 63,
        69–70
    in RNA polymerase III promoters,
        61–2, 70
    TBP binding, 68–71, 155
TBP, 67–73
    activator interactions, 151–2, 162
    binding to non-TATA box promoters,
        69
    repressor interactions, 198–9
    structure, 69
    TATA box binding, 68–71, 155
    as universal transcription factor, 67–72
TBP-associated factors, *see* TAFs
TBP-like factors (TLFs), 72–3
TDF, 103
$10T\frac{1}{2}$ fibroblast cells, 187, 188, 212–15
TFIIA
    activator interactions, 149
    inhibitor interactions, 200, 201
    in transcriptional complex assembly,
        63–4, 66
TFIIB

activator interactions, 143–5, 148–9,
    150, 151, 162
as bridge between TBP and RNA
    polymerase II, 69
repressor interactions, 196, 198
in RNA polymerase holoenzyme, 66
in transcriptional complex assembly,
    63–4, 66
TFIID
    activator interactions, 138–9, 143–5,
        148–9, 150
    activator targets, 152–3
    binding sites, 166–7
    chromatin remodelling, 68
    multi-protein nature, 68–9, 151
    promoter selectivity, 156
    TBP component of, 198, 199
    in transcriptional complex assembly,
        63–4, 66
TFIIE, 65, 66, 68
    activator interactions, 151
TFIIEβ, repressor interactions, 196
TFIIF, 68
    activator interactions, 143, 145, 148–9,
        151
    in RNA polymerase holoenzyme, 66
    in transcriptional complex assembly,
        64, 65, 66
TFIIH, 68, 71
    activator interactions, 145, 149, 151,
        162
    DNA kinase and helicase activities,
        65–6, 68
    mutations, 294–5, 296
    in RNA polymerase holoenzyme, 66
    in transcriptional complex assembly,
        65, 66
TFIIJ, 65, 66
TFIIIA, 103
    in 5S rRNA gene transcription, 61,
        62
    zinc finger motif, 102
TFIIIB, 60–2, 62
    Myc interaction, 311
    Rb-1 interaction, 326
    TBP component, 70
TFIIIC, 61, 62, 70
Thymidine kinase, 322

Thymidine kinase gene, 10
Thyroid hormone, 253
    binding by v-ErbA, 304–6
Thyroid hormone receptor (ErbA), 285,
        304–7
    activation of gene expression, 205,
        206–7
    α2 form, 236–7, 306
    alternative splicing, 236
    cellular vs viral forms, 305–6
    DNA binding, 115, 116, 250
    inhibition of gene expression, 193–4,
        205, 206–7
    inhibitory domain, 196–7
    ligand-induced regulation, 248–50
    mutations, 296–7
    repression of transcription, 254
    structure, 108, 109
Thyroid response element (TRE), 108,
        193, 194
TIF1, 58–60, 62, 160
TIF2, 160
TLFs (TBP-like factors), 72–3
TPA, 299
Transcription, importance of, 1
Transcriptional complexes, see Basal
        transcriptional complex; Stable
        transcriptional complex
Transfer RNA genes, 55
Translation, regulation of, 238–40
Transretinoic acid, 108, 115
Transthyretin, 231
TRE (thyroid response element), 193,
        194
TRF2, 72
Trithorax gene, 164, 202
TTK88, 191
Tubby, 262, 263, 269
Tumour suppressor genes, see anti-
        oncogenes
Twist, 264, 265
Two-cysteine two-histidine zinc finger,
        102–7
    DNA binding, 104–7
    transcription factors, 102–4
Tyrosine kinase, 266

U937 monocytic cells, 326

UBF (upstream binding factor), 325–6
  DNA-binding motif, 59–60, 71, 124–5
Ubiquitin, 204, 265, 275, 315–16, 319
Ubiquitination, 275–81
  histones, 5, 5–7
Ultrabithorax (Ubx) gene, 79, 80, 81
Ultrabithorax (Ubx) protein, 264
  activation domain, 140
  gene repression, 192, 193
  inducing own expression, 86–7
unc-86, 93, 99
Upstream promoter elements, 8, 9, 9–10,
  15

Valine, 118
VEGF, 106, 107
Ventral neural tube, 225–7
Ventral repression element (VRE), 264,
  265
Viruses
  DNA, oncogenic, 325, 331
  see also Retroviruses
Vitamin D, 108, 115
Vitamin D receptors, mutations, 296–7
Vitamin D3 response element, 108
Von Hippel-Lindau protein (VHL), 204,
  205, 277, 278, 279, 330–1
  mutations, 330
VP16 (Vmw65)
  activation domain, 136, 137, 138, 139,
    169, 170
  activation targets, 162
  interaction with TAFs, 153–4, 160
  Oct-1 and Oct-2 interaction, 98–9
  transactivation, 96, 136

ubiquitination, 279

Waardenburg syndrome, 101
Wilms' tumour gene product, 196, 328
WNT (wingless) proteins, 328, 329

Xenopus, 72
  RNA polymerase III transcription, 62
  zinc finger proteins, 103
Xenopus hsp70 gene, 14
Xeroderma pigmentosum, 294–5, 295
Xfin, 103

Yeast
  activation domains, 140
  a and α genes, 89–90
    homeobox-like region, 81
  heat-shock factor (HSF), 261, 262
  mating types, 89–90
  RNA polymerases, 56
  TAFs, 154, 155
  zinc finger proteins, 103, 104

Zen, 264, 265
Zerknult protein (zen), 184
Zeste, 140
Zinc, 102, 103, 109
Zinc finger motifs, 102–16, 127
  Drosophila proteins, 79, 103–4, 105
  multi-cysteine, 107–16, 126
  structure, 102
  transcriptional proteins containing,
    103
  two-cysteine two-histidine, 102–7, 126